Translational Bioinformatics

Series Editor

Xiangdong Wang, MD, PhD
Professor of Clinical Bioinformatics, Lund University, Sweden
Professor of Medicine, Fudan University, China

Aims and Scope

The Book Series in Translational Bioinformatics is a powerful and integrative resource for understanding and translating discoveries and advances of genomic, transcriptomic, proteomic and bioinformatic technologies into the study of human diseases. The Series represents leading global opinions on the translation of bioinformatics sciences into both the clinical setting and descriptions to medical informatics. It presents the critical evidence to further understand the molecular mechanisms underlying organ or cell dysfunctions in human diseases, the results of genomic, transcriptomic, proteomic and bioinformatic studies from human tissues dedicated to the discovery and validation of diagnostic and prognostic disease biomarkers, essential information on the identification and validation of novel drug targets and the application of tissue genomics, transcriptomics, proteomics and bioinformatics in drug efficacy and toxicity in clinical research.

The Book Series in Translational Bioinformatics focuses on outstanding articles/chapters presenting significant recent works in genomic, transcriptomic, proteomic and bioinformatic profiles related to human organ or cell dysfunctions and clinical findings. The Series includes bioinformatics-driven molecular and cellular disease mechanisms, the understanding of human diseases and the improvement of patient prognoses. Additionally, it provides practical and useful study insights into and protocols of design and methodology.

Series Description

Translational bioinformatics is defined as the development of storage-related, analytic, and interpretive methods to optimize the transformation of increasingly voluminous biomedical data, and genomic data in particular, into proactive, predictive, preventive, and participatory health. Translational bioinformatics includes research on the development of novel techniques for the integration of biological and clinical data and the evolution of clinical informatics methodology to encompass biological observations. The end product of translational bioinformatics is the newly found knowledge from these integrative efforts that can be disseminated to a variety of stakeholders including biomedical scientists, clinicians, and patients. Issues related to database management, administration, or policy will be coordinated through the clinical research informatics domain. Analytic, storage-related, and interpretive methods should be used to improve predictions, early diagnostics, severity monitoring, therapeutic effects, and the prognosis of human diseases.

Translational Bioinformatics

Series Editor: Xiangdong Wang, MD, PhD, Professor of Clinical Bioinformatics,
Lund University, Sweden;
Professor of Medicine, Fudan University, China

Recently Published and Forthcoming Volumes

Applied Computational Genomics
Editor: Yin Yao Shugart
Volume 1

Pediatric Biomedical Informatics
Editor: John J. Hutton
Volume 2

Bioinformatics of Human Proteomics
Editor: Xiangdong Wang
Volume 3

For further volumes:
http://www.springer.com/series/11057

Translational Bioinformatics
Volume 2

Pediatric Biomedical Informatics

Computer Applications in Pediatric Research

Editor: John J. Hutton

Springer

Editor
John J. Hutton
Cincinnati Children's Hospital
Cincinnati Ohio
USA

ISSN 2213-2775 ISSN 2213-2783 (electronic)
ISBN 978-94-017-8220-3 ISBN 978-94-007-5149-1 (eBook)
DOI 10.1007/978-94-007-5149-1
Springer Dordrecht Heidelberg New York London

Printed on acid-free paper

Springer is part of Springer Science+Business Media (www.springer.com)

Series Foreword

Translating biomedical discoveries into better health for people continues to pose major and well-recognized challenges, although the pace of those discoveries over past decades has been truly startling. Among these challenges is analyzing the huge amounts of data generated by high throughput technologies and integrating it with data about patients that is now accessible through electronic health records. Converting data into knowledge is one aim of biomedical informatics.

In recognition of the major role informatics plays in accruing, integrating, and analyzing data in the biomedical sciences and translating it into clinical practice, a series of books on *Translational Bioinformatics* is being created by united scientific forces of the *International Society for Translational Medicine* (ISTM, www.istmed.org), *Journal of Clinical Bioinformatics* (JCBi, www.jclinbioinformatics.com), *Journal of Clinical and Translational Medicine* (CTM, www.clintransmed.com), and Springer Publisher. These will cover topics such as genomics, proteomics, metabolomics, systems immunology, and biomarkers. *Pediatric Biomedical Informatics: Computer Applications in Pediatric Research* is an important volume in this series and focuses on core resources in informatics that are necessary to support translational research in a research-intensive children's medical center. One key challenge is implementing interoperable research and clinical IT systems so that data can be exchanged to support translational research.

I, as the Editor of Series Books, am privileged and honored to have Professor John J. Hutton as the Editor of this special volume. I knew John by his outstanding contributions to biomedical informatics in clinical and translational medicine and his reputation as an academic manager and leader, as well as international and national opinion leader. The most important issue for me and other CTM fellows is his friendship, professional supervision, or endless support.

The first part of this book discusses important technical challenges that underlie most types of computer-based biomedical research, e.g. storing, mining, and transmitting huge quantities of data (giga-, tera-, and sometimes petabytes), while maintaining strict cybersecurity with identity management and access control. The second part discusses a broad range of research programs and demonstrates the wide variety of software applications that are necessary to support relevant types of research. While different applications are needed to support different types of translational research, they all have in common the need to be interfaced with a well-designed hardware infrastructure. One of the most outstanding parts is that the book emphasizes new opportunities to investigate clinical phenomena through the use of data in electronic health records.

As has been emphasized by the US National Institutes of Health, the bench-to-bedside approach to translational research is really a two-way street. While translational researchers provide clinicians with new resources for prevention, diagnosis, and treatment of disease, it is the clinicians themselves who interface directly with patients and observe the onset and course of disease. Feedback from the clinicians provides essential information to the researchers and can provide new insights into ill understood elements of disease that are topics for future research.

Pediatric Biomedical Informatics: Computer Applications in Pediatric Research, as part of the Springer Series on Translational Bioinformatics, provides excellent and useful guidance to the implementation of an informatics infrastructure to support translational research in a research-intensive clinical environment. It shows how a wide variety of active research programs use this infrastructure to generate knowledge to improve human health.

Professor of Respiratory Medicine Xiangdong Wang M.D., Ph.D.
and Director of Biomedical Research Center
Fudan University, Zhongshan Hospital, Shanghai, China;
Professor (adj) of Clinical Bioinformatics,
Lund University Clinical Science, Lund, Sweden

Foreword

The Revolution in Biomedical Informatics

I was born before electronic computers existed. Therefore, I have witnessed the entire Revolution in Biomedical Informatics over the last 40 years. As a pediatric physician-scientist and a practicing pediatric cardiologist, for many years my goal was to generate data: experimental results in the laboratory to understand the structure and functions of proteins and genes, and observational and laboratory information to optimize care of patients. Generating sufficient data, cataloging the data rationally, and analyzing the limited data were my daily activities. And, there was never enough data. Until now.

Now, all of us working in medicine and science are deluged with data. Other descriptors of this Revolution of enormous information that I have heard are "avalanche" and "tsunami", as we are inundated with data. The goals for pediatricians and scientists have changed, moving beyond data creation. Although we must continue to generate data, the Revolution in Biomedical Informatics has created major challenges in the acquisition, management, storage, transmission, and analysis of data essential to accomplish our ultimate purposes of optimizing patient care, determining mechanisms underlying disease, and defining the best treatments for children.

I was fortunate to become the Chief Medical Officer at Cincinnati Children's Hospital Medical Center and Director of our Research Foundation in 2007, just as our leadership initiated the implementation of a single electronic medical or health record for all clinical care. We recognized the incredible potential of using the EMR data to assess and improve patient safety, to ascertain outcomes of our patients, and to serve as the basis for patient-oriented, clinical, translational, and health services research on a global scale. Under the guidance and tutelage of John Hutton, the Editor of this volume, and with the recruitment and expertise of the faculty and their colleagues who have contributed chapters, we have begun the journey toward maximizing this potential. The topics discussed include the major core informatics resources needed, from the EMR itself, to transmission of information in it for storage and management, to security required for patient protection, creation of usable patient data warehouses, and integration of patient information with biobanked tissue and DNA for research, all critical infrastructure to optimize care and research. In addition, some intriguing applications in both patient-oriented research and basic science are provided, to illustrate how population-based studies, assessment of language, support for decisions, and generation of networks can be done.

Biomedical informatics is the key to our future, as we integrate clinical care, genomics, and basic science to improve outcomes and discover new therapeutics. With careful design and acquisition of information, we can tame the avalanche of data. We can link data across institutions to achieve greater power of analysis and increase the speed of discovery and evaluation of treatments. This book provides insight into how to use data generated during The Revolution in Biomedical Informatics to benefit children around our world.

Arnold W. Strauss, M.D.
Rachford Professor and Chair
Department of Pediatrics
University of Cincinnati College of Medicine
Director, Cincinnati Children's Research Foundation
Chief Medical Officer
Cincinnati Children's Hospital Medical Center

Preface

Cincinnati Children's Hospital Medical Center (CCHMC) is a major teaching affiliate of the University of Cincinnati College of Medicine. Its clinical and research programs in pediatrics rank among the largest and most highly respected in the United States. A decade ago, the Board of Trustees and leadership of CCHMC envisioned a future in which advances in information technology (IT) would radically change the infrastructure needed to support clinical care and biomedical research. The faculty developed and the leadership supported a strategic plan for information technology. Since then, Hospital Information Services has implemented an electronic health record to support clinical care throughout the medical center and a Division of Biomedical Informatics has been created to support research. The two units (Information Services and Biomedical Informatics) work very closely together to assure seamless support of medical center clinical and translational research programs.

Moving the institution from the past to the present has been an adventure. We have continuously assessed what core resources in informatics are necessary to support research programs and how these can best be integrated with hospital systems to receive clinical information that is necessary to conduct translational research in pediatrics. This book focuses on the core informatics resources that are now available to our faculty and provides examples of how they support specific types of research. While the focus is upon core resources in IT and computer applications in one research-intensive pediatric medical center, CCHMC is large and diverse so that all or portions of the lessons learned and services developed will be applicable to other institutions. Most of the contributors either are experts in the planning and implementation of IT services or are active users who participate in the assessment of needs of researchers now and in the future.

The first part of the book discusses core resources in informatics, both technical and knowledge-based, that are relevant to most translational research projects. Important technical challenges underlying computer-based biomedical research are storing, mining, and transmitting huge quantities of data (terabytes and sometimes petabytes), while maintaining strict cybersecurity with identity management and access control. Another topic of broad importance is the electronic health record

(EHR) and how information captured in the EHR, during the process of clinical care, can be moved into a research patient data warehouse. Data in the warehouse can then be mined to support research, including the identification of cohorts of patients for study of a wide range of diseases. These important topics are discussed in considerable detail.

Biosamples, such as blood and tissues, are key ingredients in biomedical research and can be collected in enterprise biobanks. The information management system of the biobank can be interfaced with the research patient data warehouse so that phenotypes of patients, who contributed the biosamples, can be derived. This is critically important when conducting genetic studies, where genomes of patients with specific phenotypes will be sequenced to identify variant genes associated with disease. Chapters devoted to biobanking, data warehousing, and privacy also discuss the very important topics of patient consent and ethical and legal issues surrounding human subjects research.

The second part of the book discusses a broad range of research programs supported by core informatics resources. The first example focuses on EHR-linked registries for studies of populations and how such registries can be federated to permit data sharing among members of a distributed research network. This is particularly important in pediatrics because most pediatric chronic conditions are rare. Frequently, no single center has a sufficient number of patients to support outcomes, comparative effectiveness, and other types of clinical research. Natural language processing (NLP) is a topic of increasing relevance and is the subject of two chapters. It is apparent that much of the information in EHRs is in free text, such as clinical notes by physicians and other health care professionals. While NLP is an active field of research, we have found it essential as a core capability to support patient phenotyping, prospective identification of safety events, and mining of publications for specific types of information to be incorporated into decision support applications. Perinatal and neonatal research are unique and critically important to pediatrics. They pose challenges in information management, as presented in a chapter addressing support of perinatal and neonatal research. Neonatal terminologies are not well developed. It is generally difficult to collect and integrate information, as infants transition from obstetrical to pediatric care. Other important topics in clinical research that require computer support and are discussed include decision support, calculation of outcomes measures, and automated detection and reporting of safety events.

Six chapters provide examples of informatics applications in translational research at the interface between basic science and the clinic. The first topic is network analysis of complex molecular interactions and biological processes to identify the underlying causes of pediatric disease and to develop innovative therapies. Such analyses are being applied, for example, to drug discovery and drug repurposing, i.e., discovering new targets for old drugs developed for some other purpose.

Developmental biology is particularly relevant to pediatrics and is a discipline that is well represented among research programs in our institution. One chapter is devoted to transcriptional networks that control lung maturation. Maturation of the lung is a late event during fetal development, yet is key to survival of the newborn infant. Several decades ago, respiratory distress syndrome (RDS) was the major cause of morbidity and mortality of premature infants. Death from RDS is rare today because of research on lung maturation, which led to the development of antenatal glucocorticoid treatment and surfactant replacement therapy to prevent and treat it. A second chapter discusses how software applications and other tools can be applied to studies of the sequential action of genes during kidney and craniofacial development, i.e., define their genetic blueprint. Both chapters illustrate how software applications and computer-support permit studies of gene regulation in networks with consequent cell and organ development. Without tools of biomedical informatics, these complex, data-intensive studies could not be done.

The final chapters discuss informatics tools and applications that support genetics and genomics. Focus is upon studies of genetic variation and its effects upon human health. One chapter discusses single nucleotide polymorphism (SNP) genotyping and its integration into clinical information systems; the other discusses genome and exome sequencing to identify sequence variants that cause disease. In both cases, management of huge data sets, frequently requiring use of high performance computing, is a major informatics challenge.

Support of biomedical research in a pediatric medical center requires both a sophisticated IT infrastructure and a staff of knowledgeable professionals. One of the most important lessons we have learned during the process of building informatics support of clinical and translational research is that the information systems that support clinical care and those that support research must constantly communicate and share information. This is both a political and a technological challenge (staff from different organizational cultures must cooperate and to the extent possible, IT systems must be semantically and syntactically interoperable). These challenges can be overcome, as we hope our book demonstrates.

John J. Hutton, M.D.

Contents

Contributors

Bruce J. Aronow Ph.D. Department Pediatrics, University of Cincinnati College of Medicine, Cincinnati, OH, USA

Genome Informatics Core Laboratory, Division of Biomedical Informatics, Cincinnati Children's Hospital Medical Center, Cincinnati, OH, USA

Dominik Aronsky M.D., Ph.D., FACMI Departments of Biomedical Informatics and Emergency Medicine, Vanderbilt University School of Medicine, Nashville, TN, USA

Michael G. Barnes Ph.D. Division of Rheumatology, Cincinnati BioBank Core Facility, Cincinnati Children's Hospital Medical Center, Cincinnati, OH, USA

Jacek W. Biesiada Ph.D. Division of Biomedical Informatics, Cincinnati Children's Hospital Medical Center, Cincinnati, OH, USA

Division of Management and Informatics, Technical University of Silesia, Gliwice, Poland

Eric W. Brunskill Ph.D. Division of Developmental Biology, Cincinnati Children's Hospital Medical Center, Cincinnati, OH, USA

Ye Chen School of Electronics and Computing Systems, University of Cincinnati College of Engineering & Applied Science, Cincinnati, OH, USA

Division of Biomedical Informatics, Cincinnati Children's Hospital Medical Center, Cincinnati, OH, USA

Jeremy J. Corsmo MPH, CIP, CHRC Research Compliance and Regulatory Affairs, Cincinnati Children's Hospital Medical Center, Cincinnati, OH, USA

Louise Deleger Ph.D. Division of Biomedical Informatics, Cincinnati Children's Hospital Medical Center, Cincinnati, OH, USA

Jingyuan Deng School of Computing Sciences and Informatics, University of Cincinnati College of Engineering & Applied Science, Cincinnati, OH, USA

Division of Biomedical Informatics, Cincinnati Children's Hospital Medical Center, Cincinnati, OH, USA

Judith W. Dexheimer Ph.D. Department of Pediatrics, University of Cincinnati College of Medicine, Cincinnati, OH, USA

Divisions of Emergency Medicine and Biomedical Informatics, Cincinnati Children's Hospital Medical Center, Cincinnati, OH, USA

Phillip Dexheimer M.S. Divisions of Biomedical Informatics and Human Genetics, Cincinnati Children's Hospital Medical Center, Cincinnati, OH, USA

Gregory A. Grabowski M.D. Department of Pediatrics, University of Cincinnati College of Medicine, Cincinnati, OH, USA

Division of Human Genetics, Cincinnati Children's Hospital Medical Center, Cincinnati, OH, USA

Eric S. Hall Ph.D. Department of Pediatrics, University of Cincinnati College of Medicine, Cincinnati, OH, USA

Perinatal Institute & Division of Biomedical Informatics, Cincinnati Children's Hospital Medical Center, Cincinnati, OH, USA

Nicholas Hunt MSIA, CISSP Department of Pediatrics, University of Cincinnati College of Medicine, Cincinnati, OH, USA

Division of Biomedical Informatics, Cincinnati Children's Hospital Medical Center, Cincinnati, OH, USA

John J. Hutton M.D., FACP Department of Pediatrics, University of Cincinnati College of Medicine, Cincinnati, OH, USA

Division of Biomedical Informatics, Cincinnati Children's Hospital Medical Center, Cincinnati, OH, USA

Anil G. Jegga DVM, M.S. Department of Pediatrics, University of Cincinnati College of Medicine, Cincinnati, OH, USA

Division of Biomedical Informatics, Cincinnati Children's Hospital Medical Center, Cincinnati, OH, USA

Laurie H. Johnson M.D. Department of Pediatrics, University of Cincinnati College of Medicine, Cincinnati, OH, USA

Division of Emergency Medicine, Cincinnati Children's Hospital Medical Center, Cincinnati, OH, USA

Eric S. Kirkendall M.D., MBI, FAAP Department of Pediatrics, University of Cincinnati College of Medicine, Cincinnati, OH, USA

Divisions of Hospital Medicine and Biomedical Informatics, Cincinnati Children's Hospital Medical Center, Cincinnati, OH, USA

Michal Kouril Ph.D. Department of Pediatrics, University of Cincinnati College of Medicine, Cincinnati, OH, USA

Division of Biomedical Informatics, Cincinnati Children's Hospital Medical Center, Cincinnati, OH, USA

Long Jason Lu Ph.D. Department of Pediatrics, University of Cincinnati College of Medicine, Cincinnati, OH, USA

Division of Biomedical Informatics, Cincinnati Children's Hospital Medical Center, Cincinnati, OH, USA

John A. Lynch Ph.D. Department of Communication, McMicken College of Arts & Sciences, University of Cincinnati, Cincinnati, OH, USA

Center for Clinical and Translational Science and Training, University of Cincinnati College of Medicine, Cincinnati, OH, USA

Keith Marsolo Ph.D. Department of Pediatrics, University of Cincinnati College of Medicine, Cincinnati, OH, USA

Jaroslaw Meller Ph.D. Department of Environmental Health, University of Cincinnati College of Medicine, Cincinnati, OH, USA

Division of Biomedical Informatics, Cincinnati Children's Hospital Medical Center, Cincinnati, OH, USA

John P. Pestian Ph.D. Department of Pediatrics, University of Cincinnati College of Medicine, Cincinnati, OH, USA

Division of Biomedical Informatics, Cincinnati Children's Hospital Medical Center, Cincinnati, OH, USA

Andrew S. Potter M.S. Division of Developmental Biology, Cincinnati Children's Hospital Medical Center, Cincinnati, OH, USA

S. Steven Potter Ph.D. Department of Pediatrics, University of Cincinnati College of Medicine, Cincinnati, OH, USA

Division of Developmental Biology, Cincinnati Children's Hospital Medical Center, Cincinnati, OH, USA

Senthilkumar Sadhasivam M.D., MPH Department of Pediatrics, University of Cincinnati College of Medicine, Cincinnati, OH, USA

Perioperative and Acute Pain Service, Division of Pediatric Anesthesia, Cincinnati Children's Hospital Medical Center, Cincinnati, OH, USA

Guergana K. Savova Ph.D. Boston Children's Hospital and Harvard Medical School, Harvard University, Boston, MA, USA

Imre Solti M.D., Ph.D., M.A. Department of Pediatrics, University of Cincinnati College of Medicine, Cincinnati, OH, USA

Division of Biomedical Informatics, Cincinnati Children's Hospital Medical Center, Cincinnati, OH, USA

S. Andrew Spooner M.D., M.S., FAAP Department of Pediatrics, University of Cincinnati College of Medicine, Cincinnati, OH, USA

Division of Biomedical Informatics, Cincinnati Children's Hospital Medical Center, Cincinnati, OH, USA

Paul E. Steele M.D. Departments of Pathology and Pediatrics, University of Cincinnati College of Medicine, Cincinnati, OH, USA

Clinical Laboratory, Cincinnati Children's Hospital Medical Center, Cincinnati, OH, USA

Arnold W. Strauss M.D. Department of Pediatrics, University of Cincinnati College of Medicine, Cincinnati, OH, USA

Cincinnati Children's Research Foundation, Cincinnati Children's Hospital Medical Center, Cincinnati, OH, USA

Lirong Tan School of Computing Sciences and Informatics, University of Cincinnati College of Engineering & Applied Science, Cincinnati, OH, USA

Division of Biomedical Informatics, Cincinnati Children's Hospital Medical Center, Cincinnati, OH, USA

Michael Wagner Ph.D. Department of Pediatrics, University of Cincinnati College of Medicine, Cincinnati, OH, USA

Division of Biomedical Informatics, Cincinnati Children's Hospital Medical Center, Cincinnati, OH, USA

Jeffrey A. Whitsett M.D. Department of Pediatrics, University of Cincinnati College of Medicine, Cincinnati, OH, USA

Perinatal Institute, Cincinnati Children's Hospital Medical Center, Cincinnati, OH, USA

David P. Witte M.D. Departments of Pathology and Pediatrics, University of Cincinnati College of Medicine, Cincinnati, OH, USA

Division of Pathology and Laboratory Medicine, Cincinnati Children's Hospital Medical Center, Cincinnati, OH, USA

Yan Xu Ph.D. Department of Pediatrics, University of Cincinnati College of Medicine, Cincinnati, OH, USA

Divisions of Pulmonary Biology and Biomedical Informatics, Cincinnati Children's Hospital Medical Center, Cincinnati, OH, USA

Minlu Zhang Ph.D. School of Computing Sciences and Informatics, University of Cincinnati College of Engineering & Applied Science, Cincinnati, OH, USA

Division of Biomedical Informatics, Cincinnati Children's Hospital Medical Center, Cincinnati, OH, USA

Cheng Zhu M.S. School of Computing Sciences and Informatics, University of Cincinnati College of Engineering & Applied Science, Cincinnati, OH, USA

Part I
Core Informatics Resources

Chapter 1
Pediatric Electronic Health Records and Research

S. Andrew Spooner and Eric S. Kirkendall

Abstract Pediatric providers use electronic health record systems to review patient information, to document care, to order clinical interventions, and to perform related administrative tasks. All of these activities create data that might be useful in research, although research is seldom the objective of EHR-related data entry. Providers may use other information systems (e.g., specialized systems for analyzing electrocardiograms), but the EHR is the central application for clinical and administrative clinical activities. While there are a few EHR systems designed specifically for care of pediatric patients, most pediatric providers adopt general-purpose EHRs that must be customized for specialized pediatric environments. In this chapter we outline the special functional requirements of EHRs (e.g., growth monitoring, medication dosing, and immunization management), the relative difficulty of meeting these requirements with EHRs that are currently available in the marketplace, and current adoption trends. We discuss workflows that present special challenges to EHR implementation. We discuss the typical workflow phenomena that affect the use of data for research and other secondary uses. We also discuss special aspects of terminology systems employed by EHRs that have implications for pediatric usability. Lastly, we address special issues in the use of EHR data for the extraction of care quality measures.

S.A. Spooner, M.D., M.S., FAAP (✉)
Department of Pediatrics, University of Cincinnati College of Medicine, Cincinnati, OH, USA

Division of Biomedical Informatics, Cincinnati Children's Hospital Medical Center, 3333 Burnet Avenue, ML-9016, Cincinnati, OH 45229-3039, USA
e-mail: andrew.spooner@cchmc.org

E.S. Kirkendall, M.D., MBI, FAAP
Department of Pediatrics, University of Cincinnati College of Medicine, Cincinnati, OH, USA

Divisions of Hospital Medicine and Biomedical Informatics, Cincinnati Children's Hospital Medical Center, 3333 Burnet Avenue, ML-9016, Cincinnati, OH 45229-3039, USA

J.J. Hutton (ed.), *Pediatric Biomedical Informatics: Computer Applications in Pediatric Research*, Translational Bioinformatics 2, DOI 10.1007/978-94-007-5149-1_1,

1.1 Current State

We begin this chapter by providing an overview of the current state of electronic health records (EHRs) in pediatrics, primarily focusing on adoption rates, the contemporary pediatric EHR market, the considerations involved in use by primary care versus subspecialty care providers, and the differences in obtaining data from natural workflow versus research. The intent of describing the current state is to provide the reader with a brief background of the inherent challenges and potential of using pediatric EHRs for research. We start by first examining data regarding adoption rates of pediatric EHRs in the United States.

1.1.1 Adoption Rates

Child health providers—specifically, pediatricians—are thought to be slower than general practice providers in adopting electronic health record technology (DesRoches et al. 2008; Kemper et al. 2006; Menachemi et al. 2006; Nakamura et al. 2010). Children's hospitals, because they tend to be urban and academic, are often ahead of the adoption curve as are institutions of larger size (Nakamura et al. 2010; Jha et al. 2010). The reason for the slower adoption in pediatric practices probably relates to the difficulty of fitting child health needs into a product designed for general, adult care. In this way, current EHRs violate the pediatric dictum that "children are not small adults." If EHRs are not designed or cannot accommodate the unique needs of the pediatric population, healthcare providers are not likely to be quick adopters of such systems. Estimates of pediatric adoption of fully-functional EHRs in ambulatory practice in 2012 in the United States are at about 6% (Leu 2012).

The U.S. Meaningful Use program of the HITECH Act (HHS et al. 2009) of the American Recovery and Reinvestment Act of 2009 intends to provide financial stimulus to physicians and hospitals to adopt EHR technology. In this program, there is a Medicare program and a Medicaid program; since pediatricians do not generally see Medicare patients, child health providers and hospitals would qualify for the Medicaid program. In this program, individual providers may qualify for an incentive payment, if they have a minimum of 30% Medicaid patient volume, or, if they are pediatricians, 20% Medicaid volume. This criterion does covers about half of the office-based pediatricians in the United States (Finnegan et al. 2009), but does leave out a significant number with very low Medicaid volumes. These providers tend to practice in more affluent areas, but pediatrics is not a specialty with very high margins under the best of circumstances, so Meaningful Use will not directly affect the adoption rates for this large group. Member survey data from the American Academy of Pediatrics estimate that up to 2/3 of U.S. pediatricians may be eligible for some incentive payment (Kressly et al. 2009), so the next few years may be a time of rapidly increasing pediatric deployment of EHRs.

1.1.2 The Pediatric EHR Market

The pediatric EHR market includes small pediatric practices of one or two practitioners all the way up to large, hospital-owned practices of hundreds of pediatricians. There is similar variability in the crowded EHR vendor market, where a given company specializes in offering its product to practices of a certain type or specialty area. In the early 1990s, almost all electronic medical record systems were of the home-grown variety. Today, several hundred companies offer EHR software and the services that accompany its deployment, customization, and maintenance. While there has been some vendor dropout and consolidation, the EHR marketplace is far from the point where only a few major companies service the majority of customers. Because of the small size of the pediatric EHR market, there have been very few companies that have succeeded in marketing a product that is specific to pediatrics.

1.1.3 Vendor Systems

EHR systems today are sold by software vendors attempting to gain a large enough customer base to sustain a business. While this model provides a more sustainable source for software than the home-grown model, it creates a problem for child health providers: Most customers are not pediatric, so most EHRs are not designed specifically for pediatric care. A further problem for child health researchers is that practically none of these systems are designed with research in mind. Instead, they are designed for patient care and the administrative tasks that support patient care. Figure 1.1 is a mock-up of an EHR screen that highlights these assumptions.

While these assumptions are not truly prohibitive of these systems' use in a pediatric environment, they often force workarounds that affect the quality of data in the system and, therefore, can affect EHR-dependent research. For example, when faced with the unavailability of an adequate field to capture a concept, one may feel forced to use a free-text field intended for some other purpose to store the information. In this case the data loses integrity (such as a conversion from structured to unstructured data) and it becomes impossible to apply computational methods to the data.

Child health professional groups have attempted to promulgate catalogs of functionality necessary for the care of infants and children (CCHIT 2010; Spooner 2007; Spooner and Classen 2009). Fortunately, vendors who sell systems to pediatric practices and children's hospitals are gradually creating mature systems that respond to their customers' pediatric-specific needs.

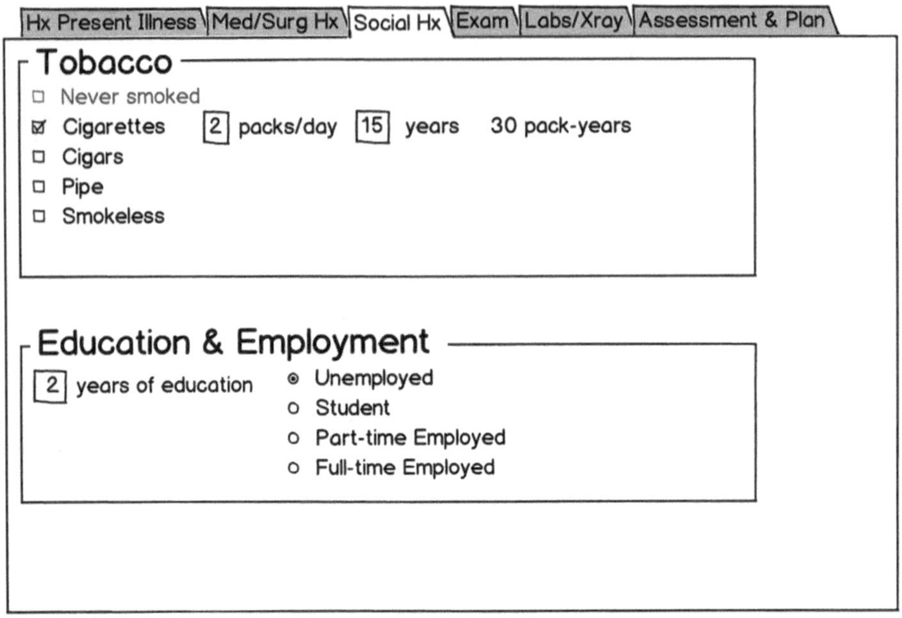

Fig. 1.1 Elements of an EHR user interface that imply an exclusive orientation to adult patients. In the case of tobacco history for an infant, one would be interested in recording passive smoke exposure, which is not included in this display. In the education section, it is implied that one's years of education are fixed and in the past, as they would be for most adults

1.1.4 Homegrown Systems and Publication Bias

Despite the prevalence of vendor systems in the marketplace, the bulk of reported literature on the use of EHRs is based on experience with homegrown systems (Friedlin et al. 2007; Gardner et al. 1999; Miller et al. 2005). The result is that the evidence on which to guide the implementation of EHRs is only partially applicable to most installed systems. Add to this the complexities of systems customized for pediatric care and the connection between the results of the adult-oriented, home-grown software and installed, vendor-provided systems is even more tenuous. This phenomenon makes the pediatric EHR environment ripe for research to be conducted on the systems themselves, but it also makes it hard to definitively answer questions about what works best. As such, reports in the informatics literature should be critically analyzed to determine the external validity of published results, in particular whether the system being tested or described is a vendor solution or a home-grown application.

1.1.5 Pediatric Versus General Environments

The main features that differentiate the pediatric environment from that of general adult care are:

- Diseases and conditions more prevalent in the young. Congenital disease and conditions related to abnormal growth and development are not usually part of adult care. Templates, data fields, and other clinical content in an EHR may therefore require customization to meet different clinical needs.
- The pediatric environment has different caretaker roles from most adult care. Parents (or guardians) are always involved in encounters and in care decisions. Siblings may receive care at the same encounter.
- The pediatric patient is growing and developing physically and mentally at a fast clip. Weights change rapidly, especially in the first year of life. Developmental capability to participate in self-care increases with age. Because of children's dependent status, social situation has a much greater impact on health than in most adult care.

1.1.6 Pediatric Subspecialties Versus the General Purpose EHR

If it were not difficult enough to apply pediatric assumptions to general-purpose systems, the difficulty is compounded in the case of pediatric specialties. Specialty care entails more detailed, less common, and oftentimes more granular, special requirements. There is also more variation of care practices at the subspecialty level because there tends to be less evidence available to standardize procedures and protocols. It is not uncommon for several physicians within the same group to have differing opinions on best practices, when little evidence exists to guide the way. In many cases there may also be a paucity of pediatric research (as compared to adults), further complicating the issues of standardization.

In pediatric specialties, the clinical content of the practice may be quite different from their adult counterparts. Pediatric cardiology, for example, is chiefly concerned with congenital disease, whereas adult cardiology focuses more on acquired cardiovascular disease. This shifting of focus on disease etiology and pathology disallows any loose extrapolation and adoption of adult data to the pediatric population.

1.1.7 Data from Natural Workflow Versus Research

Because EHRs are designed to support clinical care, data that makes its way from the EHR into a data repository is of lower quality than what one might find in data specifically collected for research. Data validation, completeness, and standard processes are very much secondary to successful completion of clinical work. It is for this reason that most research from clinical environments is based on claims data, where data are discrete rather than free text and some energy is expended to ensure its accuracy and completeness. Of course, claims data is at least one step removed from the important clinical details of the patient encounter.

1.2 Workflow and the EHR

The function of an EHR is not primarily to serve as a data-entry tool. Its purpose is to facilitate patient care for individual patients. In doing so it offers some opportunities for data extraction for other purposes (operations analysis, research, quality measurement).

1.2.1 Data Entry

Since EHRs are not designed for research, analytics, or population management, there will always be a need to input research-specific EHR data into the data repository, as well as methods to extract it. The value of that data is directly related to the quality of the data entry. Missing values threaten the validity of any measures based on the data. Data cleansing is a time and resource-consuming endeavor. For this reason, it is best to use data that are already captured reliably (like orders for tests) or to make workflow changes to increase reliable data entry. In a busy clinical environment where clinicians are already working at capacity to meet documentation guidelines for billing, there is little opportunity to make these changes. Clinicians will often ask for a "hard stop" reminder to enter data, but the effectiveness of alerts is very limited (Strom et al. 2010) and hard stops are usually abandoned as annoying. Any effort to make sense of EHR data must take into account some knowledge of how the clinical work produced it.

1.2.2 Multiple Job Roles and Their Interaction with the Record

Like the paper record, the electronic record accepts input from people in multiple job roles: physician, advanced-practice nurse, physician assistant, nurse, medical assistant, and multiple clerical roles, among others. Effective data organization depends on clear job roles related to the record. For example, if it not clear who is responsible for the accuracy of the patient's medication list, the data extracted will be of low quality. When one puts together a plan for the use of EHR data, part of the workflow analysis should include the establishment of how clear the job roles are. Job roles, or "profiles" as EHR systems refer to them, usually define how data is viewed and input in the user interface. When this variation occurs, it is not unusual for data to be entered (or not entered) in multiple ways. Great attention should be paid in designing or customizing these screens and standardization of entry and viewing carried out whenever possible.

1.2.3 Special Issues in Pediatric Workflow

***Multiple patients per encounter*:** Siblings within the same family are often seen together, especially for well-child care. In no other area of medicine is this type of multi-encounter a common experience. EHRs can be configured to allow access to multiple patient records at once, but data sharing between patients is not typically supported. In the pediatrics, there are areas of EHR data that ought to be shared between patients, like family history and social history, or guarantor information, but typically this must be entered separately for each patient.

One example where linking of records could be helpful in both the adult and pediatric EHR would be to provide the capability to update and/or duplicate the family history section in related patients' charts. For instance, if two siblings are taken care of by the same practice, family history in their direct ancestors would be identical. If the records were linked through an EHR social network, updated data in one sibling's chart could offer a prompt in the other sibling's chart that useful information needs to be verified and inserted into the record. This form of networking could also prove helpful in generation of genograms. In a more practical fashion, duplication of pregnancy circumstances and perinatal events in the charts of twins would reduce large amounts of manual data entry. There are a variety of medico-legal and ethical concerns with these kind of linkages that will not be addressed here, but the reader should be aware of the current paucity of this functionality and its implications in research data obtained from EHRs.

***Multidisciplinary clinics*:** The large number of rare disorders seen in pediatrics, coupled with the relative rarity of pediatric specialists with expertise in these disorders, creates the need to bring multiple specialists together into a single patient encounter. Arranging visits this way is a great convenience to the family, but also allows specialists to work together on difficult multi-organ system problems that might otherwise take months to coordinate. In children's hospitals, numerous clinics of these types are created or the constituents modified every year. EHR systems should support this kind of workflow, but since it is not typical in adult or non-specialty care, it is not a smoothly implemented, standard feature of most EHRs.

1.3 Special Functional Requirements and Associated Data

The following section describes some of the functionality and associated data requirements that are, for the most part, unique to pediatric EHRs. We discuss both basic functionality that should be required, as well as optimal, ideal functionality that would greatly increase the quality of data captured in EHRs.

1.3.1 Monitoring of Growth

Basic growth-chart functionality**:** Perhaps the one clinical activity that distinguishes pediatric care from adult care is growth monitoring. While weights, skinfold measurements, and even height are measured in adult care and tracked, the assumption is that these are stable measures. Growth and development are fundamental processes in pediatrics, especially in the ambulatory setting. The rapid progression in both are carefully tracked in the longitudinal health records and constantly evaluated for normalcy or deviation from expected patterns. As such, it is expected that optimal EHRs have the ability to robustly track and identify both healthy and pathologic growth. In children, of course, there are growth patterns that constitute a wide range of normal, and growth patterns that signify disease. Some diseases, like growth hormone deficiency, affect growth directly. Others, such as inflammatory bowel disease, affect growth negatively through catabolic and energy-consuming inflammatory processes. Other abnormal growth patterns are part of inherited conditions like Prader-Willi syndrome (obesity) or Turner syndrome (short stature). In routine, well-child care, examination of the growth chart is standard practice. In the ongoing management of specific, growth-affecting conditions, growth chart analysis is similarly routine. EHRs that intend to support pediatric care must support display of these data in a way that goes beyond a simple time plot of the values. Critical to the functioning of a growth chart display is concomitant display of growth norms, in order to allow interpretation of patterns (Rosenbloom et al. 2006).

Data found in growth chart**:** Weight and stature are the very basic data tracked in growth charts, but the concept of height for patients who cannot stand (or stand cooperatively) is usually conceptualized as length. Norms for young children (less than 2 years old) are typically separated from those of older children in this respect. In a typical EHR, there are growth charts for children 0–36 months old and for those over 2 years old. Data storage for the points that are plotted on the stature curves may therefore vary as to which is a height and which is a length. See Fig. 1.2 for examples of typical growth charts inuse in an EHR. The essential function of the growth chart is to give the user a sense for where the patient falls within a similar age population, expressed as the percentile at that age. Values higher than 95% or so or below 5% or so are considered abnormal, but must of course be interpreted in the context of the child's overall growth. For example, if a normal child happens to be small, owing to their genetic predisposition, they may never rise to a particular predetermined percentile. Their growth may be considered normal as it hovers around the 2nd percentile throughout life. Such tendencies are referred to as "following their own curve"; in fact, departures from that curve into "normal" range may indicate an abnormal state for that patient. It is this complexity that makes growth charts irreplaceable by number-driven decision support. There does not appear to be a substitute for a clinician viewing the curve graphically against a set of norms.

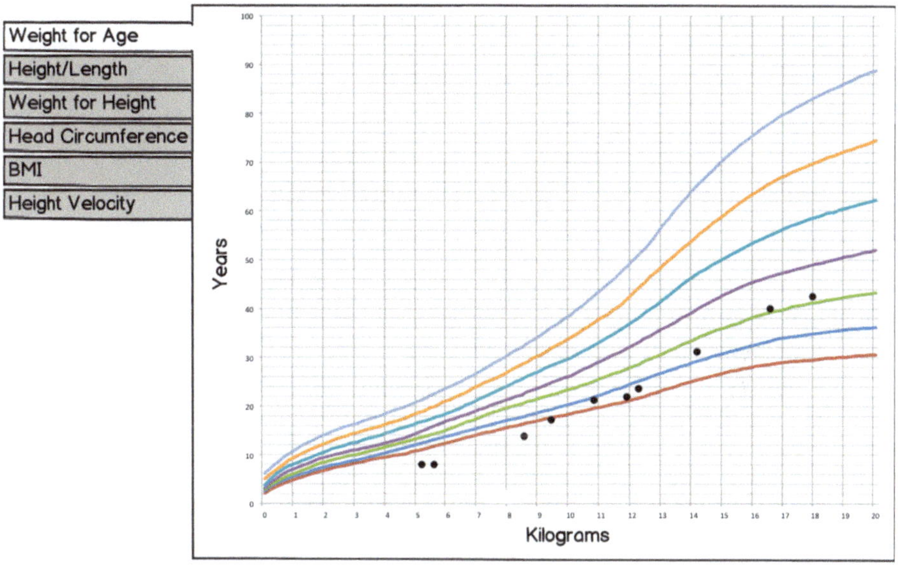

Fig. 1.2 Mockup of a growth chart as deployed in an electronic health record system. The isobars represent constant age-specific percentile for the metric (in this case, weight). In this case the patient has crossed the 3rd, 10th, and 25th percentile isobars. This might represent an abnormal growth pattern (gaining too much weight) or recovery from chronic illness to a normal weight, depending on the clinical situation

Head circumference is also essential for basic growth chart functionality. In standard growth charts used in general pediatric care, these charts go up through age 36 months. There are norms for older children and young adults (Nellhaus 1968; Rollins et al. 2010) but these are used only in specialty practices like oncology or neurosurgery to monitor head growth after radiation or tumor removal.

Body mass index, calculated from weight and stature, is also becoming a standard growth chart in pediatric practice. In adults, when BMI is used as an index of the severity of either obesity or malnutrition, the cutoff values to indicate abnormal body mass index are the same for all ages. In children, interpretation of BMI rests on the percentile value within the child's current age. The U.S. Centers for Disease control publishes these norms (CDC 2012) so that graphs can be made and percentiles calculated.

Height velocity. In specialized areas of pediatrics where growth is the focus (e.g., endocrinology) there are normative curves, implemented much like the curves for primary anthropometrics, for the rate at which height changes over time. These curves are used to evaluate the severity of growth impairment and to monitor the use of drugs that might affect growth one way or the other. There are no published curves for weight velocity, although the current interest and prevalence of obesity in the U.S. may change that.

Other anthropometric values. Norms for chest circumference, skinfold thickness, and leg length have been developed, but are used infrequently. In any case, the techniques for display, where data are displayed against normative curves, remain the same.

Percentile/Z-score calculations**:** While plotting primary data against norms makes for an effective visual display to support clinical decisions, information systems can compute the applicable percentiles given a measurement and an age, provided the proper normative data are available for the calculation. The U.S. CDC provides tables for this data for the datasets they publish, and a process for computing the percentiles (CDC 2012) (see the WHO vs CDC subsection below). Most growth charts are published merely in graphical form, and the data required to perform the computation are not provided. The computation process calculates a z-score (number of standard deviations above or below the mean for an age) and then applies assumptions about the distribution to come up with a percentile within that distribution. For extremes of growth, the z-score itself may be more useful, since the difference between a weight at the 99.96th percentile may be hard to distinguish from a weight at the 99.99th percentile otherwise. Few EHRs provide the z-score directly, but it is a desired functionality for pediatric specialists who focus on growth.

1.3.2 Data for Special Populations

Up until now, we have discussed EHR functionality associated with normal growth. In this subsection, we address the topics of collecting and managing special population data.

Congenital conditions**:** Disordered growth is a major feature of a variety of congenital conditions such as Noonan syndrome (Ranke et al. 1988), Laron dwarfism (Laron et al. 1993), and Williams syndrome (Martin et al. 2007). The measurements are the same, and the growth charts work the same way, but the normative data are different. EHR systems generally provide some of these normative curves that can be selected manually or automatically depending on clinical conditions.

Extremes of growth**:** In conditions causing extreme failure to thrive or in obesity, the usual normative curves that express curves close to the 99th and 1st percentile may not be adequate. In these cases, the data points are so far removed from the highest and lowest curves that users find it difficult to describe patterns based on the curves. In these cases it is better to create normative curves based on z-scores, so that users can express the patient's growth state relative to the position of these curves far outside the normal range.

Intrauterine growth**:** Similar to post-natal curves, intrauterine curves, based on gestational age, combined with parameters measurable via ultrasound (crown-rump length for stature or biparietal diameter for head size) are useful for expressing growth patterns of fetuses. These sorts of curves are more often found in system designed for obstetric use, but may be useful in the immediate post-natal age.

***Growth Charts: WHO versus CDC*:** The World Health Organization has published a set of growth charts for infants that are based on a sample of healthy, breast-fed infants (Grummer-Strawn et al. 2010) The motivation for creating these charts is to present a more physiologically accurate view of how normal infants should grow. Because the U.S. CDC growth data has been in use much longer, EHR system vendors have had to deal with the ambiguity of two widely accepted growth charts for normal infants. Researchers using percentile growth data from EHRs should be aware and take note of the source in order to make accurate comparisons.

***Specialized growth analysis*:** Growth chart data must sometimes be temporally related to other physiologic events. For example, one may want to indicate on the growth chart a child's sexual maturity rating, since advanced stages of sexual maturation are associated with cessation of normal growth. One might also want to indicate the "bone age" (an estimate of age based on the appearance of bones on plain-film radiography) on the growth chart in cases where the age of the patient is uncertain, as in some cases of international adoption. There are no standard ways of displaying these data within a growth chart, but practitioners who focus on growth cite this function as essential to full growth chart functioning (Rosenbloom et al. 2006).

***Correction for gestational age*:** Infants born prematurely, because of their smaller size, require special growth charts (Fenton 2003). Outside the immediate post-natal period, though, practitioners generally use regular growth charts, and graphically indicate a correction factor for prematurity. The expectation is that premature infants will eventually catch up to other infants of the same postnatal age. Part of the analysis of growth charts in premature infants is the time it takes them to achieve this catch-up growth.

1.4 Drug Dosing

Given the inherently changing growth of children, prescribing the appropriate dose of medications can be difficult. What follows is a discussion of the practical and research implications of prescribing medications through an EHR.

1.4.1 Weight-Based Calculations

Medications in infants and small children are generally dosed by body weight. As body weight increases with age, children grow into the adult dose; the weight at which one can receive an adult dose varies by medication. Such weight dependence makes the act of prescribing medications to young people more complex. In addition to the act of prescribing, there are complexities related to the storage of the prescription and the decision support that might be provided to the user. EHRs used in the care of children should, at a minimum, provide the ability to calculate a given dose of a medication based on the actual weight (Spooner 2007; Shiffman et al. 2000). More advanced functionality includes providing standard weight based doses, offering dose range checking, and providing dose ranges dependent on clinical factors, like diagnosis.

***Changes in weight*:** As infants grow, their body weight changes rapidly enough that they may "grow out of" a medication at a given dose. Providers who care for infants on chronic medications know to re-prescribe when body weight changes, but a sufficiently sophisticated information system can help to support the decision to re-prescribe, or at least to make it easier by carrying forward weight-based dosages to subsequent prescriptions. Data structures used to store prescriptions must therefore retain the weight-based dose (e.g., 40 mg/kg/day) as data.

***Dosing weight*:** It is not always the case that actual body weight is the best datum to use in calculating weight-based prescriptions. In very young neonates, who lose significant amounts of weight as they adjust to life outside the womb in the postnatal period, one may prefer to use a "dosing weight" to keep prescriptions consistent. Similarly, patients who gain weight due to edema will need a more appropriate weight upon which to base dosing decisions. EHR systems need to take into account different methods of storing and using weight in support of prescribing.

***Daily versus single-dose reasoning*:** In dose-range decision support, there are limits for single doses and for total daily doses, both of which must be accounted for in decision support. Pediatric prescribing guidelines are usually written in mg/kg/day divided into a certain number of doses. This format takes into account the per-dose and daily dose parameters, although EHR dosing rules may provide these two parameters separately.

***Power-of-ten errors*:** In providing small doses to small people one of the most common and most dangerous dosing errors is the situation where the dose is higher by a factor of 10 or 100, due to confusion between the volume to administer (e.g. 2 mL) and the mass to be administered (20 mg), faulty multiplication, or the migration of a decimal point (Dart and Rumack 2012). In adult care, doses tend to be standard, so there is no way for practitioners to recognize the "usual" dose, since there is no usual dose. Dosing decision support in EHRs is mainly intended to mitigate these kinds of errors.

1.4.2 Physiologic Variation with Development

A subtle factor that affects some pediatric prescribing, whether in routine clinical care or research, is the effect of maturation of organ systems in the immediate post-natal period that affect drug clearance rates. In order to provide adequate decision support for these dose variations, the ideal system would need to be able to compute different ideal doses at ages measured in days or even hours, and to take into account prematurity. For example, for the antibiotic gentamicin, which is commonly prescribed to neonates in the intensive care setting, one common guideline is that a premature neonate weighing less than 1,000 g at birth would get 3.5 mg/kg/dose every 24 h, but a term neonate less than 7 days old and weighing more than 1,200 g would get 2.5 mg/kg/dose every 12 h, but the same infant over 7 days old (but less than 2,000 g at birth) would get the same dose every 8–12 h (Taketomo et al. 2011). It's easy to see how such complex rules can be difficult to model in a prescribing system, and difficult to present to users in an intelligible way.

1.4.3 Off-Label Use

Vendors of drug-dosing decision support databases, commonly used in EHR and e-prescribing products, offer guidelines for dosing that are used in decision support. Because many of the drugs used in pediatrics are not actually approved for use in children under a certain age, it can be seen as controversial for a vendor to provide a non-FDA-approved dose range. Because of the absence of FDA-approved dose ranges, local variation in these ranges is common. Such variation makes it even more difficult for drug-dosing decision support database vendors to provide this decision support confidently. The result is a database with incomplete data, where users of EHRs that use these data must fill in the blanks. Across data from multiple institutions, tremendous variation is seen in the dosing rules that are brought to bear on these prescriptions.

1.4.4 Metric Versus English Controversy in the U.S.

Because of the dependency of changing body size on therapies, pediatric clinicians are in the habit of using metric-system measures for weight, height, and other measurements. Dosing guidelines are typically in mg per kilogram per day, and liquid concentrations are expressed in milligrams per milliliter. The American public, however, has not taken up the metric system, so child health providers find themselves converting weights from kilograms to pounds, and doses of liquid medicines from milliliters to teaspoons. This conversion offers opportunity for error in the communication between physicians and families. It also offers a source of error in the data that is recorded for prescriptions. Systems in an academic medical center will typically adhere carefully to metric units, but systems in community settings are more likely to store dosing guidelines and prescription records in terms of these imperial units. Merging data across sources must therefore take into account this conversion.

1.4.5 Compounded and Combined Forms of Medication

Owing to the inability of young children to swallow pills and the impracticality of the pharmaceutical market or researchers to provide liquid forms for all conceivable drugs, a small but significant number of medications must be converted to liquid form by a compounding pharmacy. Because the formulas for these compounded medications are not standard, the records for these drugs embedded in the EHR are not standard. Even within an institution, there can be multiple instances of compounding of the same medication that make comparison of prescribing data complex. Combination preparations, where more than one drug is in a preparation, are particularly common among compounded medications. Decision support aimed at one component of a combination medication may not be appropriate for the other components of the preparation, and users may be uncertain as to which component is the target of the decision support. The data model for the data in any analysis has to take these complexities into account.

1.5 Special Requirements for Decision Support in Pediatric EHRs

Rules put in place to guide prescribing decisions in child health need to take body mass and age into account. As with any factor that makes decision rules more complex, the maintainability of the corpus of rules quickly outstrips the ability of any organization to maintain these rules. An effective general strategy for managing this complexity remains an unsolved conundrum (Conroy 2007).

1.5.1 Management of Immunizations

***Decision support to determine currency of immunizations*:** While adults receive immunizations according to schedules and risk factors, the complexity of the decision-making about which immunizations to give at what time is an order of magnitude greater. This is partly due to the higher number of targeted pathogens in immunization preparations, but also to the greater number of vaccine products on the market, the changing nature of vaccine guidelines, and the complexity of the temporal reasoning required for effective immunization administration.

***Guidelines for administering the rotavirus vaccine*:** These rules illustrate the complexity that must be supported in an information system designed to give decision support for the administration of these medications. It is also illustrates the common observation that some of the concepts used in the decision support are not computable from available data ("whenever possible," "clinically stable").

Table 1.1 Rotavirus vaccine administration rules (Cortese et al. 2011)

- For Product 1, there are three required doses, to be given at 2, 4, and 6 months.
- For Product 2, there are two required doses, to be given at 2 and 4 months.
- For Product 2, there are two required doses, to be given at 2 and 4 months.
- The series should be completed with the same product whenever possible.
- The minimum age for administration of the first dose of either product is 6 weeks 0 days.
- The maximum age for administration of the first dose of either product is 14 weeks 6 days; if this deadline is missed, the vaccine should not be administered.
- The minimum interval between administrations of either product is 4 weeks.
- There is no maximum interval between doses, but the maximum age for administration of the final dose of either product is 8 months and 0 days
- Rotavirus vaccine may be give concomitantly with DTaP vaccine, HiB vaccine, IPV, hepatitis B vaccine, and pneumococcal conjugate vaccine
- Preterm infants can be immunized at the same schedule as term infants, based on the preterm infant's chronological (post-delivery) age. The preterm infant should be clinically stable, and should be given no earlier than the time of discharge from the neonatal intensive care unit or nursery.

***Decision support to schedule immunizations*:** The decision about what immunizations to deliver today is different from the decision about when the next doses are due. A convenient way to simplify this decision-making has evolved by way of the recommended schedule of well-child visits (Haggerty 1987). When new vaccines are introduced, their schedule conforms to this schedule in order to make it easier to administer. The rotavirus vaccine cited in Table 1.1, for example, conforms to the standard pattern of "well-baby checks" at 2, 4, and 6 months familiar to most parents. If a patient is able to stick to the prescribed schedule, there is little need for decision support to for what ought to be given at the visit and when to return for the next immunizations. Unfortunately, the real-life ability to adhere to this schedule is low (Selden 2006) so child-health providers are left with a great deal of decision making that they expect their information systems to help with. Only a minority of current EHRs do so (Kemper et al. 2006). This deficiency is due to the complexity of the logic required for sensible recommendations and the dependency on that logic on manually entered data.

***Immunization Registries and information exchange*:** One solution to the problem of missing immunization data and recommendations for current or future doses lies in the technology of immunization registries, now more widely known as immunization information systems (IIS) (Yasuda and C.o.P.a.A. Medicine 2006). Because of the state-based organization of public health programs, and the state-based organization of the Medicaid program, these IIS are usually established to operate within a specific state or a region within a state. The state-based nature of these systems presents challenges to usability in populations that live near state borders. It also means that resources available to administrators of these systems are as constrained as any state-funded program (Blavin et al. 2011). The case for IIS is that since each patient typically receives immunizations in a variety of settings (doctors' offices, public health clinics, immunization fairs, schools, pharmacies) a unifying system will allow all providers to make decisions about who needs immunizations, and public resources can be directed to improve immunization rates in the highest risk areas. Standards exists for the transmission of immunization information (ref: HL7 v. 2.5.1 Implementation Guide for Immunization Messaging) but are usually customized in a way that makes interoperability between systems difficult. The U.S. federal Meaningful Use program (HHS et al. 2009) at Stage 1 requires providers to conduct a test of data transmission to a public health agency; one choice in the menu of items aimed at this goal is the state immunization registry. While this is a far cry from requiring full interoperability, it is a step in the right direction toward encouraging the development of mature data sharing. In the meantime, providers who implement EHRs are faced with the dual challenge of getting local logic set up to support improvements in immunization rates while providing data to state IISs using manual methods and batch uploads.

1.6 Challenges to Identification of Patients in Pediatric EHRs

1.6.1 Newborn-Infant Transition

Newborn infants must be registered with a new medical record number (MRN) and a suitably distinctive name at the time of birth to allow normal clinical workflows. Typically these infants are assigned a name based on the mother's name, as is "Boy Jane Smith" or "Jane Smith Twin A." While the infant retains his or her MRN after this temporary name is changed to the child's given name, the MRN remains unchanged, but clinicians tend to assign more salience to a name than to a number. This change can make it challenging to integrate information across the perinatal period, especially when the venue of care changes (e.g. from the nursery to the doctor's office). EHRs used in the care of newborns must allow users to track patients based on either name. This is functionally similar to the tracking of adults who change their names after marriage, but in this case it happens to practically all individuals. At stake in this identification process is the newborn screen results, sent in for analysis in the first few days of life, but whose results come back to the outpatient provider well after the baby's given name is established.

1.6.2 Fetal-Newborn Transition

The rise of techniques useful in the diagnosis and treatment of fetal problems presents special problems for healthcare record-keeping. Typically, patients do not receive a distinct record in the system until after they are born. Records maintained on fetal procedures are usually stored in the mother's chart, as is appropriate at the time of the procedure. Tying the information about fetal procedures to the record of the infant after birth requires at least a linkage between the mother's chart and the baby's. Ideally, there should be a way to split out information from the mother's chart into the infant's chart, in a way that preserves the usual privacy standards (just because it is clinically appropriate for one to access one person's chart does not mean it is appropriate to access another's). Currently in EHR systems this kind of fine access control is not technically possible. The clinician is left with the task of manually extracting information from the mother's chart into the baby's, pending development of systems that take fetal care into account.

1.6.3 Maternal-Fetal/Infant Linking

There is one circumstance where access to one person's chart entails some access to another's. In newborn care, there are elements of the mother's perinatal chart that are just as clinically relevant to the baby as the mother: Group-B Strep colonization status of the mother at birth, HIV status of the mother, medications given to the mother around the time of delivery, and so forth. While one can tie charts together in some EHR modules specifically designed for perinatal care, the movement of this information into the baby's chart in a way that would make this information extractable for analysis or available for decision support does not exist in general purpose EHR systems. Manual abstraction or a workaround using an external system is usually what is needed to support these data functions.

1.6.4 Linking Pediatric-Adult Care

Another instance where charts need to be linked across venues is when a pediatric patient "graduates" to adult care (Cooley (Conroy 2011)). Currently, from the information system perspective, the best practice is to transmitting a care summary using a Continuity of Care Document (CCD). In current technology there is no practical way to transmit the entire corpus of information on a particular patient.

1.7 Supervision of Health Care of Children

1.7.1 Developmental Monitoring

A central feature of health supervision is screening for developmental delays. These delays in speech, motor function, or other abilities may indicate primary neurodevelopmental disease or be secondary to systemic disorders or socioeconomic factors. In any case, it is in the domain of the primary-care child health provider to screen for delays and to refer to needed services, like audiology, speech pathology, or neurology. The best practice for this activity is to use standardized developmental evaluation instruments—questionnaires—that can be filled out by the clinician or in some cases the parent. From the data perspective, the problems to be solved include how to marry parent-entered data into the medical record of the child, how to share developmental screening data for public health purposes, and how to track a very diffuse referral and follow-up process to ensure that no patients' needs go unaddressed.

1.7.2 Newborn Screening

Another screening process performed on practically all newborns in industrialized countries is newborn screening for congenital disorders. Often called "metabolic" screening, because of the emphasis on such metabolic diseases as phenylketonuria and hypothyroidism, most newborn screening programs now include screening for hearing loss (albeit not via a blood sample). Since these programs are state-run in the U.S., each state has a slightly different panel of disorders that it screens for. Challenges in the management of data for newborn screening include correct identification of the infant as he or she changes names, correct identification of the follow-up provider, presentation of the data in a way that can be stored and acted upon, and interoperability between state databases and provider EHRs. In the current environment, there is not widely implemented technical solution for any of these problems.

1.7.3 Well-Child Care

***Applicable guidelines*:** The American Academy of Pediatrics has published recommendations for well-child care according to a schedule for many years (AAP Committee on Practice and Ambulatory Medicine and Bright Futures Steering Committee 2007). State Medicaid programs expect that this schedule of visits (typically at birth, 1–2 weeks, 2, 4, 6, 9 and 12 months, then 15 and 18 months, then at 2 years and annually thereafter) be provided to their beneficiaries. The immunization schedule is arranged around the assumption of this schedule of visit, as are other guidelines for screening (anemia, developmental delay, lead poisoning, etc.). In addition to the timing of these visits, the AAP recommends what clinical events should occur at these visits: measurements, sensory screening, lab tests, and advice for parents. The Bright Futures guidelines (http://brightfutures.aap.org) provide more detail on the content of these visits.

***Required data*:** Insurers and State Medicaid agencies set standards for the content of these well-child visits. In Medicaid programs, these requirements are embodied in the Early and Periodic Screening, Diagnosis, and Treatment program (EPSDT). Audits enforce these standards, but some agencies are requiring reporting of the actual data collected in these visits, such as the records of immunizations given, screening test results, and diagnoses. A messaging standard has been developed for this reporting (ref California EPSDT format). Quality measures for pediatric primary care are built on the assumption that this schedule of well-child visits is occurring.

1.8 Terminology Issues in Pediatric EHRs

Like all specialized areas of health care, pediatrics uses terminology differently from other areas. While there are special terms used in patient histories and exams, those terms generally live in free-text portions of the record. Terminology specializations are most obvious in the regions of the EHR that focus on diagnoses (such as a problem list, past medical/family/social history, or billing section). For a diagnostic terminology system to be usable in child health, it must:

- Allow detailed descriptions of family historical risk factors
- Be descriptive of specific congenital syndromes and their subtypes
- Have detailed descriptors of anatomic anomalies that may lie outside of syndromes
- Allow description of chromosomal anomalies
- Describe patient behaviors that represent risk factors for undiagnosed behavioral disorders
- Describe family stressors that may affect child health and development
- Describe maternal conditions that affect the infant (e.g., "Maternal Group B Strep colonization")
- Describe developmental, educational, or anthropometric states (none of which may be a disorder *per se*) throughout the lifespan

1.9 Pediatric Quality Measurements and the EHR

Quality measurement has been an important part of EHR implementation for adult care providers for many years (McColm and Karcz 2010) but the maturity of quality measures applicable to children is far behind that of adults. Part of this is due to the fact that outside of asthma and attention deficit hyperactivity disorder, chronic disease in children consists of many small populations, in contrast to the large adult populations entailing diabetes, coronary artery disease, congestive heart failure, osteoporosis, and other high-prevalence adult conditions. For these numerous smaller populations, there are few simple proxy measures available to measure outcome, as one can do with diabetes (through the value of the hemoglobin A1c blood test) or process (as one can do with osteoporosis with data on the timing of bone-density studies). Nevertheless, there is increasing interest in pediatric measures that can be extracted from EHR data (Fairbrother and Simpson 2011; Kallem 2011) and the U.S. Federal Meaningful Use program is attempting to include more pediatric measures in its Stage 2 program (CMS 2012). As with any other measure, reliable data entry is prerequisite, so any data entry outside the normal workflow of ordering procedures, receiving lab test results, or prescribing medications is apt to be invalid. Special improvement programs aimed at data collection quality would usually be necessary to get to an acceptable level of validity if clinician data collection is expected.

1.10 EHR Systems as Challenges in Translational Research

Electronic health records represent a major advance over paper-based records in support of patient care. Clinical data about an individual patient can be captured longitudinally and stored in a well-architected clinical data repository. This includes medical history, physical examinations, progress notes, laboratory data, diagnoses, medications, etc. In theory, electronic records can be easily shared among health care providers and information can be made available whenever and wherever it is needed to care for a patient. It is no longer necessary to chase paper records. The EHR also has the ability to provide decision support, for example, through use of computerized reminders to ensure regular preventive and screening practices, alerts to improve compliance with best practices and to identify drug interactions, and suggestions that facilitate diagnosis and treatment.

EHRs are designed to support clinical care within an institution. Data from the EHR can also be mined to measure outcomes of care and to support quality improvement initiatives within the institution. However, EHRs have characteristics that impede many types of clinical research, especially research which requires that data be aggregated from several different institutions. Participation in a network of multiple institutions is commonly required in pediatric translational research to accrue sufficient numbers of patients with a specific condition to reach statistically significant conclusions. Almost all EHRs in the United States are products of commercial vendors, with a few companies dominating the marketplace. They are referred to here as EHR systems rather than EHRs because they generally consist of a series of modules that interact with one another to support patient care broadly. Institutions can select and implement those modules that meet the institution's clinical needs. Examples include modules that support scheduling, general inpatient care, specialized inpatient care (e.g. oncology, cardiology), ambulatory care, emergency department, pharmacy, operating room, laboratory, radiology, and billing.

While the designs appear to be modular, they are proprietary, interoperable, and details of their internal structure and coding are not publically available.

Vendors do not design their EHR systems with application programming interfaces to integrate third-party applications (Mandl and Kohane 2012), which has reduced innovation in the field. The philosophy of EHR vendors is entirely different from the philosophy that dominates the smartphone industry, where vendors build a basic mobile computing platform with advanced application programming interfaces for running third party applications. Health care institutions are generally identified as a "Vendor A" or a "Vendor B" institution because most or all components of the EHR are products of the same vendor. Unfortunately, the EHR systems of different vendors are not designed to be semantically and syntactically interoperable, e.g. vendors do not share common data structures, vocabularies, or metadata. This means that systems of different vendors have a very limited ability to communicate with one another and to share data in a meaningful way. This impedes translational research that depends upon data sharing. Workaround solutions are expensive, but available, to create clinical data warehouses that can be federated to aggregate data to support research (Chaps. 6 and 8).

A second major challenge that EHRs pose to research is the difficulty of modifying systems for data capture. A research study is designed to answer a specific question. To do this requires capture of specific data, which may or not be captured by the EHR in a routine patient visit. To capture the data in the EHR, a form containing necessary data elements must be built into the clinical workflow and be completed by health care providers during the process of patient care. To build and support the form requires technical expertise and authorization to modify the EHR of the clinic where the data are to be collected. If multiple institutions are participating in the research study, then the EHR of each institution must be modified. If a vendor agrees to incorporate the form in their version of the EHR, it can be made available to all of its customers. However, members of a network generally use EHRs of different vendors, so there is no one solution that will permit all participants to adopt a common solution. Because vendor systems are proprietary, there are generally legal restrictions around sharing details about how a vendor's system works or sharing with others software code that was developed by investigators, but depends on a working knowledge of a vendor's internal IT systems.

1.11 Conclusions

EHRs used in the care of infants, children, and adolescents must support different functionalities than systems designed for adult care. Adaptations to these systems to accommodate pediatric clinical work will affect the type of data available. Those who seek to use data from pediatric EHRs in research should examine how well the specialty workflow is supported by the EHR in order to be able to interpret the system's output.

References

American Academy of Pediatrics Committee on Practice and Ambulatory Medicine and Bright Futures Steering Committee. Recommendations for preventive pediatric health care. Pediatrics. 2007;120(6):1376.

Blavin F, Ormond B. HITECH, meaningful use, and public health: funding opportunities for state immunization registries, C.f.M.a.M. Services, editor, Washington, DC; 2011.

CCHIT. CCHIT Child Health Workgroup. 2010 [cited 2012 Apr 1]. Available from: http://www.cchit.org/workgroups/child-health

CDC. Percentile data files with LMS values. 2012 [cited 2012 Mar 14]. Available from: http://www.cdc.gov/growthcharts/percentile_data_files.htm

CMS. Medicare and Medicaid Electronic Health Records (EHR) Incentive Programs. 2012 [cited 2012 Apr 1]. Available from: https://http://www.cms.gov/Regulations-and-Guidance/Legislation/EHRIncentivePrograms/index.html?redirect=/EHRIncentivePrograms/

Conroy S, Sweis D, et al. Interventions to reduce dosing errors in children: a systematic review of the literature. Drug Saf. 2007;30(12):1111–25.

Cooley WC, Sagerman PJ. Supporting the health care transition from adolescence to adulthood in the medical home. Pediatrics. 2011;128(1):182–200.

Cortese MM, et al. Leveraging state immunization information systems to measure the effectiveness of rotavirus vaccine. Pediatrics. 2011;128(6):e1474–81.

Dart RC, Rumack BH. Intravenous acetaminophen in the United States: iatrogenic dosing errors. Pediatrics. 2012;129(2):349–53.

DesRoches CM, et al. Electronic health records in ambulatory care – a national survey of physicians. N Engl J Med. 2008;359(1):50–60.

Fairbrother G, Simpson LA. Measuring and reporting quality of health care for children: CHIPRA and beyond. Acad Pediatr. 2011;11(3 Suppl):S77–84.

Fenton TR. A new growth chart for preterm babies: Babson and Benda's chart updated with recent data and a new format. BMC Pediatr. 2003;3:13.

Finnegan B, et al. Policy Research Brief No. 9: boosting health information technology in medicaid: the potential effect of the American Recovery and Reinvestment Act, in Geiger Gibson/RCHN Comunity Health Foundation Research Collaborative, George Washington University; 2009.

Friedlin J, Dexter PR, Overhage JM. Details of a successful clinical decision support system. AMIA Annu Symp Proc. 2007;2007:254–8.

Gardner RM, Pryor TA, Warner HR. The HELP hospital information system: update 1998. Int J Med Inform. 1999;54(3):169–82.

Grummer-Strawn LM, et al. Use of World Health Organization and CDC growth charts for children aged 0–59 months in the United States. MMWR Recomm Rep. 2010;59(RR-9):1–15.

Haggerty RJ. Health supervision visits: should the immunization schedule drive the system? Pediatrics. 1987;79(4):581–2.

HHS. Health Information Technology for Economic and Clinical Health (HITECH) Act, Title XIII of Division A and Title IV of Division B of the American Recovery and Reinvestment Act of 2009 (ARRA), in Pub. L. No. 111-5 (Feb. 17, 2009), U.S.D.o.H.a.H. Services, Editor, Washington, DC; 2009.

Jha AK, et al. A progress report on electronic health records in U.S. hospitals. Health Aff (Millwood). 2010;29(10):1951–7.

Kallem C. Transforming clinical quality measures for EHR use. NQF refines emeasures for use in EHRs and meaningful use program. J AHIMA. 2011;82(11):52–3.

Kemper AR, Uren RL, Clark SJ. Adoption of electronic health records in primary care pediatric practices. Pediatrics. 2006;118(1):e20–4.

Kressly S. Testimony from the Hearing on Health IT Adoption and the New Challenges Faced by Solo and Small Group Health Care Practices, June 24, 2009. Small Business Committee Document Number 111–032. Available via the GPO Website: http://www.access.gpo.gov/congress/house. Accessed 1 June 2012.

Laron Z, Lilos P, Klinger B. Growth curves for Laron syndrome. Arch Dis Child. 1993;68(6):768–70.
Leu MG, O'Connor K, Marshall R, Price D, Klein D. Pediatricians' use of health information technology: a national survey. Pediatrics. 2012; in press.
Mandl KD, Kohane IS. Escaping the EHR trap – the future of health IT. N Engl J Med. 2012;366:2240–2.
Martin ND, et al. New height, weight and head circumference charts for British children with Williams syndrome. Arch Dis Child. 2007;92(7):598–601.
McColm D, Karcz A. Comparing manual and automated coding of physicians quality reporting initiative measures in an ambulatory EHR. J Med Pract Manage. 2010;26(1):6–12.
Menachemi N, et al. Charting the use of electronic health records and other information technologies among child health providers. BMC Pediatr. 2006;6:21.
Miller RA, et al. The anatomy of decision support during inpatient care provider order entry (CPOE): empirical observations from a decade of CPOE experience at Vanderbilt. J Biomed Inform. 2005;38(6):469–85.
Nakamura MM, et al. Electronic health record adoption by children's hospitals in the United States. Arch Pediatr Adolesc Med. 2010;164(12):1145–51.
Nellhaus G. Head circumference from birth to eighteen years. Practical composite international and interracial graphs. Pediatrics. 1968;41(1):106–14.
Ranke MB, et al. Noonan syndrome: growth and clinical manifestations in 144 cases. Eur J Pediatr. 1988;148(3):220–7.
Rollins JD, Collins JS, Holden KR. United States head circumference growth reference charts: birth to 21 years. J Pediatr. 2010;156(6):907–13, 913.e1-2.
Rosenbloom ST, et al. Implementing pediatric growth charts into an electronic health record system. J Am Med Inform Assoc. 2006;13(3):302–8.
Selden TM. Compliance with well-child visit recommendations: evidence from the Medical Expenditure Panel Survey, 2000-2002. Pediatrics. 2006;118(6):e1766–78.
Shiffman R, et al. Information technology for children's health and health care: report on the information technology in children's health care expert meeting, September 21–22. J Am Med Inform Assoc. 2000;8(6):546–51.
Spooner SA. Special requirements of electronic health record systems in pediatrics. Pediatrics. 2007;119(3):631–7.
Spooner SA, Classen DC. Data standards and improvement of quality and safety in child health care. Pediatrics. 2009;123 Suppl 2:S74–9.
Strom BL, et al. Unintended effects of a computerized physician order entry nearly hard-stop alert to prevent a drug interaction: a randomized controlled trial. Arch Intern Med. 2010;170(17):1578–83.
Taketomo CK, Hodding JH, Kraus DM. Lexi-Comp's pediatric & neonatal dosage handbook: a comprehensive resource for all clinicians treating pediatric and neonatal patients (Pediatric dosage handbook). 18th ed. Hudson: Lexi-Comp; 2011.
Yasuda K, C.o.P.a.A. Medicine. Immunization information systems. Pediatrics. 2006;118(3): 1293–5.

Chapter 2
Interfaces and Data Transmission Standards

S. Andrew Spooner and Judith W. Dexheimer

Abstract Electronic health records in pediatric care must connect to other systems to integrate information and to communicate to providers in other systems. In order to perform this data transfer, there must be formats for the construction of messages, and standard terminologies that communicate the clinical concepts. Messaging standards are usually in the form of delimited character strings that follow the HL7 version 2.X standard. We will review the state of the art of HL7 version 2 messaging types and describe the best known example of this kind of messaging in the CDC Implementation Guide for Immunization Messaging. HL7 version 3, built on XML, has the potential for richer semantics, but is not yet widely used. We will use the Continuity of Care Document as a pediatric example of the use of this standard in real systems. Terminology systems, both open and proprietary, are used to encode clinical and administrative concepts in pediatrics. We will review terminology systems in current use and their pediatric-specific limitations.

2.1 Introduction

Data interchange and analysis would certainly be easier if all information systems used the same data model. Despite the development of the Health Level 7 Reference Information Model (Rocca et al. 2006; Shakir 1997) realities of

S.A. Spooner, M.D., M.S., FAAP (✉)
Department of Pediatrics, University of Cincinnati College of Medicine, Cincinnati, OH, USA

Division of Biomedical Informatics, Cincinnati Children's Hospital Medical Center, 3333 Burnet Avenue, ML-900, Cincinnati, OH 45229-3039, USA
e-mail: andrew.spooner@cchmc.org

J.W. Dexheimer, Ph.D.
Department of Pediatrics, University of Cincinnati College of Medicine, Cincinnati, OH, USA

Divisions of Emergency Medicine and Biomedical Informatics, Cincinnati Children's Hospital Medical Center, 3333 Burnet Avenue, ML-900, Cincinnati, OH 45229-3039, USA

J.J. Hutton (ed.), *Pediatric Biomedical Informatics: Computer Applications in Pediatric Research*, Translational Bioinformatics 2, DOI 10.1007/978-94-007-5149-1_2,

software implementation preclude such uniformity, so the informatics community must concentrate its standards-development activities on the format of data transmissions between systems (messaging) and the semantic payload contained in the transmission (terminology). In this chapter, we will review the basics of messaging standards used in health information technology, and review terminology systems of importance to pediatric care and translational research.

2.2 Messaging Standards

2.2.1 Definitions

Standards Development Organizations: Voluntary standards development organizations aim to gather consensus on the structure of messaging standards.

- ***ANSI***: The American National Standards Institute accredits standards and the organizations that produce them. It coordinates the efforts of voluntary standards-development organizations and sets standards for behavior to assure openness in processes and robustness of product. There is no pediatric-specific subgroup within ANSI.
- ***HL7***: Health Level 7 (after level 7 of the Open System Interconnection model (Zimmermann 1980)) is the most widely recognized health-related standards development organization in the world. Its messaging standards are used commonly in the data interchange between clinical systems, such as laboratory and electronic health record systems. There is a Child Health Workgroup within HL7 (HL7 2012). HL7 produces a wide variety of standards, including some that are not messaging standards. For example, the Arden Syntax for Medical Logic Systems (Downs et al. 2006; Samwald et al. 2012) is a standard way to represent medical decision-making rules, including the required inputs, the expected outputs, the text of alerts, and embedded orders. The aforementioned Reference Information Model is not itself a messaging standard, but it guides the data structure of all artifacts generated within the HL7 universe of standards.
  - HL7 messages are complete units of data intended to be transmitted between systems in a single transmission. Segments within the message contain logical groupings of data fields. A field contains up to a defined number of characters that comply with the definition of a data type (string, integer, etc.). Depending on the version of HL7 you are using, fields can be delimited as logical units within an XML message (version 3) or merely delimited by the pipe ("|") character in a character string (version 2) (Table 2.1)
  - An implementation guide is a user manual intended to instruct system programmers on how to construct compliant messages. HL7 International publishes these guides in conjunction with subject matter experts and

Table 2.1 Contrast between versions 2 and 3 of HL7 messaging standards. In each case below, the portion of the message shown is intended to convey a blood glucose value (182 mg/dL) obtained 12 h after a stress test. The message includes the normal expected range of 70–105 mg/dL. The version 3 fragment is much longer, but embeds much richer semantics. The object identifiers (e.g., "2.16.840.1.113883.19.1122.4") are standard index numbers of other standards embedded in the message, like specific terminology systems

HL7 Version 2 Message Segment

```
  OBX|1|SN|1554-5^GLUCOSE^POST 12H
CFST:MCNC:PT:SER/PLAS:QN||^182|mg/dl|70_105|H|||F<cr>
```

HL7 Version 3 XML Fragment

```
  <observationEvent>
   <id root=''2.16.840.1.113883.19.1122.4''
extension=''1045813''
      assigningAuthorityName=''GHH LAB Filler Orders''/>
   <code code=''1554-5'' codeSystemName=''LN''
          codeSystem=''2.16.840.1.113883.6.1''
          displayName=''GLUCOSE^POST 12H
CFST:MCNC:PT:SER/PLAS:QN''/>
   <statusCode code=''completed''/>
   <effectiveTime value=''200202150730''/>
   <priorityCode code=''R''/>
   <confidentialityCode code=''N''
     codeSystem=''2.16.840.1.113883.5.25''/>
   <value xsi:type=''PQ'' value=''182'' unit=''mg/dL''/>
   <interpretationCode code=''H''/>
   <referenceRange>
     <interpretationRange>
        <value xsi:type=''IVL_PQ''>
        <low value=''70'' unit=''mg/dL''/>
        <high value=''105'' unit=''mg/dL''/>
      </value>
      <interpretationCode code=''N''/>
     </interpretationRange>
   </referenceRange>
```

organizations. For example, the HL7 Version 2.5.1 Implementation Guide for Immunization Messaging describes (Savage and Williams 2011) message formats for requesting immunization records, for receiving information on candidate patient patches from a query, immunization administration records, and administrative reporting related to vaccine adverse events.

- HL7 Version 2: This version is widely implemented in all healthcare settings. It is a much simpler (but more loosely defined) model for messaging than version 3, consisting of definitions of strings that can be assembled on the sending end and parsed on the receiving end into semantically appropriate components. The HL7 version 2 messaging standard allows for

locally-defined "Z" segments not defined by any HL7 standard. Use of Z segments leads to localizations of messages that make them impossible to use without special customizations on the receiving end.
  - HL7 Version 3 and the Clinical Document Architecture: A more modern messaging standard developed by HL7 on 2000 is the Clinical Document Architecture (CDA) (Dolin et al. 2001; Ferranti et al. 2006). It is "more modern" in that it is based on a reference information model—specifically the HL7 Reference Information Model—and development of document definitions follows a standard modeling process. CDA document definitions are expressed in XML, so specimens that conform to these definitions can be represented in multiple views, including human-readable ones. In a sense, a CDA document is always more than a message, since it is, by definition a complete clinical document that exists in a clinical context. One advantage of CDA is that it supports incremental interoperability, insofar as implementers can start with a simple document, with appropriate semantic tags, and still exchange data while waiting to implement more complex semantics as the application matures. Later in this chapter we will examine an example of CDA, the Continuity of Care Document.

- ***Committee E31 of ASTM International*** (formerly American Society for Testing and Materials) promulgates standards relevant to health care. The guides produced by ASTM tend to be less clinical and more administrative, as in standards for passing transcribed text or securing transmitted health information. There is no pediatric-specific subgroup within ASTM International. ASTM and HL7 collaborated on the Continuity of Care Document, described below.
- ***Accredited Standards Committee (ASC) X12.*** This group develops messaging standards for electronic data interchange of administrative data in a variety of industries, including healthcare (http://www.x12.org/). ASC X12 standards are not very clinical and have no particular pediatric flavor, but researchers may see these standard messages used in conjunction with clinical systems for transmitting insurance and payment information.
- ***The National Council for Prescription Drug Programs (NCPDP).*** Computer systems have proven patient safety utility in the area of medication prescribing (Bates 2007). For electronic prescribing to reach its full potential, there needs to be a standard way of exchanging prescriptions between providers and pharmacies (among other actors). The not-for-profit ANSI-Accredited Standards Development Organization NCPDP produces a suite of standards that allow transmission of data related to pharmacy operations, including the SCRIPT standard for prescriptions, a federally endorsed standard (Centers for Medicare & Medicaid Services (CMS) and H.H.S 2006, 2010). While the EHR may store its medication orders and prescription data in its own data model, the required standard for transmitting the prescription to the pharmacy is the SCRIPT standard, which forces the EHR data model to include data necessary to satisfy the requirements for constructing a valid SCRIPT message.

2.3 Trends that Will Encourage Interoperability

2.3.1 Meaningful Use

The Health Information Technology for Economic and Clinical Health (HITECH) Act within the 2009 American Recovery and Reinvestment Act (HHS. 2009) calls for the appropriation of federal funds to stimulate the adoption of electronic health records among U.S. health care providers. To ensure that the systems that providers implement are actually fully-functional EHRs, there is a requirement that these systems be certified that they can perform certain functions. There is another requirement that providers be able to demonstrate that they are using the systems "meaningfully"—in other words, in a way that might conceivably benefit the patient. Several provisions of this "meaningful use" (MU) rule entail the use of standards, like electronic prescribing and exchanging data with other providers. An unfortunate side effect of MU is that compliance with MU, which arguably voluntary, absorbs most of the resources of providers, vendors, and project implementers. Penalties for not complying with MU (part of the Medicare program that adult providers, but not pediatric providers, would participate in) raise legitimate questions as to whether the program can be said to be truly voluntary. In any case, the initiative will definitely increase the adoption of health information technology and the concomitant use of data standards.

2.3.2 Electronic Prescribing

In Stage 1 of MU, providers are required to send at least 30% of prescriptions via electronic prescribing. While it does not affect pediatricians, the Medicare Improvements for Patients and Providers Act of 2008 (MIPPA) (Centers for Medicare & Medicaid Services (CMS) and H.H.S 2008) requires most adult providers to adopt e-prescribing as well. Because most providers are adult providers, this will ensure rapid and robust constriction of the infrastructure needed to spread e-prescribing as the most common method of providing prescriptions within a few years. The opportunities for rich datasets for analysis under these conditions will be revolutionary.

2.3.3 Other Quality Reporting Programs

The rise of electronic health records comes with an increased interest in direct, electronic reporting of quality measures. While it does not affect child health providers (since they see no Medicare patients outside the narrow context of end-stage renal disease) the Physician Quality Reporting Initiative (PQRI) incentive program as part of Medicare has already stimulated the creation of techniques for data extraction from health information systems. Between MU and the Child Health Insurance Program Reauthorization Act (CHIPRA) of 2009, child health providers will soon be in the business of direct quality measure reporting as well. As of this writing, the proposed Stage 2 of the MU program already includes a significant number of new pediatric-relevant quality measures for electronic reporting. The net effect of these programs is that providers will have to adopt systems that have more robust data management capabilities than before. As a secondary result, they will be more likely to want to use those capabilities for analysis outside of mandated reporting.

2.3.4 Personal Health Records and Consumer Empowerment

In the "old days," information, stored on paper, was difficult to move. It was laborious to package up copies of papers for other providers, so, in some cases, providers wrote letters to each other, summarizing care. Patients or families who wanted access to their health information were similarly hobbled by the process of photocopying, but were often willing to create an independent summary record of health care, in such systems as the now defunct Google Health or the currently active Microsoft Health Vault (Do et al. 2011). Such untethered personal health record systems are themselves difficult to maintain without unusual levels of dedication. As providers use computers more and more, patients will value summary information from these systems. Health care providers will provide information from their office systems to share with personal health records, or at least provide a patient portal into a subset of the provider's system. The need to combine records across very diverse health care providers will encourage the adoption of standards for both moving the data and retaining the meaning of these records as they move into different contexts.

- ***Example HL7 V2: Immunization Data Messaging (CDC Implementation Guide)***: Within child health, the task of maintaining statewide immunization information systems (formerly known as registries) is the most important inter-institutional data exchange task. The implementation guide for immunization messaging (Savage and Williams 2011) written by the American Immunization Registry Association (AIRA) and the Centers for Disease Control and Prevention (CDC) is therefore the most clinically relevant example of real-world pediatric data exchange. This implementation guide lays out how messages containing information on patients, their immunization history, and immunization events (administrations, reactions, forecasts), without constraining how the EHR handles immunization data storage or decision support internally. Table 2.2 illustrates the structure of a version 2 message.
- ***Example HL7 v3: Continuity of Care Document***: Fundamental to consumer empowerment in health care is consumers' access to summary information like

Table 2.2 In this fragment of an HL7 version 2 message, an immunization registry is sending a record of an immunization. In real life, this message would be "wrapped" in data indicating the identity of the patient, the sender, and the receiver. RXA is the segment that contains a record of the administration of the vaccine. Within the RXA segment there are items such as RXA-5, which identified the substance administered (measles-mumps-rubella vaccine, in this case, encoded from a standard terminology system), and RXA-11, the location of the organization where the administration occurred. Other data include the identity of the person who administered the vaccine, the lot number, and the expiration date. The RXR segment describes the anatomical site to which the vaccine was delivered, in this case the left arm

```
RXA|0|1|20120401|20120401|03^MMR^CVX|.5|ML^^ISO+|||987654321^DOE^JOHN^G^
   ^^^^^^^^VEI~123123123^O=DOKES^JOSEPH^A^^DR^MD^^^^^^OEI|^^^PRIMARY
   CARE CLINIC^^^^^100 SPRING
   STREET^^CINCINNATI^OH||||W321321321|20120603|MSD^MERCK^MVX
RXR|SC^SUBCUTANEOUS^HL70162|LA^LEFT ARM^HL70163
```

diagnoses, medications, allergies, immunizations, and past medical events. While the record contains much more that just these summary data, much of the remaining data has limited usefulness beyond its original context. These data are also useful to other providers, of course. Base on a paper form used in inter-hospital transfer in Massachusetts, the Continuity of Care Record project was born, and grew into an electronic message format defined by ASTM (Ferranti et al. 2006). Eventually this message standard was harmonized with the HL7 CDA and became the Continuity of Care Document (CCD). CCD is the basis for exchanging summary records in at least some electronic systems. The expectation is that all electronic systems will be able to read and write CCDs. Desire for accurate CCDs (accurate because providers will get feedback from patients and other providers if there are errors) will drive processes for accurate data on which to base them, which can only be good for the state of patient data in electronic systems. Table 2.3 illustrates the structure of a portion of a CCD.

2.4 Terminology Standards

2.4.1 Definitions

- ***Terminology Systems***: Terminology systems are collections of related concepts defined within a discipline—are intended to convey unambiguous meaning within a defined context. Many terminology systems exist as defined standards, maintained by standards-development organizations, which specify the applicable domain, the meaning of the words, the relationships between the words, the mapping to codes, and use cases within the clinical domain. Unfortunately, in real clinical systems, system implementers react to user dissatisfaction with the rigidity of using a tightly defined list of terms by opening the list up to local customizations. Those who must grapple with data extracts from EHRs

Table 2.3 Fragment of XML from a Continuity of Care Document. The above XML expresses a resolved problem of hyaline membrane disease, which you might find in a Continuity of Care Document from a newborn intensive care admission. SNOMED-CT is indicated by the use of the object identifier (*OID*) 2.16.840.1.113883.6.96

```
problemObservation{
        id(root:'9d3d416d-45ab-4da1-912f-4583e0632000')
        code(code:'ASSERTION', codeSystem:'2.16.840.1.113883.5.4')
        value(
          builder.build{
cd(code:'494627013',codeSystem:'2.16.840.1.113883.6.96',displayName:
   'Hyaline Membrane Disease')
          }
        )
        problemStatus{
          value(code:'413322009',
    codeSystem:'2.16.840.1.113883.6.96',displayName:'Resolved')
        }
        problemHealthstatus{
          value(code:'162467007',
    codeSystem:'2.16.840.1.113883.6.96',displayName:'Symptom Free')
        }
```

need to clarify to what extent a terminology standard is being used in its pure form, or even within its intended domain. The most important reason to adhere to the pure form of a terminology is to support interoperability—both between people and between machines. Terminology systems exist for a variety of clinical domains, like diagnoses, procedures, exam findings, organisms, anatomic terms, psychosocial problems, and other clinical concepts. To the extent that system implementers use the appropriate terminology for the right part of the application, terminologies can facilitate *semantic interoperability*; that is, the meaning of the record can be conveyed across organizational boundaries. This kind of data exchange is a major goal of the U.S. effort to create a National Health Information Network (Dixon et al. 2010).

- ***Post-coordination Versus Atomic Concepts***: When designing the terms for a terminology system that a human user will see, developers have the dilemma of how much coordination of concepts to do. If one does no pre-coordination (bringing together of separate concepts into a more complex concept, as in "fracture of the humerus," where "fracture" and "humerus" are the atomic [indivisible] concepts) then users will be left with the laborious task of bringing them together themselves. If one does too much pre-coordination, then users may not be able to find the exact concept they want to express (e.g., "exacerbation of mild intermittent asthma" may be an inappropriate term for a child with exercise-induced asthma with exacerbation). For terminology systems that are attached to common, real-world processes, the level of coordination may be straightforward (e.g., in LOINC, the term 57698-2 refers to "lipid panel with direct LDL," a heavily pre-coordinated term, but one that represents a laboratory test that is actually obtained in practice). Clearly, pre-coordination imposes a tremendous maintenance burden on the terminology developer, since there are new, complex concepts created every day in clinical medicine. It also creates an issue with historical data, since a given pre-coordinated term might not have been available at the time the data were collected. A classification system that groups related concepts is often necessary when aggregating terms that were collected via clinical processes over many years. All terminology systems, no matter how pre-coordinated, undergo periodic updates, of course, but the more clinically relevant the terminology is, the more rapidly it will be updated. These updates present special challenges for data aggregation, as mentioned, but in pediatric applications there are apt to be more local terminology customizations to accommodate special pediatric terms not found in standard terminologies.

- ***Types of Terminology Systems***: Sets of terms are usually designed with a specific purpose in mind. An *interface terminology system* is made to support data entry (the "interface" here is the human-computer interface). As such, an interface terminology system emphasizes the use of natural language terms with synonyms. Terms in these systems are heavily pre-coordinated, in order that they may be selected rapidly for real clinical situations. SNOMED-CT is an example of an interface terminology system. In contrast, a *reference terminology system* is designed for data processing like retrieval, aggregation, and determination of equivalent data points. Reference terminology systems may appear to a clinical user to be overly rigid, but this is the result of a reference terminology system's main functional requirement to eliminate ambiguity. A *classification* is a set of terms intended to divide concepts into groups for a defined purpose. The most familiar classification system is the International Classification of Diseases (e.g., ICD-9, Clinical Modification, used to encode diagnoses for all U.S. healthcare claims), maintained by the World Health Organization. A common blunder in system implementation is to use a classification like ICD-9-CM when an interface terminology system would much better capture the concepts that clinical users are trying to express. Interface terminology systems typically map their terms to an appropriate classification so that it can be captured for administrative purposes, allowing the user to use the more appropriate reference information system.
- ***Pediatric Aspects of Terminology Systems***: Terminology used to describe pediatric concepts must match the findings, conditions, and treatments found in the pediatric environment, of course. But there are some requirements of pediatric terminology systems that go beyond the simple description of clinical phenomena. In addition to very specific descriptions of diseases seen only in pediatrics, terminology systems must also accommodate the capture of data where terms may need to apply to the patient and the parent separately. For example, a term like "38-week gestation" may refer to the mother's pregnancy or the child's time of delivery. Likewise, terminologies must capture the fluid nature of children's growth and development, and social situations that apply only to the young. For example, descriptors of a patient's capability of self care may represent an abnormal state in adults, but a normal state in infants.

- ***Diagnoses and Conditions***: The most familiar use of a standard terminology system in pediatric medicine is for diagnoses, including descriptions of conditions for which no diagnosis is (yet) known. Physicians are usually very familiar with the assignment of diagnosis codes in the billing process. Since the terminology system used for billing (the classification ICD-9-CM) is not especially rich, physicians will often incorrectly assume that computer systems cannot adequately capture clinical concepts with adequate specificity. In fact, there are several terminology systems that clinicians can use that do an excellent job of capturing clinically relevant concepts. The dilemma the system implementer has is which one to use, and whether to try to ask users to use a reference terminology system directly, or a more clinically relevant interface terminology.

2.4.2 SNOMED-CT

SNOMED-CT is a extensive clinical terminology standard formed in 2002 when the College of American Pathologists merged the England and Wales National Health Service's Clinical Terms (a UK-based terminology for primary care previously known as the Read Codes)(Harding and Stuart-Buttle 1998) with the reference terminology subset of SNOMED known as SNOMED-RT (Cornet and de Keizer 2008). The resulting system, called SNOMED-CT (for Clinical Terms) is intended to be used as a general-purpose reference terminology system, including the domains of diagnoses, conditions, historical findings, exam findings, and test results. SNOMED began as a pathology terminology system, but has expanded its domain to the point that the National Library of Medicine (NLM) licensed it for use throughout the United States in 2003 (Cornet and de Keizer 2008). Since 2007, SNOMED-CT has been maintained by the International Health Terminology Standards Development Organisation (IHTSDO) with the goal of promoting international adoption of its use. One can obtain direct access to SNOMED-CT by obtaining a license to use the NLM's Metathesaurus (National Library of Medicine 2012).

The goal of any terminology system, including SNOMED-CT, is to reduce variability in data capture and encoding in order to facilitate better clinical care and research (Ruch et al. 2008). Unfortunately, no terminology system has perfect coverage of all conceivable concepts, so the natural evolution of terminology systems in clinical care is to customize the content, or develop home-grown terminologies. Mapping such homegrown systems to standard terminologies usually shows that the reference terminology system fails to cover all the needed concepts by a significant margin (Wade et al. 2008; Wade and Rosenbloom 2008). The U.S. National Library of Medicine's Unified Medical Language System (UMLS) Metathesaurus is an attempt to draw lines of semantic equivalence between terms of various systems, but is not itself used as a clinical terminology system. UMLS Metathesaurus is, as the name implies, more of a thesaurus than a terminology system. Via the UMLS Terminology Services web site (https://uts.nlm.nih.gov/), one can access SNOMED-CT and other terminologies that are mapped within UMLS.

SNOMED-CT is a good example of a terminology system that includes a semantic network, indicating relationships between concepts. The simplest romantic relationship is the "is a" relationship, as in "von Gierke disease **is a** glycogen storage disease" (a network of terms connected by "is a" relationships is also known as a *taxonomy*). In this relationship, "glycogen storage disease" is the parent of the child concept, "von Gierke disease." Semantic networks can be useful in EHRs, but their navigation adds complexity to the user interface that limits their use. For example, a user may want to take a concept that is in the record and express it with more specificity. Following "is a" links to child concepts can achieve this specificity, provided that the EHR embeds those links. Semantic relationships can be more sophisticated, and can, for instance, connect diseases to the symptoms that it causes or a drug to its common side effects. Other relationships include "is caused by," "site of finding," and "manifests as." An EHR equipped with semantic links between terms can give significant decision support to a user who, for example, may want to generate a list of symptoms commonly associated with a disease "on the fly" for documentation purposes. The structure of SNOMED CT contains the current concept, parents and children of that concept and descriptors of the concept, including defining relationships, qualifiers, and descriptions or synonyms. It is freely available for research use and can be accessed both online and through the Internet (http://snomed.dataline.co.uk/). SNOMED-CT's terms can be used within pediatrics; adaptation and extension of the terms have been performed in a research setting to help improve pediatric care and concepts (James et al. 2009).

2.4.3 ICD (International Classification of Diseases)

ICD (International Classification of Diseases) is the standard diagnostic tool for epidemiology, health management and clinical purposes. ICD classifications (ICD-9-CM, ICD-10) encompass many of the same domains as SNOMED-CT (diagnoses, symptoms, and procedures), but are not intended to be an interface terminology system. ICD-10 represents an order-of-magnitude increase in the number of terms available, so it may seem to users that ICD is becoming adequate for use as an interface terminology system for diagnoses. While it is true that the higher granularity of ICD-10 will allow better use of administrative (claims) data in clinical decision support, ICD is still a classification system, so system implementers should continue to use a true interface terminology system to capture these data.

2.4.4 LOINC (Logical Observation Identifiers Names and Codes)

Logical Observation Identifiers Names and Codes (LOINC) is a freely available "universal code system for laboratory and clinical observations" developed in 1994 by the Regenstrief Institute and Indiana University (Huff et al. 1998). The database includes medical and laboratory codes, nursing diagnoses, nursing interventions, outcome classifications, and patient care data. The laboratory coding in LOINC includes chemistry, hematology, serology, microbiology, and toxicology. The clinical portion of the coding includes vital signs, intake/output, procedures, selected survey instruments and other clinical observations. LOINC's primary use is to identify laboratory tests. It is endorsed by the American Clinical Laboratory Association as well as the College of American Pathologists. Since laboratory systems used in child health settings use certified clinical labs, LOINC supports the coding of tests used in pediatrics.

2.4.5 Other Terminology Systems

There are many other terminology systems in use in pediatric applications, like terms from the American Psychiatric Association's Diagnostic and Statistical Manual of Mental Disorders, Fourth edition (DSM-IV) (American Psychiatric Association 2000) or the American Medical Association's Current Procedural Terminology (Abraham et al. 2011). Outside the United States, ICD-10 has been used for several years to classify diagnoses, and SNOMED-CT has gained traction as a reference terminology because it incorporated the Read codes and is maintained by an international agency. Of course, interface terminologies, which must reflect the language and clinical practice of real end users, will continue to be especially heterogeneous across countries.

2.4.6 Proprietary Systems

- ***Terminologies***:
  - **MEDCIN**: MEDCIN® is a proprietary system of standardized vocabulary developed by Medicomp Systems, Inc. (Chantilly, VA, http://www.medicomp.com/products/medcinengine/) for use as an interface terminology system for electronic medical records. Introduced in 1978, it now includes more than 280,000 clinical concepts, linked to each other by a rich semantic network. MEDCIN terms describe chief complaints, symptoms, exam findings, diagnostic test findings, and diagnoses, among others. The semantic network can be implemented within an EHR to allow one to generate a template of symptoms associated with a possible disease on the fly, for example. The terms in MEDCIN are mapped to terms in reference terminology systems and classifications like SNOMED-CT, LOINC, and ICD.
  - **Intelligent Medical Objects**: Intelligent Medical Objects (IMO, Northbrook, IL, http://www.e-imo.com/) develops and maintains medical terminology

systems for problems (symptoms, diagnoses), procedures, and medications. IMO's emphasis is on providing pre-coordinated terms that a clinician in practice would normally use, allowing clinicians to use clinically relevant descriptors rather than the exact code necessary to achieve some administrative goal. The result of this approach is there are many synonymous terms, which vary from each other only in minor lexicographic features. As with any heavily pre-coordinated terminology system, there is much ongoing development, and the terms are updated continuously. Customers can request new codes based on new clinical phenomena. Using a dataset coded in IMO will necessarily involve referring to the crosswalk to standard terminology systems, or consulting with a clinician about nuances of meaning between similar terms. IMO is particularly useful in pediatrics due the terminology's ability to cover rare disease found primarily in pediatric practice.

- **Other terminology systems**: There are other terminology systems available, like the commercial products offered by Health Language (Denver, CO, http://www.healthlanguage.com/) and the public domain Omaha System, used chiefly by nurses in classifying patient problems, interventions, and outcomes. In all cases, expertise about the intended use of the codes is required in order to use these systems in data analysis. Appropriateness for pediatric applications varies among these systems and in certain clinical domains within each one.

• ***Medications***

- **RxNorm**: RxNorm (http://www.nlm.nih.gov/research/umls/rxnorm/index.html) was developed by the U.S. National Library of Medicine to provide normalized names for clinical drugs and to link its names to many of the drug vocabularies commonly used in pharmacy management and drug interaction software. By providing links between these vocabularies, RxNorm can mediate messages between systems not using the same software and vocabulary. Standardized vocabularies also exist for medication-specific uses. While one could use straightforward chemical descriptions to name medications (like 2-acetoxybenzoic acid for aspirin) these standard names are not used in real clinical practice, and are absurdly complex to boot (e.g., (2S,5R,6R)-6-([(2R)-2-amino-2-(4-hydroxyphenyl)-acetyl]amino)-3,3-dimethyl-7-oxo- 4-thia-1-azabicyclo[3.2.0]heptane-2-carboxylic acid for amoxicillin). RxNorm attempts to assign a unique concept identifier for each synonymous drug term complied from several drug databases. The databases from which RxNorm draws terms are sold commercially or published by the U.S. government. The avowed purpose of RxNorm is to provide drug names that describe medications that patients actually receive, which explains the approach of aggregating real-world drug databases. The technique RxNorm uses is to identify non-conflicting codes from within the combined data, and to assign a concept identifier. RxNorm's data model includes semantic relationships between concepts like "has-ingredient" for combination preparations or "has-trade name" to identify brands. Because RxNorm is derived from databases that primarily comprise products that are sold, it is lacking in content for drugs that are prepared ad hoc by compounding pharmacies. This deficit has profound implications for pediatrics, where many children are treated with medications that are not available commercially, like a compounded liquid preparation derived from a crushed pill ordinarily used in adults. Another challenge to pediatric coding of medications is in compounded combination preparations for topical drugs. These recipes are not standardized. RxNorm is free and there is a publically available browser (http://rxnav.nlm.nih.gov/).

- **Medi-Span**: Medi-Span is maintained by Wolters Kluwer Health (Medi-Span, Indianapolis, IN). Like other commercial drug databases, Medi-Span includes not just a coding system for medications, but also clinical content regarding drug interactions, allergens existing in drug products, acceptable dose ranges, and other clinical information needed for appropriate prescribing. It is this information that represents the main value that these databases offer. None of these databases are specific to pediatrics.
- **First Databank**: First Databank (FDB) is a clinical drug and drug interaction database designed and maintained by First Databank, Inc (South San Francisco, CA). In the realm of pediatric-specific dosing and medications, companies are acknowledging the special needs to pediatric practice in the software that they offer, including support for compounded medications (First Databank). In some EHR systems, implementers can choose which commercial drug database to embed in the system. In others, the EHR application is tied to a particular database. EHRs that are otherwise identical, then, can be fundamentally non-interoperable based on the variance in the coding scheme used to represent drug concepts. It is this variance that RxNorm is intending to rectify, but the technical overhead of mapping to standard terms is not yet commonly undertaken.
- **Multum**: Multum is a drug database maintained by Cerner Corporation (Kansas City, MO) and is embedded with the electronic medical record systems and healthcare services they provide, but can also be implemented in non-Cerner systems.
- **Other databases**: There are other commercial databases (e.g., Gold Standard' Elsevier, Netherlands) that implementers might choose. The wide diversity of these coding systems presents challenges to data aggregation, but RxNorm promises to help.

2.5 Challenges in Pediatrics

There are other standards operating in healthcare, like the imaging standards set by the Digital Imaging and Communications in Medicine (DICOM) for diagnostic imaging. To the extent that data standards for terminology and messaging achieve national scope, they will necessarily be focused on the needs of large, national populations—namely, adults. It is unlikely that any pediatric-specific data standard would gain enough traction to achieve widespread use. For this reason, pediatric informaticians recognize that advocating for the needs of children within larger data standards is more productive than trying to make systems that specialize in pediatric needs. Researchers who work with data that employ these standards need to be aware that standards implemented in the pediatric healthcare environment are at high risk for non-standard implementations because of the differences between pediatric and adult care.

References

Abraham M, et al. CPT 2012 (Cpt/Current Procedural Terminology (Professional Edition)). Chicago: American Medical Association Press; 2011.

American Psychiatric Association. Diagnostic and statistical manual of mental disorders. 4th ed., text rev. Washington, DC: American Psychiatric Association; 2000.

Bates D. Preventing medication errors: a summary. Am J Health Syst Pharm. 2007;64(14 Suppl 9):S3–9, quiz S24-6.

Centers for Medicare & Medicaid Services (CMS), H.H.S. Medicare program; identification of backward compatible version of adopted standard for e-prescribing and the Medicare prescription drug program (version 8.1). Interim final rule with comment period. Fed Regist. 2006;71(121):36020–4.

Centers for Medicare & Medicaid Services (CMS), H.H.S. Medicare program; revisions to the Medicare advantage and prescription drug benefit programs. Interim final rule with comment period. Fed Regist. 2008;73(182):54225–54.

Centers for Medicare & Medicaid Services (CMS), H.H.S. Medicare program; identification of backward compatible version of adopted standard for e-prescribing and the Medicare prescription drug program (NCPDP SCRIPT 10.6). Interim final rule with comment period. Fed Regist. 2010;75(126):38026–30.

Cornet R, de Keizer N. Forty years of SNOMED: a literature review. BMC Med Inform Decis Mak. 2008;8 Suppl 1:S2.

Dixon BE, Zafar A, Overhage JM. A framework for evaluating the costs, effort, and value of nationwide health information exchange. J Am Med Inform Assoc. 2010;17(3):295–301.

Do NV, et al. The military health system's personal health record pilot with Microsoft HealthVault and Google Health. J Am Med Inform Assoc. 2011;18(2):118–24.

Dolin RH, et al. The HL7 clinical document architecture. J Am Med Inform Assoc. 2001;8(6):552–69.

Downs SM, et al. Using Arden Syntax and adaptive turnaround documents to evaluate clinical guidelines. AMIA Annu Symp Proc. 2006;2006:214–8.

Ferranti JM, et al. The clinical document architecture and the continuity of care record: a critical analysis. J Am Med Inform Assoc. 2006;13(3):245–52.

First Databank. FDB MedKnowledge framework: integrated software components for building drug informations systems quickly. [cited 2012 May 30]; Available from: http://www.fdbhealth.com/~/media/downloads/form%20required/us/datasheet%20-%20fdb%20medknowledge%20framework.ashx

Harding A, Stuart-Buttle C. The development and role of the Read Codes. J AHIMA. 1998;69(5):34–8.

HHS. Health Information Technology for Economic and Clinical Health (HITECH) Act, Title XIII of Division A and Title IV of Division B of the American Recovery and Reinvestment Act of 2009 (ARRA), in Pub. L. No. 111–5 (Feb. 17, 2009). U.S.D.o.H.a.H. Services, editor, Washington, DC; 2009.

HL7. Health Level Seven International: child health. 2012 [cited 2012 Apr 1]. Available from: http://www.hl7.org/Special/committees/pedsdata/index.cfm

Huff SM, et al. Development of the Logical Observation Identifier Names and Codes (LOINC) vocabulary. J Am Med Inform Assoc. 1998;5(3):276–92.

James AG, et al. A reference set of SNOMED terms for the structured representation of respiratory disorders of the newborn infant. Stud Health Technol Inform. 2009;150:243–7.

National Library of Medicine. UMLS terminology services. 2012 [cited 2012 Apr 1]. Available from: https://uts.nlm.nih.gov/home.html

Rocca MA, et al. Development of a domain model for the pediatric growth charting process by mapping to the HL7 reference information model. AMIA Annu Symp Proc. 2006;2006:1077.

Ruch P, et al. Automatic medical encoding with SNOMED categories. BMC Med Inform Decis Mak. 2008;8 Suppl 1:S6.

Samwald M, et al. The Arden Syntax standard for clinical decision support: experiences and directions. J Biomed Inform. 2012;45(4):711–8.

Savage R, Williams W. HL7 version 2.5.1 implementation guide: immunization messaging (Release 1.3). C.f.D.C.a. Prevention, editor, Atlanta; 2011.

Shakir AM. HL7 Reference information model. More robust and stable standards. Healthc Inform. 1997;14(7):68.

Wade G, Rosenbloom ST. Experiences mapping a legacy interface terminology to SNOMED CT. BMC Med Inform Decis Mak. 2008;8 Suppl 1:S3.

Wade G, et al. Assessing voids in SNOMED CT for pediatric concepts. AMIA Annu Symp Proc. 2008;2008:1164.

Zimmermann H. OSI reference model – the ISO model of architecture for open systems interconnection. IEEE Trans Commun. 1980;28(4):425–32.

Chapter 3
Data Storage and Access Management

Michal Kouril, Nicholas Hunt, and Michael Wagner

Abstract One of the most basic challenges arising from the increasing use of electronic data in both clinical practice and research lies in the design and implementation of storage solutions capable of accommodating modern demands. Complex organizational structures, which often cross clinical care and basic research, the often sensitive nature of data, and the ever growing volume of both structured or annotated and unstructured data all motivate innovations, e.g., in identity management, audit trails, monitoring and security, and permissions management. Clinical, translational, and health services research generate very large amounts of data and take place within a complex regulatory environment. Data management, including long-term preservation of value, requires careful attention to security, ease of use and access, transparency, and compliance with numerous state and federal laws. It also demands close collaboration and mutual trust between of the IT group responsible for support of research and the IT group responsible for support of clinical care and business operations in a complex research intensive medical center.

3.1 Introduction

Many research-intensive medical centers struggle with an ever-increasing deluge of electronic data. While the per-unit cost of storage infrastructure has steadily and rapidly declined, there are significant costs (which can trump the hardware expenses) associated with the technical and managerial challenges of how to keep up

M. Kouril, Ph.D. (✉) • N. Hunt, MSIA, CISSP • M. Wagner, Ph.D.
Department of Pediatrics, University of Cincinnati College of Medicine, Cincinnati, OH, USA

Division of Biomedical Informatics, Cincinnati Children's Hospital Medical Center, 3333 Burnet Avenue, ML-7024, Cincinnati, OH 45229-3039, USA
e-mail: michal.kouril@cchmc.org

J.J. Hutton (ed.), *Pediatric Biomedical Informatics: Computer Applications in Pediatric Research*, Translational Bioinformatics 2, DOI 10.1007/978-94-007-5149-1_3,

with the growing storage demands. Questions are inevitable about the value of data and whether its long-term preservation is justified. At the same time, the increasing volumes of data increase demands on processes to ensure data integrity and security, including the need for audit trails to record all access requests. This chapter relays several of the challenges our Research IT group has faced around the broad topic of data storage and access management, as well as some solutions that have worked well. Much of the content is written purely from the vantage point of a centralized Research IT Services group, i.e., with a focus on a global view of an institution's strategic IT needs rather than from the point of view of an individual investigator or research group. Rather than focus on specific technologies or software solutions, we instead describe the challenges and solutions in general terms. Finally, we write this chapter almost entirely based on our subjective experience and our unique circumstances. Nevertheless we believe that many of our conclusions and strategies described herein are applicable in other settings.

3.2 Classification of Data Types

Before delving into any specific strategies for data storage, we will first describe the types of electronic data typically encountered in (but not specific to) a pediatric medical research institution, their categorization, and how the various types of data can be leveraged to maximize the long-term value of the data. From our vantage point as Research IT Service providers and with our goal of efficient management and long-term data value preservation in mind, we have found it useful to distinguish two broad categories of electronic data, the main criterion being whether and how well the stored data (we will simply call the stored data a file) is described by metadata.

"Unstructured" Data – In traditional file storage systems, an electronic file can only be identified by its file path (the folder it resides in), its name (including, if applicable, its extension), date of its creation, access or modification date, its size and possibly by permissions. These attributes are typically not sufficiently informative to allow someone to understand the context and purpose of the data in the file, say, years after it was generated. This includes many cases of plain-text files, but is especially problematic for binary data, which typically require specialized software for processing. Unless the team that originally generated the data is involved and has, say, sufficiently well-kept laboratory notebooks, it could well be impossible to re-process the contents of the file. Unstructured data files clearly present a challenge when it comes to long-term preservation of value, as they are unlikely to be easily interpreted and re-processed long after they were initially generated.

"Structured" Data – We call data "structured" if they are described by standardized, consistent, and well-defined metadata (structured information about the data in the file in question). This metadata must contain the minimum necessary information about the data file in order for it to be "useful" in the long term, e.g., amenable to re-processing. Note that the detailed requirements for minimum necessary information differ for each data type, and that much care needs to be taken to define sufficient standards for different data types. The structure of the data and metadata can then be exploited to help, for example, with storage, retrieval, and access. Examples of structured data include well-annotated databases, where specific information is stored in well defined, fixed fields within a record, but also, say, a DNA microarray primary data file that is described by sufficiently rich metadata. Note that many unstructured data files can rather easily be turned into structured data, if processed for example, by a simple parser. Conversely, data that may be amenable to being categorized as structured (i.e., where the required minimum metadata can easily be parsed out directly from the file) may not be identified as such and simply stored on an ordinary network storage device. The latter case represents a missed opportunity of increasing the ease of retrieval and interpretability of this data file and is something to be avoided and/or detected with automated methods.

It is clear that in order to maximize the long-term value of data, an institution should strive to maximize the proportion of structured to unstructured data. This, however, comes with significant challenges, especially in a large pediatric research setting where individual investigators enjoy a great deal of autonomy in terms of deciding on which processes and workflows to implement in their daily operations. The definition and capture of metadata can be burdensome and involve significant changes in processes. End-users who see few or no short-term benefits will likely resist change. Our approach toward overcoming these challenges has been threefold. We encourage the use of user-friendly tools such as electronic laboratory notebooks and web-based data portals that can to some extent automate the generation of metadata (see the following sub-section). Secondly, we made the decision to discourage unstructured data storage by levying charge-backs for networked storage drive, which are not conducive to retaining metadata. Thirdly, we have worked closely with the various institutional research cores that generate large amounts of data (e.g., flow cytometry, DNA sequencing, etc.) in order to ensure that most of the data generated is appropriately categorized, annotated and stored, e.g., in web-based data portals (see the following section). As a consequence, a large fraction of data can be captured and categorized at the source, and this minimizes the chances of the same data ending up as unstructured files.

3.3 Sample Systems for Structured Data

In this section we will provide examples of ways to capture, store, and share structured data. The requirement that metadata accompany each primary data file makes the use of a (typically relational) database along with a file storage suitable for these data types. We discuss a variety of different options for different types of data in the following subsections. Clearly all require significant investments, whether they are commercially available solutions or home-spun implementations, and there is significant overhead associated with the addition of a database system on top of the file storage solution. However, we believe these are the best ways to maximize the long-term value of data, and we strongly advocate incentivizing their use by investigators.

A database, almost by definition, holds well-structured data, which is decomposed in the case of a relational database into tables, columns and rows. Typically, end-users do not interact with databases directly, instead they usually enter and retrieve data from a database through a middleware software layer. While well-designed databases almost naturally fall into the category of structured data storage systems, care must still be taken to ensure the long-term value of the data, e.g., by ensuring that properly curated data dictionaries are included in the database's documentation. Databases provide many advantages and features to help with fast storage and retrieval of information, such as indexing of values for fast search, optimized underlying data storage, etc. Many database engines also have the capability to control access to the data up to individual tables and rows. To optimize the storage and retrieval mechanisms, especially when working with a large homogenous datasets, databases might not hold the data itself, but merely contain metadata and links to the large data files. Other considerations include the fact that not all data are amenable to being stored in databases, and that there is a significant overhead associated with the *de-novo* design and implementation of a database.

We have implemented several web-based data portals for data storage for specific data file types, and we briefly outline their functionality in the following subsection. There also are several off-the-shelf commercial or open-source data capture and storage solutions that rely heavily on databases and have found widespread use in medical research centers. A number of these will be discussed in subsequent sections of this chapter.

3.3.1 Web-Based Data Portals

Typical sources of large data streams in a biomedical research institution are the research cores such as flow cytometry, DNA sequencing, RNA expression, imaging and microscopy. These data-generating laboratories are typically centralized facilities used by large numbers of investigators, and they often generate very large numbers of large data files that are inherently amenable to what we term structured data storage. Typical metadata for these primary research data files should include details about the instrumentation that generated the data, details about the experimental setup, the sample sources etc., much of which can be parsed out of the file or extracted from the software that controls the instrument that acquires the data. In an effort to capture these data soon after it is generated, we have opted to develop (separate) web-accessible "data portals" for each data type. These user-friendly, accessible software packages let users upload, store, search for, retrieve and share their primary data securely and efficiently. Advantages include their web accessibility (zero footprint on client computers!), a powerful, flexible search engine which allows searching through a large amount of metadata to retrieve a dataset, the security provided by enterprise authentication (including single sign-on) and user-driven permissions management. Depending on their level of access, end-users can provision others with controlled access to their data. This minimizes the need for sharing data by creating copies on file systems, which represents an unnecessary, wasteful and costly burden on the storage systems. Finally, the web-based portal software can be enhanced with processing pipelines, e.g., to move data to a high-performance computational cluster, thus relieving end-users of the need to tax their desktop computers with memory- and CPU-intensive computations. We have implemented such systems for genotype calling, Genome-Wide Association Studies, microarray processing, and next-generation DNA sequencing. End-users can launch analysis jobs using state-of-the-art processing software on a high-performance computational cluster without requiring in-depth familiarity with Linux-based commands or interactions with cluster batch scheduling systems. Challenges with these systems include the need to help end-users overcome the change in workflow that inevitably comes with the use of any new software system, as well as to convince them to provide and enter sufficiently rich metadata. From an institutional point of view, there is an up-front cost and effort to develop these systems and to define standards for minimal metadata that must accompany any file of a given type. Open-source solutions like the iRODS system (http://www.irods.org) (Rajasekar et al. 2006, 2009) are beginning to mitigate the need for home-grown solutions.

3.3.2 Databases for Clinical and Translational Data Capture

Clinical research data, while quite different in nature than the data types described in the previous subsection, are also very amenable to being stored in systems based on databases. The electronic health record (EHR) can serve as a source of data to support translational research as discussed in Chaps. 6, 8, and 10. To complement the EHR and support research projects by direct data entry, we briefly mention two applications that are used by a large number of investigators in our institution. Many other solutions are available.

- ***REDCap*:** REDCap (Research Electronic Data Capture, http://project-redcap.org/) (Harris et al. 2009) is a widely used, almost self-service, secure web-based application designed to support data capture for translational research studies. It is installed locally on web and database servers and provides an intuitive interface for validated data entry, audit trails for tracking data manipulation and export procedures, automated export procedures for seamless data downloads to common statistical packages, and procedures for importing data from external sources. It has been most widely used for studies that do not need to comply with extensive regulatory requirements.

- ***OpenClinica***: OpenClinica (OpenClinica) is a widely used, web-based open source (LGPL – Lesser General Public License) software for managing clinical trials (https://www.openclinica.com/). It is modular in design and users are permitted to review, modify, and share its underlying source code. Investigators can build studies, design electronic Case Report Forms (eCRFs), and conduct a range of clinical data capture and clinical data management functions. OpenClinica is designed to permit compliance with Good Clinical Practice (GCP) and regulatory guidelines such as 21 CFR Part 11 via differentiated user roles and privileges, password and user authentication security, electronic signatures, SSL encryption, de-identification of Protected Health Information (PHI), and comprehensive auditing to record and monitor access and data changes. There are tools for data cleaning, clinical data management, and site monitoring. Datasets can be extracted in real time and in a variety of formats for analyses.

3.3.3 Electronic Laboratory Notebooks (ELNs)

Just as software such as Microsoft Office has provided a user-friendly replacement for paper-based documents, one can reasonably expect that electronic laboratory notebooks (ELNs) will eventually replace paper-based laboratory notebooks. ELNs accept input of data from laboratory personnel via computers. Times of data acquisition and changes made in the data can be tracked to meet regulatory requirements. Data is stored in a standard database (e.g., Oracle) and therefore is fully searchable. Modern ELNs will also have ontology-based semantic search features.

ELNs are not easily implemented in a research environment because they require changes to the way researchers traditionally collect and record data. Paper notebooks provide almost unlimited freedom in data collection and in use of free-text. Although the electronic version can be very flexible, it requires a mind-shift, together with the availability of appropriate software tools. Numerous commercial solutions are available.

Our institution chose a particular ELN product in 2007 (CERF by ELN Technologies, Inc.) and has been incentivizing the use of this tool by covering all associated software license fees. Nevertheless, because of the significant changes in workflow required when ELNs are used, adoption rates have been rather low and only a relatively small percentage of eligible investigators have chosen to forgo paper-based solutions. Upcoming release of ELN clients for tablet computers (including Apple iPad) are expected to increase the adoption rate significantly.

3.3.4 Example Custom Data Capture Systems

It is often the case that no single off-the-shelf software completely satisfies the data capture, storage and processing demands of a particular project. If sufficient funding and local IT expertise are available, then the best option may be to design and implement custom software to support the project. We will describe two examples of complex, grant-supported, research projects that have required extensive resources to plan and develop appropriate data management systems for them.

3.3.4.1 Genome-Wide Association – Data Management Support

Genome-wide association studies (GWAS) use genome-scale information about single-nucleotide variants in large cohorts and correlate the allele frequencies with phenotypic data that is generally derived from electronic health records. Phenotyping is complex and may require use of Natural Language Processing (NLP) of text in the EHR. Phenotyping is discussed in Chap. 10, Natural Language Processing – Applications in Pediatric Research. The desired outcome of a GWAS study is a list of genomic markers that are (statistically) significantly associated with a phenotype of interest, thus suggesting they be followed up for functional studies. The data required for these kinds of analyses consists of (typically very large) genome-wide Single Nucleotide Polymorphism (SNP) variants assayed on human samples (typically blood or saliva) in conjunction with demographic, clinical and phenotypic data for the individuals in the study. At the present time, SNP data are increasingly being replaced by data from genomic sequencing, which adds to the complexity of data management. The SNP genotypic data typically simply consists of information about whether a sample is homozygous for the major allele, heterozygous or homozygous for the minor allele, which in turn can be easily encoded in a very compressed two-bit format as 00, 01 or 10. The value 11 can thus be reserved for the cases where the data is missing, e.g., for the case where the software that interprets the acquired fluorescence images fails to make a confident call. This leads to a very compressed representation of SNP data and implies that data from chips with millions of genotypes can be stored easily in a relational database as a binary string.

While several excellent packages for GWAS data analysis are readily available (e.g., PLINK, http://pngu.mgh.harvard.edu/purcell/plink/) (Purcell et al. 2007) and systems such a research patient data warehouse that extracts, transforms, and loads data from the institutions EHR (Chap. 6, Research Patient Data Warehousing). Software such as RedCap or OpenClinica can also be used to capture and store the clinical data. It was critical for our purposes to have a system in place that would integrate the clinical phenotypes and the genotypes in one single database and thus facilitate integrated analyses and flexible queries by end-users, who are not necessarily well-trained. Hence, we implemented a web-accessible database using open-source tools such as PHP and MySQL. A relational database stores all demographic and phenotypic information and also holds the genotypes in binary strings. The query interface then permits users to assemble, for example, combined phenotypic and genotypic data for case–control studies, which can be further analyzed using standard open-source software (See Chap. 19, From SNP Genotyping to Improved Pediatric Healthcare).

From the perspective of data storage, which is the primary topic of this chapter, however, it is important to stress that this represents a solution that clearly falls into the "structured data" category: the primary data (genome-scale variant data) is well annotated by individual demographics and clinical data, making it useful for further downstream analysis. If the same data were stored as disjoint flat data files and spreadsheets on a network drive, in all likelihood it would be of little value to researchers in the future who would be unfamiliar with details of data accrual. The web interface and the highly optimized database, while certainly not inexpensive because of the need to repeatedly involve programmers and designers, do provide structure and preserve the value of these data in the longer term.

3.3.4.2 MRI Imaging Data Management Support

A customized data management system was developed to support a specific project that requires integration of neuropsychological test data on normally developing children ages 0–18 years with data from neuroimaging (functional Magnetic Resonance Imaging (fMRI)). The purpose of the study is to establish a normative database of brain development in children that can be used as a reference in studies of diseases. Similar to the GWAS database described above, this study requires storing very disparate data types (clinical, neuropsychological, and imaging) which would be very difficult to link together and ensure their integrity, if they were stored as spreadsheets, text files and images. A relational database, however, together with its web-based interface was developed to provide a uniform, user-friendly gateway for entering, storing and processing all data pertaining to the study. As with the GWAS database, role-access is provisioned via an enterprise identity management system, all data changes are logged. In contrast to the GWAS database, data can be entered directly by study coordinators through the web gateway, using double-data entry where appropriate (if transcribing from paper-based forms). fMRI data is even larger and more unwieldy than genome-wide genotyping data, and so in this case the data are stored in protected files on the file system and only references to the files are stored in the database. The web interface supports a very general search function that permits users to query the more than 500 variables stored in the data and to retrieve images satisfying search criteria. The data can be downloaded and processed offline, or it can be submitted directly from the web interface to the LONI Pipeline (http://pipeline.loni.ucla.edu/) (Rex et al. 2003) workflows, which run on the local Linux cluster. For more information the interested reader is directed to the project web site: http://research.cchmc.org/c-mind.

To summarize this section, both of the examples provided are instances of large, well-funded projects that warranted custom data storage and management solutions. In both cases the databases along with the respective web interface facilitate access to diverse data types and very large, unwieldy files in a central repository. The ability of end-users to directly access, search (using rich meta-data!), download, and process complex data sets, the built-in security, audit and quality control features as well as the ability to access the data easily with only a web browser have made these tools indispensable. The initial design and development costs are certainly high, but these can be amortized by offshoot projects that only require relatively minimal modifications to the initial designs (e.g., other fMRI research studies).

3.4 Unstructured Data

As examples of unstructured data storage options, we briefly describe network attached shared (NAS) disk data storage and the newly emerging online cloud storage technology.

3.4.1 Personal and Shared Network Storage

Personal and shared network storage referred to as NAS (Network Attached Storage) is generally set up to mimic a local hard drive. Similarly to local drives, NAS solutions allow users to store flat files without imposing any requirements on metadata or other structure. NAS servers typically provide added benefits including sharing stored files and data with multiple people or devices via a network. NAS devices are in general centrally managed and backed up by an IT group, thus eliminating the need to manage backups on individual workstations. Often NAS devices also support audit functions, which means that a history of all operations on the file system, including who performed them and from where, is stored. This facilitates (but is certainly not sufficient for) compliance with regulations such as HIPAA (United States 1996). The resiliency of the unit is often favorable compared to local storage. NAS units are often able to withstand the loss of one or more disks before an actual data loss occurs. To provide physical security, NAS servers are typically located in data centers, which protect them from theft and unauthorized physical tampering.

There are two leading protocols to access data over a network: CIFS/SMB (Common Internet File System/Server Message Block) – often referred to as Windows based file sharing – developed by Microsoft and NFS (Network File System) – Unix/Linux-based file sharing – originally developed by Sun Microsystems.

Network transfer speeds are improving and the times to store and retrieve data from NAS systems are becoming comparable to a local hard drive. Potential disadvantages of using NAS devices are related to the complexity of the NAS setup (which requires some in-depth technical expertise) and their maintenance (compared to local disk). Moreover, NAS systems by default generally do not support tying metadata to the primary files, and thus the associated ILM (information lifecycle management), if one exists at all, is somewhat less effective than ILM implemented over structured data. NAS storage is currently the most common mechanism to store data in medical research centers. Personal network data storage is typically restricted to files that do not need to be shared. Shared network data storage on the other hand typically contains business and research files that do need to be shared.

3.4.2 External Hard Drives and USB Flash Drives

External hard drives (sometime called USB hard drives) are popular because of their low price and portability. Their use is strongly discouraged within an academic health center, except for special purposes such as encrypted data transport. Their major drawbacks include:

- General lack of support for encryption
- Poor physical security (easy to steal)
- Prone to failure and typically do not support any kind of internal redundancy
- Can only be connected to a single computer

Flash drives (sometime called USB memory sticks) are often use for data transfers, presentations, and movement of data sets from one computer to another or possibly to another institution. This flexibility has to be accompanied by strong encryption because the stored datasets in an academic health center often contain Protected Health Information and flash drives are very easy to lose. In order to use USB flash drives properly, an institution should strongly consider deployment of an enterprise encryption solution for USB flash drives.

3.4.3 Online and Cloud Storage

To address the ubiquitous needs for sharing data and easy availability and accessibility of data anytime and anywhere, several companies offer so-called cloud storage. In reality these cloud-based storage solutions are (merely) internet-accessible servers with large amounts of storage. Cloud storage vendors often develop clients for multiple platforms so that a user can access the same file from multiple sites, e.g. a laptop at home, a workstation at work, a smart phone, or a tablet computer. The major advantage of using online or cloud storage is the users' ability to access stored data from virtually anywhere and to share data with other users.

There are major disadvantages and risks for an institution adopting an external cloud solution compared to deploying an in-house solution. First there is an issue of compliance with laws and rules governing access to and use of many types of information, such protected health information. Institutions and vendors must assure compliance before allowing general use of cloud storage. Second, transfer speeds can be a problem with bottlenecks, possibly on a vendor side or related to connectivity between the vendor and the institution. Problems arise with transfer of the huge data sets characteristic of many types of biomedical research. Last, but not least, there is the question of costs, including subscription, internet bandwidth, support, etc.

3.5 Compliance and Retention

With any type of data storage (structured or unstructured), retention, compliance, and security are always of concern, both for investigators and for host institutions. Strategies to deal with these issues differ for different types of data.

3.5.1 Disaster Recovery

Strategies for disaster recovery for different data types can take many different paths based on the criticality, sensitivity, and amount of data that must be protected. Traditional methods for backing up and recovering data (such as tape drives) may be adequate for smaller data sets, however these are becoming increasingly difficult to utilize because of the very large size and scope of datasets that are commonly generated in translational research. A number of key considerations must be taken into account before establishing a strategy that is appropriate for the environment.

- **Business Impact Assessment (BIA):** The BIA will document the criticality, impact, and sensitivity of the data in relation to other applications, systems, and data within the environment.
- **Types of recoveries:** The types of recoveries that are going to be required must be cataloged. Strategies for recovery of structured and unstructured datasets can be different. With structured data, the recoveries in all likelihood involve backing up and restoring entire databases, which can imply that very large amounts of data need to be copied on a regular basis, and maintained in the environment. With unstructured data, the files that are generated and need to be restored in the event of a recovery may be small in size, but could be large in number.
- **Implementation:** Once the BIA and types of recoveries are established, a solution needs to be implemented. Based on the criticality and impact of the data, this solution will typically be multi-faceted and have several guises. A disk-to-disk and disk-to-tape solution will work for most situations. In the disk-to-disk scenario, backups are performed using secondary disks to store the backups for rapid recovery and speed of backups. For longer-term retention and where data volume allows it, the backup set on the disk is transferred to tape for off-site storage and data retention.

3.5.2 Retention

Data retention strategies are among the most difficult to implement in the context of institutional data. Most data is generated and then stored indiscriminately because of complexities related to establishing institutional retention policies or strategies over all data generated by an institution. This leads to storage of large amounts of data that are used once or never used. This is an expensive waste of institutional resources. When establishing an institutional policy for data retention, it is important to keep the following key considerations in mind.

- **Usability:** How long does the data maintain its usability for the institution? Many data sets, once generated and analyzed, can be destroyed because they are no longer useful to investigators. On the other hand, some data must be retained to meet regulatory requirements or to document critical research or business reports. Institutions should establish policies with time frames for retention of data derived from different sources for different purposes.
- **Impact:** In the event of a breach or loss of data, what is the potential impact on the institution? If data is no longer needed, then it should be destroyed so that it is no longer a potential source of institutional liability should a breach occur.
- **Privacy/Compliance:** Protected health information (PHI) is typically stored to support translational research. Regulations require that once such data is no longer needed or if the owner of the PHI requests its removal, then the data must be deleted. Data may be subject to other types of regulations, such as those promulgated by the Food and Drug Administration for data related to clinical trials or by the Defense Department or Homeland Security.

Once an institutional policy and a strategy is developed, it must be implemented. Processes must be in place to allow for the removal of data elements, once they are no longer needed or have reached their retention lifetime. The content of files containing unstructured data is generally poorly documented. This poses major challenges in file management. Typically, the files containing unstructured data will need to be indexed before decisions about deletion can be made. For structured data, indexes exist and can guide the removal of data elements from tables/structures.

3.5.3 Encryption

Encrypting structured and unstructured data can be difficult, depending on the format of the data, how it is accessed, and how it is transported. File access protocols such as CIFS do not have encryption built-in, which can make it difficult to protect the data when being transmitted. To make it more difficult, compliance standards may require that protected information be encrypted when being transported or stored.

- **Transport (data in flight):** Encrypting the data while it is transmitted across networks and between systems.
  - This provides protection of data during transport so that malicious users cannot intercept and read it
- **Storage (data at rest):** Encrypting the data when written to systems for storage and access.
  - This provides protection against lost disks or other physical attacks against storage locations

For unstructured data, depending on how the data is accessed (CIFS, NFS, HTTP), services can be added to provide encryption. When working with protocols such as CIFS, encryption is not inherently built-in so additional tools such as Virtual Private Networking (VPN) must be implemented to provide the encryption. By moving the storage location behind a VPN device, all access to unstructured files using the CIFS protocol will be transmitted down an encrypted tunnel to the CIFS volume thus providing the level of protection required. Other protocols such as NFSv4 and http allow the addition of encryption to the protocol, using built-in standards (e.g. HTTPS, NFSv4 encryption). When working with structured data, access is generally by protocols that support encryption (HTTP(S)). Since the application is presenting the data, encryption such as SSL/TLS can be added to the protocol to encrypt the data in transport from the application.

Encrypting data storage can be achieved through different methods and must be applied to all locations (e.g., primary and backup locations). Tools such as Full Disk Encryption (FDE) and Full Database Encryption (FDbE) can provide the encryption, when written to systems. No matter what method of encryption is used, there are important items to remember when implementing encryption in an environment:

- **Key Management:** When implementing encryption, it can only be as secure as the keys that are used to encrypt the data. Use of weak keys or poor protection of keys, can allow attackers to decrypt data and bypass the encryption. It is essential to choose strong keys when encrypting data and ensure that key access is limited to those individuals who require access.
- **Key Rotation:** Encryption keys, like any other password, are susceptible to compromise. While keys may be large enough to make this difficult, with computing resources available, they may be guessed at some point. To address this, keys should be rotated at least periodically to reduce the possibility of their being guessed and used to decrypt protected information.
- **Performance:** Encryption requires additional processors to handle the encryption/decryption operations. In high-volume environments, encryption places significant overhead on servers, reducing the overall performance of the system. Where possible, hardware devices such as Hardware Security Modules (HSM) and SSL Offload should be added to the system to handle the encryption/decryption, since they have cryptographic processors and are purpose built to handle these operations.

3.6 Management of Permissions and Access

Translational research generates huge amounts of data that is sensitive and must be access-protected. While there are extensive regulatory requirements regarding the protection of data from human subjects, it is clear that other types of data must be stored securely for a variety of reasons, for example, to protect investigators' intellectual property rights or to meet requirements of state and federal agencies that support research in the United States, such as the NIH, CDC, FDA and the Departments of Defense and Homeland Security. Furthermore, investigators must be able to ensure the integrity and validity of their electronic data, which would be impossible without tight control of access to their data.

A less obvious question is that of data ownership and responsibility for data protection. While host institutions usually hold claim to all data generated by their employees, academic investigators typically retain a great deal of independence and demand that they control access to all data generated under their supervision. This is typically very different from a corporate environment, where the company retains complete control and ownership of all data and thus can decide how access to sensitive data will be provisioned and justified.

3.6.1 Permissions

The complexity of large research organizations fundamentally dictates that any data storage solution (be it for structured or unstructured data) have a well-defined process for permissions and access management. Such a process is typically anchored in an institutional identity management solution, which uniquely identifies the data every defined entity (individual, application etc.) is permitted to access. Ideally, a self-service solution is in place that allows end-users to request access to data and resources. Changes in permission for an individual or group become effective as soon as all required authorizations have been granted, that is, the process between the request for access and the authorization and implementation of change is smooth and timely. Changes in permission to access should be audited and be transparent to both end-users and data owners. In practice, the process to grant permissions is generally better defined than the process to deny or withdraw permission, as when a user leaves the institution or moves to another department. This issue can be mitigated by carefully tracking the owner of data (e.g., a protected folder, database, etc.) and requiring periodic entitlement reviews to confirm the current list of provisioned users.

A specific example of a self-service permissions management system for unstructured data storage that works well in the author's institution is the commercially available DataPrivilege package (Varonis Inc). The initial setup involves defining all protected directories on the file systems and importing them into the permissions management system. For each protected directory, a list of authorizers must be defined. They have the power to confirm or deny requests for permission to access data. Once the system is setup, users can login to a webpage to request permissions, which triggers an approval workflow potentially resulting in provisioning access. This workflow does not require involvement of IT specialists and is entirely self-service. An IT team is required only to keep the system running, monitor for unauthorized changes, and help users setup new protected directories. This system has worked especially well in our setting because investigators can fully control and oversee permissions for all folders containing their data. Furthermore, the audit trails in the system, as well as the various monitoring functions, allow staff to find rare cases of unauthorized changes in permission.

Standard database engines (such as those used for web-based storage portals and clinical research data capture system) offer role-based access for the management of permissions. If *ad-hoc* applications provide the front-ends to databases, then compliance with institutional guidelines for control of permission and access should be followed.

3.6.2 Access Auditing

In addition to processes for data protection and access, it is important to have an audit infrastructure in place to capture who is accessing data, when they are accessing it, and how they are accessing it. Audit data is important when investigating unauthorized changes in data, security breaches, and failures of compliance. Audit capabilities need to be enabled within the applications and systems hosting the structured and unstructured data. This can be as easy as enabling built-in audit features or can require additional development within applications to document access to data hosted by the application. In either case, once the audit data is enabled or generated, the audit data needs to include the following items to ensure it can be used. The fields are built-into most protocols such as the Syslog and Windows event log:

- **Date/time of event:** All audit data should include the date/time of the event. This should be synchronized with other systems to ensure consistent time across events within the environment.
- **Log source:** The location / facility that generated the event (e.g. server, application, etc.).
- **Outcome:** Whether the event was successful / failure.
- **User/system:** The users and/or system that generated the event.

Once the audit data is generated and stored it is beneficial to correlate it with other events within the environment. Tools such as Security Incident and Event Management (SIEM), Security Incident Management (SIM), or Security Event Management (SEM) can be used to collect, store and correlate event data from multiple sources. These tools will take the event data and normalize it with other sources in the environment to provide a consistent picture of the event data and can be used to alert personnel in the event of violations of pre-defined rules. Audit data are typically reviewed on a regular basis so that suspicious events are not missed and possible breaches are quickly detected. Event data should be maintained within the event correlation systems for at least a month and can then be rotated to offline storage. This permits timely investigation of unauthorized access. Offline storage should be maintained for at least a year to ensure records can be restored and reviewed when required. This will provide adequate time to perform investigations and to meet compliance requirements.

3.6.3 Sharing Data with External Collaborators

While investigators need convenient options for sharing information with colleagues, host institutions need to protect themselves and their staff from data leakage, unintentional exposure of often-sensitive data, and hacking. Data can be transmitted by email or uploaded/downloaded using many of the data transfer protocols such as FTP and HTTPS. Content management systems such as Microsoft SharePoint, or the open source Drupal and Plone content management systems (CMS) have capabilities to securely share data files, but are not often optimized for large data sets. Xythos (Blackboard, Inc.) is an example of a commercially available application to facilitate data sharing. It provides a web interface to local storage with many convenient features, such as quota management, data sharing via web tickets, automated work flows, subscription services, and automatic expiration of access provisions.

E-mail, despite being a technology that is now several decades old, is still one of the most common ways to share data. Email is transmitted as plain text unless attachments are explicitly encrypted. Email servers have built-in size limitation, which is often the reason why researchers switch from email to an online storage for file sharing. Most institutions employ so called spam filters, which remove unwanted or harmful emails. Such filters can interfere with an exchange of attachments. There are commercially available encryption devices, which ensure that the emails and attachments are delivered to the intended user over an encrypted channel. In order to address limitations inherent to email, alternatives such as ZIX (http://www.zixcorp.com/) or Xythos are often deployed on premises to complement email. Besides the on-premise alternative, cloud solutions, such as DropBox and Google Drive, can be used for certain types of data, provided that the institution enforce appropriate conditions and limitations.

3.7 Managing Growth

Translational research programs pose enormous challenges in data storage and management because of the enormous growth in the quantity of data that is being generated. For example, because next-generation DNA sequencers and confocal imaging microscopes are capable of generating terabytes of data in a matters of days, the focus is shifting from how to create petabyte size storage servers (a relatively simple task) to deciding how to effectively *manage* what is stored on these servers and how it is stored. The key is exploring the data itself – what do they contain and can we leverage metadata to create an effective strategy for information lifecycle management.

3.7.1 Managing Growth of Structured Versus Unstructured Data

Structured data, by definition, is annotated by metadata and thus can be incorporated more easily into an information lifecycle. The implication is that the decisions around what to require for specific data types in terms of their associated metadata are crucial and have a high impact on the ability to manage the data effectively. One should attempt to find a balance among the quantity of information that needs to be stored, how it should be stored, how it will be backed up with appropriate methods of disaster recovery, and associated costs. Data-generating research cores such as the flow cytometry facility, confocal imaging core, sequencing cores and others represent a large fraction of the demand for storage. Therefore, it is especially desirable to ensure that all data generated by these cores can be stored in well-structured and annotated forms, making it conducive to cost effective strategies for information lifecycle management. Typically, the growth in demand for storage by these cores is fairly predictable, using simple assumptions such as average core utilization rates.

Of greater concern is growth in unstructured data, which can arise when users download (often large) data sets from public sources without annotating them or store duplicate copies of basic research data, which also resides in a repository in structured form. Growth in unstructured data is often unpredictable, especially if data storage is made readily available, and users have little or no incentive to be selective in what they store. The challenge for IT service groups is to steer growth from storage of unstructured data to storage of structured data with accompanying metadata.

3.7.2 Chargebacks and Quotas

Chargebacks create a financial incentive to minimize data storage and to delete data when it is not longer needed. For example, research groups must decide whether to keep all intermediate data related to an experiment or only those that cannot be recreated easily. The investigator must weigh the costs of storage (through chargebacks) against the cost to re-generate the data. Users can be assigned disk space they "own", i.e. users pay for allocation of a disk, which they then manage, rather than paying for storage as needed, which is assigned and managed institutionally.

A storage quota typically refers to the maximum amount of data a user can store. Once the user crosses the pre-set threshold, the system can lock the directory, file system, etc. and prevent the user from writing additional files. Alternatively, crossing the threshold can trigger an alert to notify the users, who then negotiate either a quota increase at increased charge or file removal.

3.8 Joint Management – Hospital and Research

Translational research typically occurs in a clinical environment. IT services in support of research may be managed by a hospital or by a separate academic organization. Hospitals typically have a large IT staff to support clinical care and business operations, especially if they have implemented an electronic health record. In theory, the hospital IT group could support the less demanding needs of research programs. In practice, this poses challenges because support of clinical care and critical business functions will always be of higher priority than support of research. In addition, research programs generally use many different customized software applications and platforms, while hospital operations typically are supported by products of commercial vendors. To support research in a clinical environment, it is important to define clearly who is responsible for providing IT support to the different missions – patient care, business operations, and research. In the authors' institution, clinical and business operations are supported by hospital information services (IS), while research is supported by biomedical informatics (BMI). The two groups are separately staffed and budgeted, but must collaborate very closely to provide coherent support to the overall research and clinical enterprise. The two have developed a responsibility matrix (Table 3.1), which has served the organization well and prevented misunderstandings. A similar matrix of responsibilities should be developed by institutions conducting translational research that requires collaboration among components of the organization or with other organizations (such as a medical school, university, hospital, clinical practice group, research foundation). In our case, the research group has taken advantage of the hospital's need to have a secure 24/7/365 data center and has incorporated a research data center in the same physical facility. Research has service agreements with the hospital to take advantage of the hospital's facilities and staff for data backups and disaster recovery. Because translational research requires extraction of data from clinical systems (see for example, Chaps. 6 – Research Patient Data Warehousing and 8 – EHR-linked Registries for Studies of Populations), BMI complies with the hospital's policies and procedures related to security and protection of information.

Table 3.1 Distribution of responsibilities for support of research between Hospital Information Services (IS) and Biomedical Informatics (BMI)

Issue	IS	BMI
Centralized enterprise data storage for research	X/J	P
Specialized data storage for research	J	X/P
Backups and archives of research data	X/J	P
Security	X/P	J
Authentication	X/J	P
Research applications hosting	J	X
Research web server hosting	J	X
Open source and non-enterprise software for research	J	X
Support for research applications	J	X
Support for enterprise applications that support research administration (e.g. grants administration, Institutional Review Board, animal care)	X/P	J
Content ownership, end-user training and support for enterprise research applications	J	X/P
Software evaluation and development for research	J	X/P
Clinical software implementation	X/P	J
Enterprise software evaluation and implementation to support faculty research (e.g. Electronic Laboratory Notebook, REDCap)	J	X/P
Manage licenses for commercial enterprise software	X/P	J
Manage licenses for research non-enterprise software	J	X/P
Help desk 24/7, Desktop support	X	J

X primary responsibility for operational support, *P* primary responsibility to formulate policies, *J* participates in formulation of policy

Acknowledgements This publication was supported in part by an Institutional Clinical and Translational Science Award, NIH/NCATS Grant Number 8UL1TR000077-04. Its contents are solely the responsibility of the authors and do not necessarily represent the official views of the NIH.

References

Harris PA, et al. Research electronic data capture (REDCap) – a metadata-driven methodology and workflow process for providing translational research informatics support. J Biomed Inform. 2009;42(2):377–81.

OpenClinica. [cited 2012 June 29]. Clinical trials software for electronic data capture. Available from: http://www.openclinica.com

Purcell S, et al. PLINK: a tool set for whole-genome association and population-based linkage analyses. Am J Hum Genet. 2007;81(3):559–75.

Rajasekar A, Wan M, Moore R, Schroeder W. A prototype rule-based distributed data management system. In: HPDC workshop on "Next Generation Distributed Data Management", Paris; 2006.

Rajasekar A, Wan M, Moore R. Event processing in policy oriented data grids. In: Proceedings of intelligent event processing AAAI spring symposium, Stanford; 2009. p. 61–66.

Rex DE, Ma JQ, Toga AW. The LONI pipeline processing environment. Neuroimage. 2003;19(3):1033–48.

United States. Health Insurance Portability and Accountability Act of 1996. Washington, DC: U.S. G.P.O.; 1996. 168p.

Chapter 4
Institutional Cybersecurity in a Clinical Research Setting

Nicholas Hunt and Michal Kouril

Abstract The principal challenge facing IT groups that support research on a daily basis lies in striking a fine balance: On one hand researchers must share data and use cutting edge analytic tools on a variety of computing platforms to enhance their creativity and productivity. On the other hand, much of the data that supports translational research contains personal health information derived from patients' medical records. Hospitals are justifiably concerned about the highly sensitive nature of the data and the need to comply with a myriad of federal, state and local laws and contractual regulatory requirements that demand high levels of security and access control. In this chapter we discuss these challenges and the approaches taken at a research intensive children's hospital to put a policy framework in place that enacts standards for security, monitoring, testing and design of the IT infrastructure. These protect the institution, while enabling collaboration and innovation among researchers. We stress the organizational need for a close and collaborative relationship between IT groups that support research and those charged with support of the medical center's clinical and business operations. It is also important to recognize that technology alone cannot assure security. Institutional policies and user education also play key roles in assuring that confidential information is in fact protected.

N. Hunt, MSIA, CISSP (✉) • M. Kouril, Ph.D.
Department of Pediatrics, University of Cincinnati College of Medicine, Cincinnati, OH, USA

Division of Biomedical Informatics, Cincinnati Children's Hospital Medical Center, 3333 Burnet Avenue, ML-7024, Cincinnati, OH 45229-3039, USA
e-mail: Nicholas.Hunt@cchmc.org

J.J. Hutton (ed.), *Pediatric Biomedical Informatics: Computer Applications in Pediatric Research*, Translational Bioinformatics 2, DOI 10.1007/978-94-007-5149-1_4,

4.1 Introduction

Translational research typically requires access to data that reside in geographically distributed data warehouses that are called upon by a team of collaborators using a variety of software applications. IT support of this type of translational research requires networks to permit data sharing, while maintaining high levels of security. Without networking, investigators cannot access the applications, servers, and other resources in a distributed environment. Given this criticality to providing services, networking can be the major infrastructure that provides an environment where access to data is restricted to authorized users and the overall system is protected from malicious attacks by unauthorized individuals (data breaches). According to the Verizon 2011 Data Breach Report (Verizon Data Breach Report 2011), which outlines the incidents investigated by Verizon security response teams and publicly accessible data, the top 2 of 10 threats to systems involved direct hacking against servers. Given this threat, network security and sound network security practices can provide a large layer of protection against current threats to the security of the environment.

4.2 Secure Network Design

In this section, secure network design practices along with models for architecting secure networks will be covered including data-centric network, firewalls, intrusion prevention/detection systems (IPS/IDS), and remote access/virtual private networking (VPN'ing).

4.2.1 Data-Centric Networking

The traditional model for building enterprise networks takes into account the different trust zones within which applications must be accessed and published. These typically include: (1) the internal zone for trusted employees and machines on the Local Area Network (LAN), the Wide Area Network (WAN), and those authorized to remotely access the network; (2) the Demilitarized zone (DMZ), a subnet that provides limited connectivity to the internal network; and (3) the external untrusted Internet zone. This model has worked well for most organizations, providing a barrier between internal and external environments, as well as segregating different populations of users. However, access requirements for most applications are evolving, as technology changes so this model may not work for all situations. Medical centers are finding that users are increasingly using mobile devices, requesting access from all locations, and requiring more help with troubleshooting, when accessing applications from different locations. Given this,

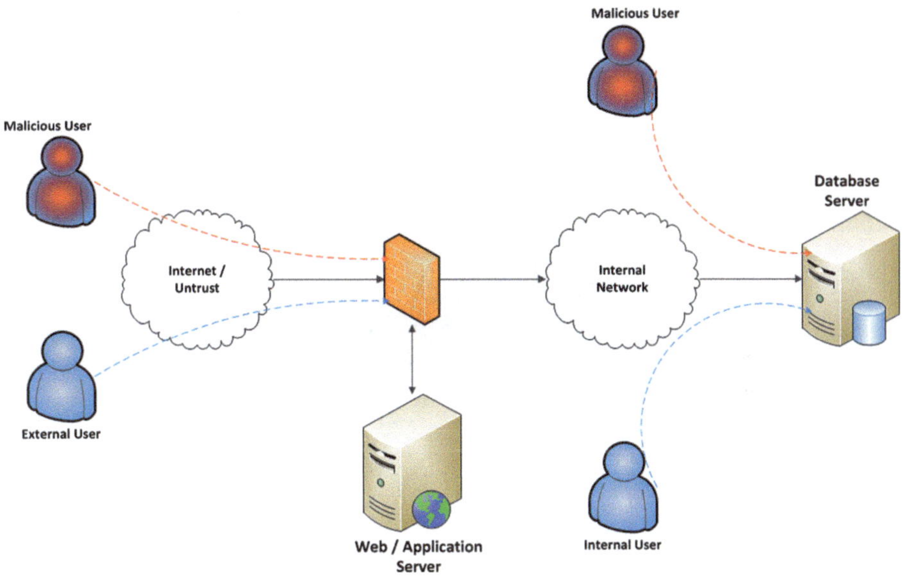

Fig. 4.1 Typical architecture of a network with different zones of trust

the barriers between the internal and external environment are becoming less clear and less stringent than they were originally intended. To address this, the concept of data-centric networks has evolved. The aim is to re-architect networks around protecting data, unify the internal/external experience, and build core protections for what must be protected; the data.

Figure 4.1 outlines the setup of a typical network with the different trust zones, using an application server that has a database backend. External users and any malicious users on the Internet will be restricted from accessing the database server directly and have limited access to the Web/Application server through firewall filtering. However, any user who is on the internal network (malicious or not) will have direct access to the database server, where the application data may be stored. Given vulnerabilities or misconfigurations that may be present within the server or database application, users may be able to bypass the application protections and pull data directly from the database, whether they have been authorized or not. Given the scale and size of most enterprise networks, the malicious user in Fig. 4.1 could be accessing via remote offices, affiliates, or other locations that may not have the same level of physical security as a main campus or office may have.

To address the risks posed by the direct access to data as shown in existing network design, data-centric networking looks at building rings of protection around data. Figure 4.2 provides a high-level overview of this design philosophy. This architecture follow similar security designs used within CPU privilege rings (so-called Ring 0 access) (Processor Privilege Levels). Starting from the center of the circle and moving outward, the rings and access become progressively less

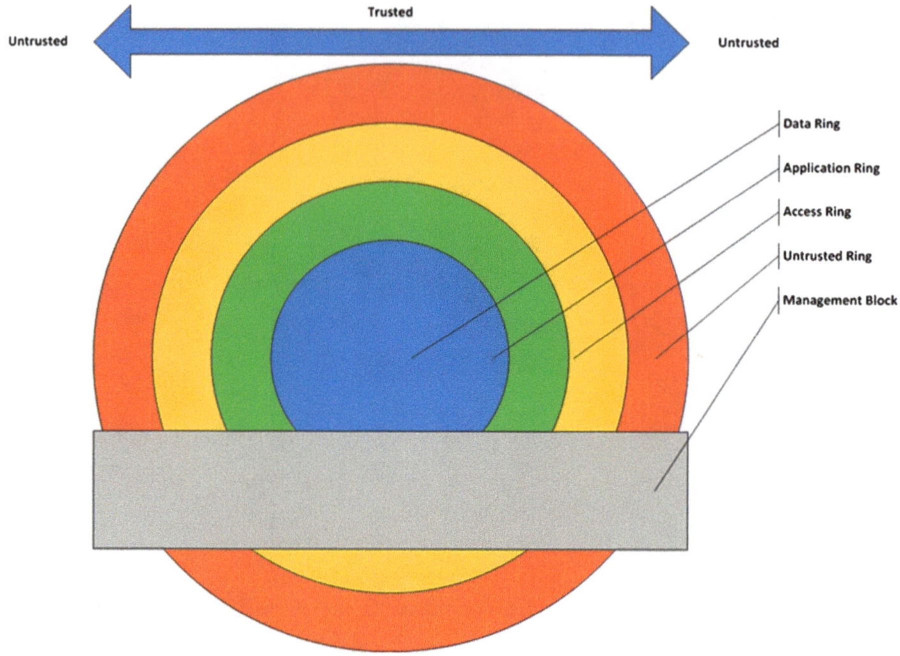

Fig. 4.2 Building rings of protection around data

trusted. No user or process can jump a ring without going through a ring above it. This forces all access to data through known applications and paths that can be hardened, monitored, and protected using internal controls. Rings include:

- Data Ring: This would include all structured and unstructured data sources such as database servers, file servers, etc.
- Application Ring: This would include all application servers serving content or publishing data that is accessed, used, or manipulated by users.
- Access Ring: This is the main ring that presents the initial login, authentication, and authorization to the users. This would typically include VPN, proxies, etc. that are used for access to the environment.
- Untrusted Ring: This is anything outside of the environment. This can include both the Internet and internal network segments.
- Management Block: This would include management systems used by system administrators to maintain the environment, such as patching, authentication, backups, etc.

Figure 4.3 shows a network with the same type of users as the network in Fig. 4.1, but has been re-architected in accord with the data-centric network model. As shown in the Figure, there is no direct access to the database server or application except from the management layers. Users can access an application only through the access server, which will present the data back to them. All access to the application

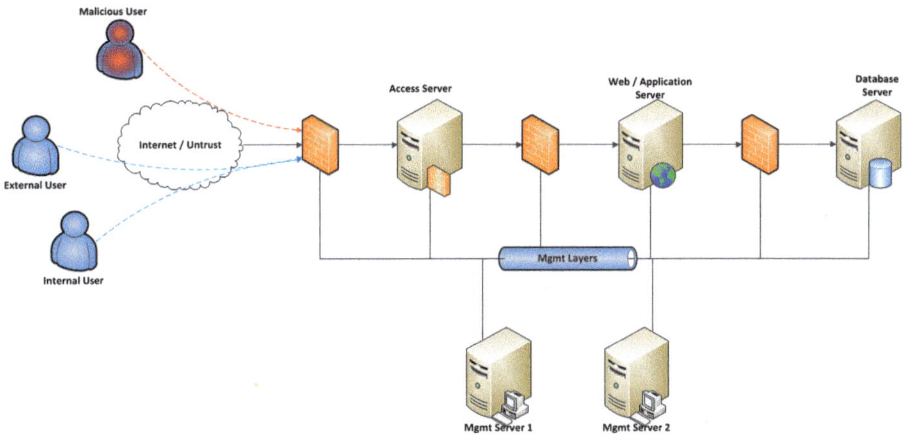

Fig. 4.3 The network shown in Fig. 4.1 has been re-architected in accord with principles of the data-centric network model

is through the external firewall, to the access server, and then to the application. At these initial rings, filtering for users, locations, etc. can be put in place to further limit the scope of access by users. None of the traffic is ever sent to the application or database server until it has been properly authorized to be legitimate traffic within the environment. The data-centric network model reduces the access footprint on the database server from the typical 50–200 ports that may be accessible in the traditional model down to zero ports because only the application server can access the database server. This reduces the risk of unauthorized internal users accessing the database server. It also decreases the possibility that malicious users can gain access to the internal network or access the server to launch attacks. Some of the other benefits of this model include:

- Define Firewall Access: Since all application and data access must go through a firewall interface, specific inventories of port usage are gathered during the deployment of applications and can be used for auditing and compliance, when reviewing servers and applications in the environment.
- Unified User Experience: A single method of access is used by all authorized users regardless of location, i.e. the process of access is the same, whether they are at an internal (e.g. at work) or external (e.g. at home, travelling).
- Enforced Standards/Monitoring: Since all access to the applications from within the internal and external zones is funneled through a limited number of access methods in the access ring, more monitoring of access and enforcement of standards can be enabled to reduce the potential of successful attacks on the environment.
- Outbound Access Protections: Since all the application and database servers are on internal limited access segments, outbound access from the servers can be denied. This will reduce the possibility of an attacker gaining access to the system and installing software to copy data to outside sources. A user who gains access

to the internal network through the firewall can only access services/resources that have been pre-authorized. This reduces the possibility of data escaping the environment.
- Reduced Horizontal Attack Surface: Since most environments are only as secure as their weakest server or application, limiting access between rings and systems within the rings reduces the possibility of a lower-security application server being compromised and then used as a pivot point to launch attacks against more secure systems.

There are a number of challenges that must be overcome in using the data-centric network model. These include:

- Performance: In a typical application/database scenario, data access is relatively simple and is over a switched network or routed across a core router (Fig. 4.1). Typically, performance is excellent. In a data-centric network, all application database access must be sent through a firewall with limited access (Fig. 4.3). This can impact database access times and cause application performance issues based on the bandwidth the firewall can handle. Large infrastructure firewalls or firewall service modules in core routers/switches can address this and allow for setup of different network and trust zones.
- Network Re-Architecture/Change: Implementing the data-centric model requires a re-design of existing environments and for example, moving servers, retraining staff, rerouting connections, and implementing additional firewalls. This is expensive and can be quite stressful to the IT staff.
- User Access: When database servers are on internal networks, users may become complacent/accustomed to accessing servers directly when they are in their office/workspace. In the new model, users at the office, as well as at home, will be required to VPN or use other remote access methods to access databases and other unstructured data within the environment. Users may require substantial training to learn new processes.
- Firewall Changes: When there are no firewalls or other protections in place, server owners and administrators may not have accurate inventories of what ports are required for an application to run. When moving to the data-centric model, issues can arise because non-standard ports are being used and may not have been added to new firewall rules, thus causing issues with application access/operation. Proper sniffing and network monitoring can help identify the ports so that firewall policies can be established before applications are migrated to the new environment.
- Application Development: Developers writing applications for the data-centric environment need to make sure they are writing their applications to use standard database ports or confined port ranges to ensure they can work through different layers of firewalls. When ephemeral ports and other dynamic ranges are used, this can cause issues in negotiating firewall policies.
- Outbound Application Access: Many applications today are built to have auto-update or other web service integration to pull in and process data used within the application. In an environment with limited to no outbound-access,

application calls to external resources must be inventoried and allowed to use the infrastructure or other proxies to access. Open source products such as Squid (Squid Cache Proxy Project) Proxy or built-in proxies in firewalls can be used to allow access to a restricted number of external resources.

When implementing the data-centric network model, it is important to assess the environment and use of applications within the environment. Given the right architecture and use of the model, the footprint of threats to applications and data within the environment can be drastically reduced with improvement in security.

4.2.2 Intrusion Prevention/Detection

Once the secure infrastructure and firewalls are properly implemented, unwanted traffic is stopped. Given the frequency of attacks and the number of potential vulnerabilities in all systems, best practices dictate that monitoring be in place to alert when possible attacks are taking place within the environment. Continuous monitoring will help respond to attacks in a timely manner, reduce the potential impact of the attack and reduce the possibility of the attack occurring again within the environment. To achieve this, Intrusion Prevention Systems (IPS) or Intrusion Detection Systems (IDS) can be configured to monitor key ingress/egress points and critical segments of the network to alert and/or block attacks as they occur. There are different types of IDS/IPS as well as different models for deployment to be considered, when planning to implement monitoring of the environment. The main types of IDS/IPS include (NIST IDS IPS Guide):

- Network-Based: monitors network traffic across particular network segments or devices and analyzes the network and application protocol activity to identify suspicious activity.
- Network Behavior Analysis (NBA): examines network traffic to identify unusual traffic flows that may represent threats, such as distributed denial of service (DDoS attacks), certain forms of malware, and policy violations.
- Host-Based: monitors the characteristics of a single host and the events occurring within that host for suspicious activity.

Network-based IDS/IPS is generally the most prevalent type of monitoring, but host-based monitoring is a rapidly improving technology and provides detection of a broader range of internal and external attacks. When planning to deploy an IDS/IPS system, the deployment model plays a key role in determining what the system can provide and how it affects the entire environment. In a traditional IDS role, the system is placed in an out-of-band mode (Fig. 4.4), whereby a monitor port or tap is copying all traffic to the monitoring system. IDS analyzes the traffic to detect an intrusion/attack. In this scenario, a possible attack generates an alert to a console or response team that responds to mitigate the attack and address the vulnerability that lead to the attack. This is the model that is used in most environments, given the

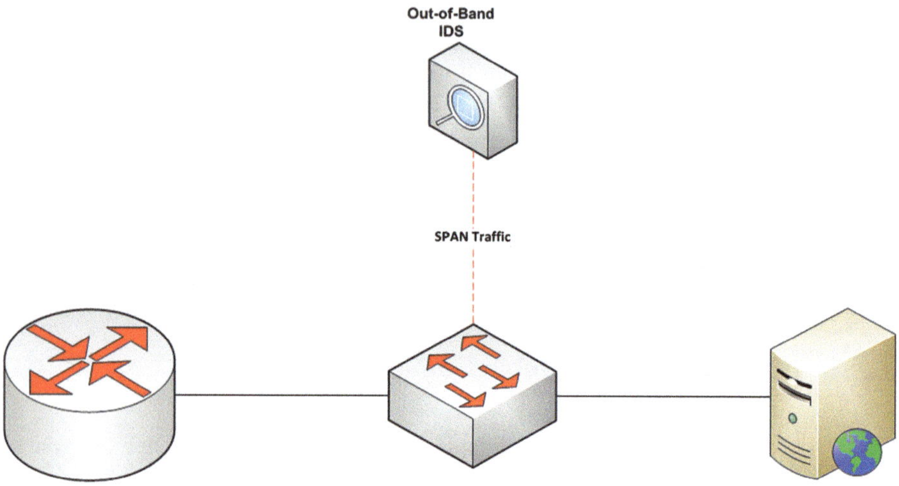

Fig. 4.4 Out-of-band intrusion detection system

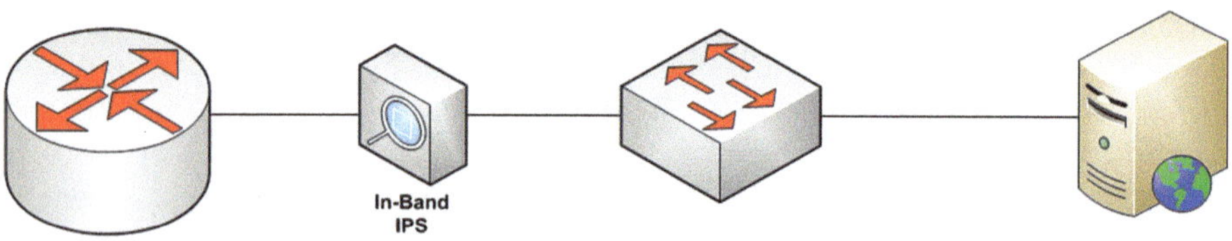

Fig. 4.5 In-band intrusion prevention system

concerns over blocking legitimate traffic and the possible impact an inline device may have on flow of legitimate traffic.

In the Intrusion Prevention Mode all traffic is routed directly through the IPS system, inspected, then sent to the destination host/network (Fig. 4.5). In this scenario, traffic designated as an attack is dropped directly inline and is not sent to the destination. This model of implementation is rarely used because of the fear that legitimate traffic could be slowed or blocked. However, in terms of attack mitigation, IPS provides the quickest response and most effective way of detecting and responding to attacks.

When planning to deploy an Intrusion Prevention System, options are available that can help address the typical concerns. The key options include:

- Performance: When purchasing/building an IPS, the network must have enough bandwidth to handle the amount of traffic that will be forwarded through the device. A number of vendors have technical specifications published for appliances, but purchasers must test the inline inspection using traffic simulators to ensure the system can handle the volume of traffic it will be inspecting in their institution. In any case, anticipate at least a 10–20% negative impact on traffic flow to allow for full inspection of sessions.

- Network taps: When placing the system inline, bypass network taps must be purchased or built so that systems can be taken offline without affecting traffic flow in the environment. In maintenance windows, bypass taps can be put in bypass mode allowing reboots, updates, etc. without affecting the traffic in the environment.
- Placement: An IPS deployment for every network segment within an environment would be very costly. When planning to implement IPS, the institution should plan to place it in critical segments or ingress/egress points to get as much coverage as possible, without being required to deploy multiple devices to get complete coverage of the network. Systems placed within access infrastructure and key uplinks to routers/switches can give coverage across the environment, while not requiring multiple expensive devices to cover all segments of the network.
- Tuning/configuration: Signatures that are included in the product and from the vendor must be used. A number of implementations fail because signatures have been disabled, tuned out of use, or removed. Most vendors have so called 'Low-Fidelity' or 'High-Accuracy' signatures, which accurately distinguish between legitimate and illegitimate traffic, thereby minimizing false-positives that may block legitimate traffic. At a minimum, high accuracy signatures must be enabled for true blocking, whereas less accurate signatures are enabled for detection only.
- Spend time tuning: When deploying the system, it is critically important to spend time developing a baseline to learn what is normal traffic in the environment, i.e. tuning the system. Ad-hoc configurations and deployments without adequate tuning can lead to a situation where signatures that have not been properly tested must be disabled and are not monitoring the environment. Adequate time spent tuning and building the signatures will lead to more effective monitoring and blocking.

Once an IPS is in place and properly tuned, it can provide a wealth of knowledge and visibility into the environment. In networks with high traffic flow, inevitably both true and false alerts will be generated. Alert fatigue may occur. True alerts of attacks may be ignored or missed in the myriad of alerts that are generated by the IPS. To address this, a plan for monitoring and responding to alerts must be developed before the IPS is implemented. Unless there is a 24/7 presence of staff within the environment, organizations should evaluate the possibility of using a Managed Security Services Provider (MSSP) to monitor the system and alert staff when an attack is detected. An MSSP augments current staff and will help them gain knowledge of current trends. Such knowledge can be helpful in identifying actual attack traffic and reduce the amount of time spent monitoring and responding to attacks.

IDS/IPS systems complement firewalls. They allow monitoring of traffic that has been authorized for possible malicious content and facilitate responding to attacks in a timely manner. Properly deployed and configured within the environment, IDS/IPS provide additional layers of protection for the institution, which handles sensitive information and is potentially liable for breaches of security.

Table 4.1 Methods of remote access by various applications

Application	Remote user access	Remote admin access
Public site	http/https	Db Server Studio, https
eCommerce site	https	https
eMail	https, eMail Client (MAPI)	RDP, https

4.2.3 Remote Access/VPN

Translational research is inherently collaborative with both researchers and data frequently located at multiple geographically separate sites. Researchers need remote access to multiple networks. An infrastructure to support remote access must be implemented to allow users to access the private environment securely and reliably. There are many methods of remote access, ranging from traditional Virtual Private Networking (VPN) to other methods such as Remote Desktop/Terminal Services. Advantages and disadvantages of each must be considered when planning the infrastructure to permit remote access. It is important to take into account all the different methods that are going to be used/required /supported for remote access to the environment. Given the data-centric model for the application infrastructure, all access to applications can be considered 'remote' since there is no internal network. Looking at remote access in this way is valuable, when deciding what is going to be supported in the environment. To begin, planners must inventory: (1) the applications within the environment that are going to be accessed; (2) how users will access the applications; and (3) how administrators will access the applications. Most recently developed applications utilize web services, or have some form of web service client that accesses the application. Given this, some form of web remote access will be required and is generally provided through web-proxies or some other form of remote access. Administrators of applications may require access to terminal services or to some other type of client/server application. Table 4.1 can guides an inventory of the type of remote access that is going to be required for the environment.

Once the applications and their methods of remote access have been established, a proper set of remote access tools can be selected and implemented. When looking at remote access tools, remember that more than one method will generally be required. Not all applications, users and security models for remote access are going to work for all applications and users in the environment. It is important to select a few options that will work for most of the environment, then work to integrate the few that do not work into what is supported. Table 4.2 lists several options/methods for remote access along with their typical use, advantages and disadvantages.

Once methods for remote access have been selected for a particular environment, administrators can focus on the tools necessary to support all users and can configure them properly. Proper configurations include:

- Limited Access: Users should be restricted to those applications and services for which they have been pre-authorized. The default 'any' or 'allow all' methods of access should not allowed in the environment. Different institutional policies are generally required for different classes of users.

Table 4.2 Targeted use, benefits, and issues with various methods of remote access

Method	Target use	Benefits	Issues
VPN (IPSEC)	Full network access to the environment	Is mature and been around a long time. Is built into most security and firewall devices.	Can be problematic with slow/unreliable connections Gives user full network access (IP access) to the environment Network ports may not be opened from some secure networks
VPN (SSL)	Full network access to the environment Proxy network access to the environment	Uses standard https protocols Reliable across slow/unreliable network connections Open from most environments	Gives user full network access (IP access) to the environment Some require concurrent user licenses which can increase cost of the solution
HTTP	Web-based applications over public and private networks	Browser-based Familiar to most users Has handlers/methods for building security onto protocol	Can be insecure in default deployments Very visible attack footprint
Terminal Services/Remote Desktop	Full console access to remote servers and devices	Traditional desktop experience Full access to remote system Built-into Windows operating systems	Can be insecure in default deployments Older tools may not support authentication/encryption requirements Can give elevated access to systems Is not built for open/public networks using default configuration
Secure Shell (SSH)	Remote user login to shell (linux) Remote user tunnel	Secure transport protocol for open/public networks Low cost	Takes more configuration from client side Requires configuration changes on some client machines

- Logging/Monitoring: Logs of access by users regardless of method must be in place. Syslog and other Security Information and Event Management (SIEM) tools can be used to collect and analyze the remote access logs so that they can be generated when needed.
- Central Authentication: Ensure the remote access infrastructure uses a common identity format so that users are not required to remember multiple user-names/passwords and/or tokens to access the environment. Inconvenient access can encourage people to share credentials or methods of remote access, which can lead to elevated access privileges and violation of institutional policies.

A well-conceived strategy for remote access with methods supported by the enterprise can provide access to internal systems and applications in a secure manner. Systems that are convenient for users permit them to perform their jobs in a secure environment, while insuring that the institution complies with federal and state laws, regulations, and policies that govern access to sensitive data.

4.3 Operating Systems and Cybersecurity

Operating systems of applications are probably second to networks in terms of implications for security of the environment. Operating systems of servers are complex and. have vulnerabilities that can lead to compromise of applications. It is important to choose operating systems that meet institutional needs, secure them by default, and continuously monitor and improve the security of the operating systems within an institution's particular environment. This section reviews challenges of securing operating systems with specific examples of how these can be addressed.

4.3.1 Configuration

Configuration standards, policies, and processes document the setup steps and procedures for configuring applications, network devices, servers, etc. within the environment. Configuration standards should be documented for each application or device within the environment. This ensures that applications and devices are consistently setup, securely built based on best practices, and are accessible in the event that primary support personnel are not available.

Initial configuration standards must be built for a particular environment. They should include:

- Best Practices: Do not re-invent the wheel. There are a number of sources of standards that have been well tested and can serve as a starting point for an institution (National Vulnerability Database Checklist Program; Center for Internet Security).

- Do Not Copy Standards: Many of the standards that are available are purpose built for specific regulatory and/or security requirements and may not be fully applicable to a particular organization and its environment. Copying the standards and logos of others may bind an organization to standards that cannot be achieved and will invite failure in an audit of compliance. Institutions may use the standards of others as a guide, but standards adopted by an institution must fit its specific needs.
- Conduct Regularly Scheduled Reviews: Once a standard is written, it needs to change over time as technology, regulations, and the environment change. Configuration standards should be reviewed at least annually to make sure they are still relevant and accurate.
- Auditable: Compliance with standards must be auditable. This can be done by manual, scripted, or automated reviews. The Security Content Automation Protocol (SCAP) provides specifications and practical guidance to enable automated vulnerability management and audits of compliance (Security Content Automation Protocol).

There are a number of good sources for configuration standards and checklists. Examples include:

- Center for Internet Security (CISecurity)
  - http://benchmarks.cisecurity.org/en-us/?route=downloads.benchmarks
- NIST National Vulnerability Database Configuration Checklist Project
  - http://web.nvd.nist.gov/view/ncp/repository

In addition to these vendor neutral databases, most vendors provide guides to setting up the operating system with proper levels of security. Important items to consider, if a vendor guide is not available or is incomplete include:

- Services/Daemons: Disable all unnecessary services, programs, applications, and daemons that are not explicitly needed by the operating system. This reduces the footprint of the server and reduces the possibility of unneeded software affecting the operating system.
- Limited Access/Authorizations: Only allow local access to those administrators and power users who must directly access the system. This can be achieved through local authentication/authorization requirements along with network level controls for limiting access to administrative ports and services.
- Secure Management: When allowing remote management, insecure management protocols such as telnet, rsh, VNC, etc. should be disabled and only secure management protocols enabled. This will reduce the possibility of individuals intercepting credentials of administrators and using them for malicious purposes.
- Auditing/Logging Events: System should be configured to log access and changes to the system. This will provide an audit trail of who made what changes and when. This can be useful in investigations of possible breaches of security and in troubleshooting.

- Patching/Updating: Build requirements into the standards to require periodic patching and updating of systems to ensure they have the latest software and any vulnerabilities that have been identified are patched as soon as possible.

When writing configuration standards, they should not be limited to operating systems and other network based devices, but should include any applications within the secure environment. This helps ensure consistent setups, knowledge transfers, and continued secure builds of applications running within the environment.

4.3.2 Patching

Given the size and complexity of applications and operating systems, it is inevitable that vulnerabilities will be found and require patching. This is highlighted in the latest Symantec Internet Threat Report (Symantec Threat Report), which noted a 30% increase in the number of vulnerabilities in software releases in 2010 compared to 2009. Patches are released to address vulnerabilities, after a period of delay while methods of correction are developed. If left unpatched, the vulnerabilities can lead to compromise of the system. To ensure these patches are applied and the systems are secure from these vulnerabilities, a patching program that includes regular scanning and updating needs to be put in place.

The first step in implementing a patching process is to regularly scan the environment for potential required patches. The scans should not be limited to server and applications, but should also include the network equipment (switches, routers, firewalls, etc.) within the environment. Vendors generally release patches and updates for their equipment. Systems such as Windows and Linux have automatic update capabilities that can download and install the available patches for the system. There are products that automate scanning to identify vulnerabilities and available patches. Examples include Windows Server Update Services (WSUS) (Windows Server Update Services), Freeware OpenVas (OpenVas), Microsoft Baseline Security Analyzer (MBSA) (Microsoft Baseline Security Analyzer), and Secunia Personal Software Inspector (PSI) (Secunia Personal Security Inspector). Once the list of required patches is developed and the available patches are downloaded, they must be installed on all the affected systems. As verification that patches have been installed properly within the environment, verification scans should be run against all the systems. This can be part of the follow-on monthly scans to ensure continuity and continued updating of systems within the environment.

The real challenge to implementing a patch process is setting a standard and regular process for installing patches. Most vendors have moved to a once a month release of patches (outside of critical updates). Organizations can follow the same schedule and regularly install patches on all systems, applications, and network devices within the environment. This will ensure the continued update and review of all the systems within the environment.

The inevitable argument that is raised is: 'This will cause issues with my [system, application, device]'. If a patch or update has been shown to have compatibility issues with a particular application or device, there must be a process to document the exception, approve it, and regularly review it. The documentation should include the source of the exception (from support, through testing, etc.), who generated the exception (owner), and the expected timeframe for resolving the problem. During the regular scanning and installing of patches, all exceptions should be reviewed to ensure they are still applicable and required within the environment. Failure to resolve exceptions can lead to accrual of systems and applications within the environment that are out of standard and create potential vulnerabilities that could adversely affect the rest of the environment.

4.4 Protecting Sensitive Information

Sensitive information such as Personal Health Information is commonly used in translational research and must be protected. Typically, it is captured and stored in three types of systems: Research Patient Data Warehouse, tools for Electronic Data Capture that may include data storage, and Research Data Storage systems.

4.4.1 Protected Health Information

Protected Health Information, also known as Personal Health Information, (PHI) does not include publicly available information that is lawfully made available to the general public from federal, state, or local government records. Protected health information that cannot be exposed to individuals without permission of the patient is defined as information that when used alone or in combination with other information can identify an individual (whether living or deceased). Table 6.1 lists these identifiers, as defined in the United States by the Health Insurance Portability and Accountably Act (HIPAA) Privacy Rule.

4.4.2 Research Patient Data Warehouse

The Research Patient Data Warehouse (RPDW) is one of the most highly protected components of the IT infrastructure in a research-intensive medical center. Typically it contains information from the electronic medical records of large numbers of patients. See Chap. 6, Research Patient Data Warehousing, for an extended discussion of data warehouses. Protection of PHI within a RPDW requires special attention to:

- Access control
- Auditing of control and data changes
- Strict change control of any alteration of the system

A RPDW typically contains huge amounts of data, e.g. patient's demographics, diagnostic and procedure codes, medical history, results of laboratory tests, images, and genomic data. Accruing, storing, and searching the data in a timely manner requires high performance systems. Security systems must be carefully designed to avoid overly compromising performance of the warehouse. It requires a balance between ability to process and deliver data in a timely manner and ability to implement security controls that adequately protect the data. Performance can be improved by moving some controls to auxiliary systems surrounding the RPDW:

- Increased physical security of the data center as a trade off for on-disk data encryption
- A dedicated network segment surrounded by a firewall as a trade off for an in-band firewall
- Accessing data through an application server with a tight control rather than a direct database connection
- Application server to database connection might not be encrypted for performance reasons putting an additional burden on security of the application server
- The audit trail might be stored externally rather than stored on the system holding the RPDW

Many of the trade-offs are specific to an institution, its environment, and infrastructure. Institutional specificity might be related to:

- Processes for change control
- Types of queries or in general how the system is utilized
- The process of loading data (nightly full or incremental refresh, continuous/online versus batch load)
- The system of data backups such nightly, online DR, virtual vs. physical tape, etc.
- Method of user access (accessing a single RPDW or accessing a separate datamart that contains an extract of the primary data in a separate database)
- Reporting utilities and the security of service accounts accessing the data
- Utilizing all database built-in security controls – row level security, encryption

4.4.3 *Tools for Electronic Data Capture*

Tools for electronic data capture are commonly provided to investigators as part of the design and implementation of translational research projects. These allow researchers to collect data to be used in studies, clinical trials, etc. From the point of view of those responsible for security, it is important to note that tools for electronic data capture are typically user facing and hence require strong identity and access management. One such system might host a few or 100s of studies making it a high level target for unauthorized users. In order to minimize the possibility of data compromise:

- Encryption between the client and the server is a must. This is typically done by SSL (https), but additional layers such as VPN or SSL VPN might be used
- Regularly scheduled changes in passwords must be enforced, typically at 90 day intervals
- Role management – distinguish between study coordinators and data entry personnel. Grant permissions based on need and role.
- Entitlement reviews – ensure that only the people who are entitled to access receive access and regularly review these entitlements.
- Comply with best practices such as regular server patches, centralized review and storage of the logs, change and configuration management.

4.4.4 Research Data Stores

Biomedical research institutions typically house many data generating groups/cores. In order to share data efficiently, network attached storage (NAS) or online (web) based storage systems are generally used. Much of the data in a medical research institution will contain Protected Health Information, as previously discussed for Research Patient Data Warehouses. There are a several basic considerations, when planning to protect information. In a NAS environment it is important to realize that:

- The data in flight might not be encrypted. The Common Internet File System (CIFS) does not have encryption built-in, whereas the Network File System (NFS) does in version 4 (NFSv4). Direct Internet access to file servers should be provided only through VPN.
- Although users can manage permissions to some degree themselves in CIFS and NFS environments, the built-in permissions management might not be ideal. Add on products and careful monitoring of permissions is encouraged in order to keep the permissions well-defined and manageable long term.
- Identity within the NAS is only as strong as the institution's identity management system. Inadvertent additions of users to groups, confusion of users with the same or similar names, unclear definition of groups and the data to which members have access are some of the issues that can plague an identity management system and must be clearly resolved. It is important to develop standardized workflows for addition and deletion of users and groups.
- Audit-trails are essential not only for compliance with laws and regulations, but as a best practice. Many users fail to realize the side effects of, for

example, moving or removing data directories. Having the ability through audit trails to quickly identify who did what, when and where helps dispel many misconceptions and improve trust in the system.
- Data classification
- Data storage guidance (what goes where)

Web based data stores typically handle internet access better that NAS, yet can be mounted as a network drive (WebDav). With regard to security, some characteristics of web based data stores are:

- Encryption using SSL (https)
- Authorization/permissions are not limited to pre-defined protocols, but are dependent on the hosting application.
- Audit trails are easy to add
- During the upload process, information can be automatically parsed from the data and hence allow more flexible searches
- Typically stores "structured" data

4.5 Role of Users in Protecting Information

The focus of this chapter on institutional cybersecurity has been on technology. In reality, breaches of security with loss of protected or confidential information usually occur because users fail to comply with institutional policies and standards rather than because of technical failures of network security and intrusion prevention systems. For example, theft, loss, and/or misuse of portable devices are one of the most common breaches of security across industries (Verizon Data Breach Report 2011). It is critically important that institutions develop and enforce policies defining acceptable and unacceptable uses of information resources. Institutional policies and standards should address such issues as:

- Behaviors of users must be consistent with the mission, vision, and core values of the institution.
- Users must comply with applicable laws, regulations, and policies.
- Users acknowledge that sharing of Confidential Information with unauthorized individuals may result in the waiver of legal protections and may place the user and employer at risk of criminal or civil liability or may damage their financial standing, employability, privacy, or reputation.
- Users are individually responsible for maintaining the confidentiality, security and integrity of their chosen passwords.
- Users must promptly and properly dispose of, sanitize, and/or destroy confidential information in any form (e.g., paper, electronic, on Portable Devices) when no longer useful
- Users consent to institutional auditing and monitoring for excessive or inappropriate personal use, criminal activity, regulatory violations.

- Users understand that violations of any policy may subject them to disciplinary action up to and including termination of employment.
- Users must ensure that: Confidential Information is not visible in an unattended work area or on an unattended workstation; they log off of systems after access; they do not post passwords or user IDs in visible areas.
- Users are responsible for ensuring optimum security of Portable Devices (regardless of device ownership) that access, transmit, or store the institution's Confidential Information. Users must immediately report any loss or theft of computing devices.
- Users are prohibited from interfering with the intended functioning of the information system, e.g. disabling security devices, inserting viruses, inserting network monitoring devices, installing unapproved software, destroying files.

To summarize, maintenance of high levels of cybersecurity to protect confidential and protected health information requires a combination of best practices in technology and thoughtful institutional policies that are enforced. One without the other precludes success.

4.6 Joint Management – Hospital and Research

As discussed in Sect. 3.8, to support research in a research oriented medical center, it is important to define clearly who is responsible for providing IT support to the different missions – patient care, business operations, and research. In the authors' institution, clinical and business operations are supported by hospital information services (IS), while research is supported by biomedical informatics (BMI). The two groups are separately staffed and budgeted, but must collaborate very closely to provide coherent support to the overall research and clinical enterprise. With regard to operation of networks and implementation of cybersecurity, IS and BMI have developed a responsibility matrix (Table 4.3). This has served the organization well and prevented misunderstandings. IS is responsible for installing and maintaining the physical networks, including firewalls. BMI works with IS to design networks that support research programs and to develop standard operating procedures for

Table 4.3 Distribution of responsibilities for support of research between Hospital Information Services (IS) and Biomedical Informatics (BMI)

Issue	IS	BMI
Physical networks	X/J	J
Security	X/J	J
Authentication	X/J	J

X primary responsibility for operational support, *P* primary responsibility to formulate policies, *J* participates in formulation of policy

identity and access management. The goal is to create an environment in research that is secure and meets regulatory requirements with regard to protection of sensitive information. With this in place, the hospital can transfer clinical information from electronic medical records to the research data center with assurance that the information will be properly protected.

Acknowledgements This publication was supported in part by an Institutional Clinical and Translational Science Award, NIH/NCATS Grant Number 8UL1TR000077-04. Its contents are solely the responsibility of the authors and do not necessarily represent the official views of the NIH.

References

Center for Internet Security. [cited 2012 June 11]. Available from: http://www.cisecurity.org
Microsoft Baseline Security Analyzer. [cited 2012 June 5]. Available from: http://technet.microsoft.com/en-us/security/cc184923
National Vulnerability Database Checklist Program. [cited 2012 June 11]. Available from: http://web.nvd.nist.gov/view/ncp/repository?tier=&product=&category=&authority=&keyword=
NFSv4: overview of new features. [cited 2012 June 11]. Available from: http://www.iaps.com/NFSv4-new-features.html
NIST IDS IPS Guide [cited 2012 June 5] Available from: http://csrc.nist.gov/publications/nistpubs/800-94/SP800-94.pdf
OpenVas. [cited 2012 June 5]. Available from: http://www.openvas.org/
Processor Privilege Levels. [cited 2012 June 5]. Available from: http://cs.usfca.edu/cruse/cs630f06/lesson07.ppt
Secunia Personal Security Inspector. [cited 2012 June 5]. Available from: http://secunia.com/vulnerability_scanning/personal/
Security Content Automation Protocol (SCAP). Available from: https://scap.nist.gov/
Squid Cache Proxy Project. [cited 2012 June 5]. Available from: http://www.squid-cache.org/
Symantec Threat Report. [cited 2012 June 5]. Available from: http://www.symantec.com/threatreport/print.jsp?id=threat_activity,vulnerabilities,malicious_code,fraud_activity
Verizon Data Breach Report. 2011 [cited 2012 June 5]. Available from: http://www.verizonbusiness.com/resources/reports/rp_data-breach-investigations-report-2011_en_xg.pdf
Windows Server Update Services. [cited 2012 June 5]. Available from: http://technet.microsoft.com/en-us/updatemanagement/default.aspx

Chapter 5
Protecting Privacy in the Child's Electronic Health Record

S. Andrew Spooner

Abstract With the increasing availability of electronic health records, translational researchers frequently wish to extract information from or add information to these records to support various research projects. It is important that investigators be familiar with laws and regulations that protect the privacy of patients and restrict who can view or extract information from their EHRs. Laws and professional standards demand that all individually identifiable health information be secured at a high level and handled as private, sensitive information. Some information (mental health, reproductive health, abuse) is generally regarded as more sensitive than others in the health care of a patient at any age. In a situation where minors are involved, these security and privacy policies become more complicated because of possible conflicts between the interests of the child and the interests of parents or guardians. Situations specific to pediatric clinical practice and research increase the difficulty of implementing these policies: adolescent care, adoption, fetal care, foster care, and genetic disease. Security policies for access to systems intended to be used by patients (personal health records and patient portals) are complex. They can become even more challenging when the child has participated in clinical research and unexpected clinically relevant results are obtained.

S.A. Spooner, M.D., M.S., FAAP (✉)
Department of Pediatrics, University of Cincinnati College of Medicine, Cincinnati, OH, USA

Division of Biomedical Informatics, Cincinnati Children's Hospital Medical Center, 3333 Burnet Avenue, ML-9009, Cincinnati, OH 45229-3039, USA
e-mail: andrew.spooner@cchmc.org

J.J. Hutton (ed.), *Pediatric Biomedical Informatics: Computer Applications in Pediatric Research*, Translational Bioinformatics 2, DOI 10.1007/978-94-007-5149-1_5,

5.1 The Information in an Electronic Health Record (EHR)

Investigators conducting translational research frequently wish to extract information from or add information to patients' EHRs to support various research projects. It is important that they be familiar with laws and regulations that protect the privacy of patients and restrict who can view or extract information from patients' EHRs.

5.1.1 Basic EHR Data Integrity

Like the research record, the electronic health record demands a high level of data integrity. While the goal of policies governing data integrity in the research record is to ensure scientific rigor, the goals of policies governing data integrity in the EHR are of a different nature and affect the use of the EHR to support research:

- The EHR is a medical record that supports clinical care of a patient
- The EHR must be clinically accurate, even in the absence of well-defined data collection processes
- The data in the EHR belong to the patient, and must be provide to the patient at any time. The patient can also request changes to the chart (although these requests do not have to be honored if they are inappropriate) and the patient/parent may also add documentation to the chart at any time.
- The EHR plays an important role in legal defense of malpractice claims. Although the medical record is classified as hearsay, one may still use it in court if one can show that the record is maintained in a businesslike way. Any evidence that the medical record is being used for purposes other than clinical care may render the record useless in legal defense. For this reason, there are usually limitations on which people in which job roles are allowed to make entries in the record.
- The medical record may be shared among many providers or even across institutions. As electronic EHRs gain in prevalence, electronic methods for doing this sharing also become more prevalent, and it becomes even more important for the records to be maintained by those in familiar job roles (physician, nurse, etc.).

5.1.2 Data Entry

Clinical care is accompanied by the recording of a large amount of free text and a small amount of discrete data. While EHRs vary in the extent to which they demand discrete data entry, it is accepted that free-text entry (in the form of dictation, text-generating macros, or typing) is necessary to capture the complexity of clinical care. One might be able to reduce very simple patient encounters to a series of check boxes, but in academic medical centers when even moderately complex disease is addressed, it is not reasonable to expect clinicians to adhere to templates that generate primarily discrete data.

There are areas of the EHR, like laboratory test results and medication orders, which contain a preponderance of discrete data. In these areas there are usually a number of regulatory agencies that govern how these data are structured. For example, U.S. clinical laboratory procedures are certified through a program defined by federal law (Kroger 1994). Under these conditions, one is not free to set up investigational clinical laboratory tests as a part of routine care and incorporate them in the EHR. Likewise, prescription data must conform to data standards that allow electronic prescribing, so investigational drugs present a challenge to represent in clinical EHRs. These regulatory hurdles, while they serve a good purpose, may make it impossible to use the EHR itself as a research record, even if the proper institutional review board assurances are obtained. "Shadow records" that parallel the clinical record for research can cause confusion in the clinical operation, especially when the research activities overlap with normal clinical activities.

Another particular challenge of maintaining research data that parallels clinical data is how to handle discrepancies between the two. It is customary to apply data quality standards to research data. For example, one may want to select a particular blood pressure, collected under certain conditions, for a data point in a research study. One might then delete all other blood pressures from the research record in order to establish the data point of interest. This kind of selection of data is not usually possible in an EHR. All data are retained, and deleting data—even if it is erroneous—must be done in a way that retains all data for future inspection. Most clinical operations that allow corrections of data in the EHR have strict policies about how the change is documented. It would be unusual to see a situation where data from a clinical research study would flow back to the clinical record as a correction, regardless of how valid the correction might be. In any case, only those personnel authorized to make entries in the clinical record can initiate those changes.

5.2 Privacy Concepts in Pediatrics

Health care information is sensitive, and as such is protected by the Administrative Simplification provisions of the Health Insurance Portability and Accountability Act of 1996 (National Committee on Vital and Health Statistics 1997), as well as state laws. Because most episodes of pediatric care involve at least two people in a patient role (the actual patient and the parent or guardian), and perhaps many more, the task of securing information while maintaining the appropriate level of access is especially challenging in pediatrics. As technology moves toward fulfilling the goal of faster information flow and higher transparency, these issues are exacerbated. Pediatric clinical research, especially in genomics, can also generate health care information that creates privacy concerns. These issues are discussed in Chap. 4, Institutional Cybersecurity in a Clinical Research Setting; Chap. 6, Research Patient Data Warehousing; and Chap. 7, Biobanking in Pediatric Research.

5.2.1 HIPAA – Health Insurance Portability and Accountability Act of 1996

The U.S. Health Insurance Portability and Accountability Act of 1996 (HIPAA) intended to provide for continuity of health insurance coverage after a change in employer. This initial goal never materialized, but the portion of the law that required electronic transmission of health care claims (Title 2) remained. This portion of the regulation, known as "Administrative Simplification," raised concerns about privacy and security of the claims information that was required to be sent. This concern spawned the HIPAA Privacy Rule and the HIPAA Security Rule, enacted in April 2003 and currently enforced by the U.S. Office of Civil Rights (HHS 2002). While the full detail of these rules is beyond the scope of this text, it is important to appreciate that HIPAA remains the main driver of how clinicians behave to protect health information (privacy rules, mostly) and how systems are designed to protect it (security). An important principle regarding the use of EHR information is the "minimum necessary" rule, which states that those who access the record see only that part of the record that is necessary for performance of their job. This principle affects (or should affect) users's behavior, but it also guides policies for who is given access to what parts of the EHR. A researcher wanting to examine records of patients solely for the purposes of research would violate this rule. The HITECH Act of the American Recovery and Reinvestment Act of 2009 (HHS 2009) strengthen HIPAA's privacy and security requirements, and imposed stiffer penalties for violations.

5.2.2 HIPAA Business Associate Agreements

Those who work with health data, unless the data are suitably rendered anonymous, may be subject to the HIPAA privacy and security rules and the attendant penalties, through business associate agreements. These agreements bind recipients of health care data to the same rules that the clinical originators of data must follow, and applies the same penalties for breaches of confidentiality.

5.2.3 Pediatric Aspects of HIPAA

The HIPAA Privacy Rule allows parents or guardians access to the child's health information in almost all situations. Exceptions include when the minor is the one who consents to care and the consent of the parent is not required under State or other applicable law; or when the minor obtains care at the direction of a court; or if the parent agrees that the minor and the health care provider may have a confidential relationship (HHS 2002). Privacy laws vary from state to state, and providers are obliged to follow the most stringent one. Since control of children's health information is sometimes a hot political topic (as in the case of minors' access to reproductive health services) these legal conflicts can make control of data very complicated (Chilton et al. 1999).

5.2.4 FERPA – Family Educational Rights and Privacy Act

A law that existed many years before HIPAA was the Family Educational Rights and Privacy Act (FERPA) (Kiel and Knoblauch 2010) which attempts to give students some control over the use of their educational records. When healthcare is provided at a school, the line between health records and educational records is blurred, and there can appear to be conflicts between HIPAA and FERPA. If one is attempting to aggregate data from both educational and healthcare settings, these specific laws may come into play. The U.S. Department of Education and the U.S. Department of Health and Human Services published joint guidance on navigating these apparently conflicting laws in 2008 (HHS 2008).

5.2.5 Release of Information

A common function of the information systems in a healthcare organization is the release of information based on a request from a patient, parent, guardian, lawyer, State agency, or other suitably approved group. Release of information (ROI) in a hospital is typically handled via a controlled process through the Health Information Management department or similar entity. Before the age of EHRs, the actual conveyance of medical records was achieved by a tedious and time-consuming process of photocopying paper records or printing images of documents from archived storage. It was considered normal for this process to take several weeks. The difficulty of this process rendered the medical record effectively inaccessible to all, but the most dedicated patients and their representatives.

In the information age, expectations about the ease by which one can get information are changing. The Continuity of Care Document project (Ferranti et al. 2006) is a manifestation of the expectation that electronic health records can produce immediate summary information for the purposes of sharing across venues of care. The expectation of immediate access has spread to all areas of the EHR (How et al. 2008). These expectations entail more sophisticated authentication methods than the typical notarized permission form that usually initiates the process of Return of Information (ROI) in health care institutions using traditional paper based systems.

ROI is important to understand in pediatric care because it means that all information in the chart (or at least that part designated the "legal medical record") is available to the guardian at all times. While in the past it may have been comforting to a child/adolescent to assume that information would be "secure" from prying parental eyes because of a 6-week wait for photocopying, that wait will eventually be reduced to practically zero through electronic methods. Parents or guardians will have contemporaneous access to all details in a child or adolescent's chart. We have not yet had the opportunity to evolve habits in practice that take this into account, or sophisticated privacy policies that balance the need to keep things truly private between a provider and a minor patient under the assumption of immediate parental electronic access.

5.2.6 Clinical Data Sharing Versus Financial Data Sharing

Regardless of privacy policies put in place, the fact that guardians receive billing information about health services provided also runs counter to the concept of keeping things private between a minor and a provider. Doctors who treat adolescents have been known to write prescriptions on paper or provide samples rather than run the risk of notifying a parent via a pharmacy claim. Regardless of how one feels about the appropriateness of such confidential care, such practices do create holes in the protections set up in the electronic record.

5.2.7 Parental Notification Versus Consent to Treat

Adolescents can consent to treatment at an age younger than the age of majority in certain clinical contexts (Weddle and Kokotailo 2002). For example, an adolescent at age 12 can, in the states of California or Illinois (as of 2003, English et al. 2003) consent to treatment for mental health services. In North Carolina, the minor can consent at any age. This varying age of consent has little impact on EHR functionality or data storage, but it is often confused with the concept of parental notification just because an adolescent can consent to treat for his or her own care does not make the record of that treatment confidential, or obviate parental notification regulations. Once again, the availability of that information in the medical record may appear threatening to both patient and provider, to the point that the provider may record data in a non-standard place (like a "sticky note" field that is not part of the legal medical record). Once again, full appreciation of the workflow used to produce health data is necessary in order to construct meaningful queries and analysis.

5.2.8 Mandated Reporting

Child health workers are obliged under the law of all U.S. states to report suspected child abuse. This obligation overrides HIPAA or other concepts of health information privacy (AAP and C.o.C.A.a.N 2010).

5.3 Health Information Privacy in Adolescent Care

5.3.1 The Nature of Adolescent Practice

The care of adolescent patients—as in the care of all patients—must address issues of particular sensitivity: reproductive health, sexually transmitted disease, substance abuse, physical abuse, eating disorders, sexual abuse, mental health, and sexual orientation. The difference with adolescents that affects EHR implementation is that the patients are more sensitive to the effects of confidentiality on their decision to seek care (Ginsburg et al. 1995). Most agree that adolescents need to share in the decision-making about their care, regardless of their inability to legally consent to their treatment. For sensitive topics, adolescents may forego care in order to hide information from parents (Britto et al. 2010; Ford et al. 2001). Since a fundamental goal of health information technology is usually to make information *easier* to share, the adolescent's prerequisite to restrict information dissemination may be impossible to accommodate without non-standard methods of information management. As a result, clinical users may resort to obfuscation of data or the use of paper to manage the information that would otherwise be contained in the EHR. Obviously, this would have major downstream effects on the interpretation of data derived from these environments.

5.3.2 Adolescent Health, Privacy, and Research

Adolescents participate as subjects in clinical research, but the process for weighing the risks and benefits of parental consent are complex. Even when parental consent is not a sensitive issue, researchers intending to engage in clinical research involving adolescents should familiarize themselves with local legal issues regarding assent and consent at various ages. The Society for Adolescent Medicine maintains guiding policies for these issues (Santelli et al. 1995). It is a basic principal of adolescent healthcare, endorsed by professional societies, that they be offered confidential care when appropriate (Ford et al. 2004; Gans Epner 1996). Since health information is already considered confidential, a promise of confidential care essentially means that information will be kept from parents or guardians, a concept that flies in the face of some state law and EHRs designed to provide information to parents or guardians in the form of printed summaries and on-line portals. As of this writing, there are no standards for adolescent privacy policies to govern such patient-accessible information, whether for clinical care or research.

5.4 Health Information Privacy and Mental Health

Mental health information was singled out in the HIPAA Administrative Simplification rules in the sense that "psychotherapy notes" do not have to be disclosed to patients or families as part of the usual release of information. These kinds of notes are usually made to record a therapist's thoughts during a patient's therapy, and, if a patient accessed these notes, they might be damaging to the patient's progress. The regulation specifies that these notes cannot contain "medication prescription and monitoring, counseling session start and stop times, the modalities and frequencies of treatment furnished, results of clinical tests, and any summary of the following items: diagnosis, functional status, the treatment plan, symptoms, prognosis, and progress to date" (HHS 2007).

This minor exception to the idea that a patient or family owns the information in the chart with complete access rights has no direct effects on data analysis. It does, however, impose requirements for more complex access control on developers of EHRs. It also has the potential to confuse clinical users, who are already struggling with how to practice medicine in the era of patients' immediate access to their information. For example, if psychotherapy notes should not be shared, are there not other classes of data in the chart that ought to be afforded this same protection, for the same reasons? HIPAA did not describe other exceptions, but clinicians' desire to document care without disrupting care may create new use cases that make data access policies even more complex than they are now.

5.5 Guardianship Issues (Adoption, Foster Care, Fetal Care)

In pediatrics, as with elder care, the patient is not assumed to be the main decision-maker in health care decisions. For most children, the parents are responsible for the child's care as well as the financial and administrative transactions involved in that care. In some cases, the guardian must be distinguished from the financial guarantor. For children whose parents have had their parental rights severed, or who have otherwise been taken from the care of their parents, other adults are designated guardians. In specific legal proceedings, a court may appoint a guardian ad litem with defined decision-making authority for the child. The only impact these complex arrangements may have on data used for research is that it may affect the consent processes associated with the study.

References

AAP, C.o.C.A.a.N. Policy statement – child abuse, confidentiality, and the health insurance portability and accountability act. Pediatrics. 2010;125(1):197–201.

Britto MT, Tivorsak TL, Slap GB. Adolescents' needs for health care privacy. Pediatrics. 2010;126(6):e1469–76.

Chilton L, et al. American Academy of Pediatrics. Pediatric Practice Action Group and Task Force on Medical Informatics. Privacy protection and health information: patient rights and pediatrician responsibilities. Pediatrics. 1999;104(4 Pt 1):973–7.

English A, et al. State minor consent laws: a summary. Chapel Hill: Center for Adolescent Health & the Law; 2003.

Ferranti JM, et al. The clinical document architecture and the continuity of care record: a critical analysis. J Am Med Inform Assoc. 2006;13(3):245–52.

Ford CA, Best D, Miller WC. The pediatric forum: confidentiality and adolescents' willingness to consent to sexually transmitted disease testing. Arch Pediatr Adolesc Med. 2001;155(9): 1072–3.

Ford C, English A, Sigman G. Confidential health care for adolescents: position paper for the society for adolescent medicine. J Adolesc Health. 2004;35(2):160–7.

Gans Epner JE. Policy compendium on reproductive health issues affecting adolescents. Chicago: American Medical Association; 1996.

Ginsburg KR, et al. Adolescents' perceptions of factors affecting their decisions to seek health care. JAMA. 1995;273(24):1913–18.

HHS. Does the HIPAA Privacy Rule allow parents the right to see their children's medical records? [Web page]. 2002 [2006 Mar 14; 2012 Apr 22]. Available from: http://www.hhs.gov/hipaafaq/personal/227.html

HHS. Title 45 – Public Welfare, Section 164.501 – definitions, in 45 C.F.R. § 164.501; 2007.

HHS. Joint guidance on the application of the Family Educational Rights and Privacy Act (FERPA) and the Health Insurance Portability and Accountability Act of 1996 (HIPAA) to student health records, E. Health and Human Services, editor, Washington, DC; 2008.

HHS. Health Information Technology for Economic and Clinical Health (HITECH) Act, Title XIII of Division A and Title IV of Division B of the American Recovery and Reinvestment Act of 2009 (ARRA), in Pub. L. No. 111–5 (Feb. 17, 2009), U.S.D.o.H.a.H. Services, editor, Washington, DC; 2009.

How SKH, et al. Public views on U.S. health system organization: a call for new directions. New York: The Commonwealth Fund; 2008.

Kiel JM, Knoblauch LM. HIPAA and FERPA: competing or collaborating? J Allied Health. 2010;39(4):e161–5.

Kroger JS. Coping with CLIA. Clinical Laboratory Improvement Amendments. JAMA. 1994;271(20):1621–2.

National Committee on Vital and Health Statistics. Publication of recommendations relating to HIPAA health data standards–HHS. Notice. Fed Regist. 1997;62(195):52563–5.

Santelli JS, et al. Guidelines for adolescent health research: a position paper of the society for adolescent medicine. J Adolesc Health. 1995;17(5):270–6.

Weddle M, Kokotailo P. Adolescent substance abuse. Confidentiality and consent. Pediatr Clin North Am. 2002;49(2):301–15.

Chapter 6
Research Patient Data Warehousing

Keith Marsolo

Abstract Recent years have seen a push to increase the adoption of electronic health records and other clinical information systems in the United States. This has led to a corresponding increase in the volume and breadth of data that are generated during the course of clinical care. Institutions have a strong desire to analyze and mine this information for quality improvement and research purposes. To accomplish this, they are turning to systems that allow them to integrate data from several systems and synthesize it into a coherent view. These systems, commonly called data warehouses or integrated data repositories, are now serving as the basis for much of the clinical and translational research that occurs in medicine today. Here we describe the different components of a research patient data warehouse, including data models and terminologies, extract, transform and load (ETL) procedures used to populate the warehouse and steps institutions need to take to ensure patient privacy. We also provide an overview of one of the more popular open-source warehousing technologies, the Informatics for Integrating Biology and the Bedside (i2b2) platform.

6.1 Introduction

The increased adoption of electronic health records (EHRs) in the United States over the past few years has led to a tremendous increase in the amount of data that are captured during the course of clinical care and has stimulated efforts to ensure that information is put to "meaningful use" (Blumenthal 2009, 2010). This increase has led to a corresponding desire to analyze and mine that information for research and quality improvement purposes (Weiner and Embi 2009). Despite the increased

K. Marsolo, Ph.D. (✉)
Department of Pediatrics, University of Cincinnati College of Medicine, Cincinnati, OH, USA
e-mail: keith.marsolo@cchmc.org

J.J. Hutton (ed.), *Pediatric Biomedical Informatics: Computer Applications in Pediatric Research*, Translational Bioinformatics 2, DOI 10.1007/978-94-007-5149-1_6,

implementations of enterprise EHRs, data often exist in multiple disparate systems. As such, there has been an increased desire for a system that supports integration of data from multiple sources and of multiple types into a coherent view, which can then be used for analysis and reporting purposes. Data warehouses (also called integrated data repositories) have been employed by the business community for decades to integrate and analyze financial data (Kimball 1998; Agosta 2000). Their initial use in the healthcare industry also had a financial purpose, to allow for the integration and analysis of billing, claims and administrative data. Recent years have seen a push to augment this financial and administrative information with clinical data from the EHR (Mackenzie et al. 2012).

6.2 Research Patient Data Warehousing

To facilitate research and quality improvement, many institutions have or are now creating research-specific data warehouses (Mackenzie et al. 2012; Murphy et al. 2010; Research Patient Data Registry – RPDR; Eggebraaten et al. 2007; Lowe et al. 2009; Zhang et al. 2010). While it is possible to use an operational data warehouse for research purposes, a research specific-warehouse is intended to support a different set of use cases and workflows and will often include additional data sources (e.g., genomic, sensor data). In order to address privacy and security concerns, particularly those related to the Health Insurance Portability and Accountably Act (HIPAA) Privacy Rule and Health Information Technology for Economic and Clinical Health (HITECH) Act (Blumenthal 2010; Title 45 Code of Federal Regulations Parts 160, 162, 164: Combined Regulation Text), they typically add additional safeguards like de-identification or obfuscated access.

6.2.1 Place in a Combined Clinical and Research Environment

Clinical information systems exist on a continuum within the healthcare delivery environment (Fig. 6.1). EHRs, by their nature, are patient-centric, allowing users to interact with the record of a single patient, one patient at a time. They are designed to be high-availability, and as such, are typically transactional in nature. This results in an application and data model that makes it very difficult to perform population or cohort-level analysis.

To offset these limitations, many EHRs are packaged with an operational data store (ODS), a database that can be used for reporting purposes. Most ODSs are still very transactional in nature, often reflecting the underlying architecture of the EHR (to simplify the process of refreshing the ODS with updated data). As the name implies, one of the primary uses of an ODS is to generate data that provides insight into the operations of an institution – clinical, financial, or otherwise. However, they are complex. For example, the ODS for the Epic EHR (Clarity) has over 15,000

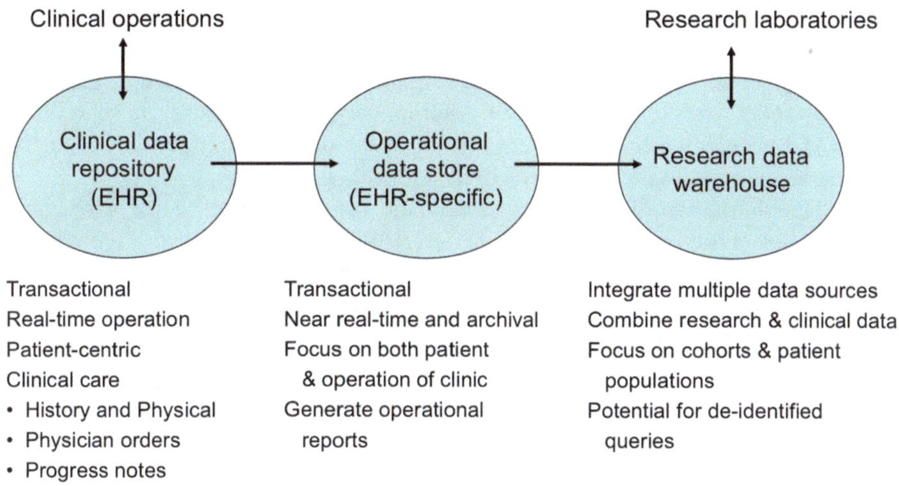

Fig. 6.1 Continuum of information systems, ranging from those focused purely on clinical care, to those geared towards research

tables. It is often difficult, if not impossible, to integrate information from outside sources into the ODS. In addition, population level analysis is still a challenge, as the relevant data may be spread out over dozens of different tables.

Data warehouses are designed to overcome the limitations of ODSs. Data are pulled from multiple tables/sources and combined into a single integrated record that can then be queried and analyzed.

6.2.2 Use Case

The overriding use case of a research data warehouse is cohort identification, i.e., to find patients who meet criteria X, Y and Z. Query criteria can include demographics (age, race, gender), diagnoses, laboratory results, procedure codes, medication orders, etc. and can be constrained by date, exclusion/absence or number of occurrences. While cohort identification may seem like a rather trivial use case, it is often the first step for any kind of biomedical analysis, whether for clinical care, research or quality improvement (e.g., find all patients of a particular doctor or clinic, all patients with a given disease/genetic anomaly, all patients who have not come to clinic during the past 6 months). If the data are housed in a single warehouse, cohort queries can be complex, but are feasible. If the data are stored in a set of siloed source systems, they can become incredibly time-consuming, if not impossible, to complete.

Once an initial cohort has been generated, a data warehouse can be used to facilitate the analysis of data about its members. This can occur through tools that have been developed to sit on top of the warehouse and interact through a middleware service (as in STRIDE, i2b2, and Physio-MIMI Murphy et al. 2006, 2010; Lowe et al. 2009; Zhang et al. 2010), allowing for self-service or mediated access by super users (example below). Alternatively, the warehouse can be a source of cohort- or project-specific datasets, serving as the source of extracts or data marts that can be analyzed with a user's favored statistical package.

While cohort identification is typically the primary use case for a research data warehouse, other use cases include: searching for and requesting biosamples available for research (another version of cohort identification); using it as a repository of patient outcome and process measures (which are stored as derived observations), and as a data source for research and quality improvement registries.

6.2.3 Data Model

There are a number of different data models that can be chosen to underlie a data warehouse. The simplest and most traditional is the star schema. A central fact (observation) table is linked to a number of different dimension tables, which are used to group related data (for example, by patient, visit, provider). In a research data warehouse, observations can be anything from the presence of a diagnosis, a patient's birthdate, a lab result or a billing record. A slightly more complex version of this model is the snowflake schema, where a single fact table is linked to a number of dimensions that are normalized into multiple tables (Agosta 2000; Kimball 1996). Other alternative data models include the Health Level 7 – Reference Information Model (HL7-RIM) and Observational Medical Outcomes Partnership (OMOP) (Eggebraaten et al. 2007; Lowe et al. 2009; Overhage et al. 2012; Lyman et al. 2003). The choice of data model is influenced by the primary use case of the warehouse. For instance, the primary focus of OMOP is to monitor drug safety, so if the main use of a warehouse is along those lines, the OMOP data model may be a good choice. If a more generalized approach is desired, then the standard star/snowflake schemas may be a better fit. It is possible to convert from one model to another, though if one model expects data or metadata that is not present in the other, data may be lost during the transformation. Later on in the chapter, we examine the open-source i2b2 framework (Murphy et al. 2006; Mendis et al. 2007; Gainer et al. 2007), which uses a star schema as its data model.

6.2.4 Terminologies

A particular challenge in research data warehousing revolves around terminologies, or how to code a particular data element. There are two major approaches: either rely on a "local" code (typically derived from the coding scheme of the source system), or transform the data so that it is consistent with one of the many terminology standards, like ICD-9/-10 or SNOMED for diagnoses, LOINC for laboratory results, or RxNorm for medications. Terminologies are further discussed in Chap. 2 (Interfaces and Data Transmission Standards).

The decision to transform the data again depends on the primary use of the warehouse. If data will be routinely shared with outside institutions, transforming an element to a standard can ease the process of ensuring semantic interoperability. If the main users of the warehouse are primarily internal customers, it may be beneficial to retain the coding used in the source systems. Local users will be familiar with local terminology that appears on the application screens of the source and can navigate more easily in the research system. In addition, retaining the source terminology in some form can help eliminate confusion for end users. If an EHR uses ICD-9 as an internal coding for diagnoses, for instance, and the data warehouse converts them to SNOMED, a query for all "asthma" patients run on the clinical ODS may not match the results returned from the research data warehouse because SNOMED codes at a different level of granularity than ICD-9. ICD-9 is a relatively coarse terminology system, and converting to a fine-grained terminology like SNOMED can lead to errors, if not done correctly. Discrepancies of this kind can lead to frustration on part of staff, as they seek reasons for the discrepancy in results obtained from searches of the two local warehouses that are supposed to be populated from the same data sources. Such discrepancies can cause users to lose faith and trust in future results.

A particular advantage to using internal terminologies is that they are easier to maintain. If the source system changes the way that items are defined, the terminology of the warehouse can simply be updated to reflect the modifications. If items are mapped to a reference terminology, the work is essentially doubled. The internal terminology will need to be monitored for modifications, as will the reference terminology. While there are tools to help with mapping (Musen et al. 2012; Kohane et al. 2012), the process almost always requires time consuming manual reconciliation.

6.2.5 *Data Provenance*

Clearly understanding the original source of a data element (data provenance) within a research patient data warehouse is essential, but a significant challenge. Within an EHR, there may be multiple diagnoses and each may come from a different source within the system (admission, discharge, problem list, encounter, billing, etc.). Depending on how the diagnosis was generated, some sources of the data element, diagnosis, may have more "credibility" than others. For instance, a diagnosis of lupus erythematosis assigned in an optometry clinic may not be as reliable as one assigned by a rheumatologist. Within an EHR or ODS, items like diagnoses are typically stored in a location that is based on how the data is entered into the system. In a data warehouse, all of this information is brought together into a single table of observations. If care is not taken to denote where a value came from, or to allow users to add filters so that they can query on multiple parameters such as clinic, provider, and billing office, then results returned to users may not mean what they think it means. This becomes a concern particularly when users are allowed to generate self-service queries.

6.2.6 Extract-Transform-Load (ETL)

One of the biggest challenges in creating a data warehouse lies in the procedures used to extract the data from the source systems, transform it to the correct format and load it into the warehouse (Kimball and Caserta 2004). This process (extract-transform-load) is often referred to as ETL. Maintaining an ETL process can be expensive in both people and software. Enterprise EHRs are generally supplied by commercial vendors and are constantly being updated with new versions. Organizations may also modify the local EHR to meet needs of internal users. Automated ETL tools like Microsoft SQL Server Integration Services (SSIS), Oracle Data Integrator, Informatica, and Talend can help monitor the quality of the source data and notify warehouse administrators of any changes. In many cases, though, manual SQL procedures must be created to optimize and customize the process.

The design of an ETL process depends on the frequency of the data refresh from the source systems, as well as the frequency of refresh to the target. Dealing with a laboratory information system (LIS) that sends results as real-time HL7 messages is a more complex endeavor than one that transmits results as a daily flat file. The closer to real-time the transmission of data from a source is, the more difficult it is to monitor and maintain the ETL process. The same rule applies when refreshing a target system. A challenge when dealing with EHR data is that the size of an ODS can be in the range of terabytes. A complete refresh of a data warehouse with data of this size will eventually take longer than the available refresh window. This can be mitigated by the creation of an incremental or delta refresh, where the only data that are loaded are the new/changed elements. A refresh like this is more challenging to architect, as one must figure out which elements have been added/deleted/modified, which can be difficult depending on the size and complexity of the source. While ETL processes to load EHR data into a warehouse can occasionally be shared among institutions, most EHR implementations include a fair degree of customization, requiring similar customization in the corresponding ETL process. If data are routinely pushed from the warehouse to other sources (data marts, registries, etc.), this adds complexity to the refresh process because a system must be in place to move data from the source to the warehouse to the users application in a timely manner.

6.2.7 Privacy

Laws governing human research vary from country to country. This discussion focuses on requirements in the United States. Issues of privacy and consent are also discussed in Chaps. 4 (Institutional Cybersecurity in a Clinical Research Setting) and 5 (Protecting Privacy in the Child's Electronic Health Record). In order to conduct research with patient identifiable patient data, users must obtain consent (Title 45 Code of Federal Regulations Parts 160, 162, 164: Combined Regulation Text; Title 45 Code of Federal Regulations; Part 46). The alternative is to work with a de-identified dataset, or a dataset that has been stripped of all personal health information (PHI). Working with de-identified data is not considered to be human subjects research and does not require an IRB protocol. This makes the creation of a research data warehouse a challenge, as the act of creating a warehouse requires the use of identifiable patient data. Users of warehouses that have been created for clinical or operational purposes do not face these challenges because the use of PHI in those contexts is covered under the HIPAA Treatment, Payment and Operations (TPO) exclusion. There are several strategies that can be employed to satisfy the requirement that consent be obtained from a patient in order to have their data included in a research warehouse. The first is simply to have each patient sign a consent form denoting whether their data can be included. This approach is relatively expensive and time-consuming, however, because of patient volume, and is almost never used. A variation of this approach is to have hospital registrars ask for patient consent as part of the registration process. Such a "self-consenting" process, i.e., a member of the research staff is not immediately available, may grow in favor as enterprise EHRs with integrated registration become more prevalent. Another strategy is to employ a passive consent, or "opt-out" strategy, where all patients and their data are included by default unless they explicitly request to have their information excluded. One of the more common approaches is simply to request that an IRB waive the consent requirement, the justification being that it is impractical to obtain consent from every single patient at a hospital. The final method is to include language in the clinical consent-to-treat document that all patients must sign stating that any of their data may be used for research and included in a research patient data warehouse.

Because many warehouses are established under an IRB protocol that waives patient consent, they typically operate under the "honest broker" principle (Dhir et al. 2008). An honest broker serves as an intermediary between the individual providing the data and a researcher (SACHRP 2010). The broker aggregates and integrates identified data from multiple sources, removes the identifiers and then provides the resulting dataset to the investigator. They provide a service, but do not function as a member of the research team (i.e., they cannot participate in the analysis of the data). The same principle can be applied to biorepositories and biospecimens (SACHRP 2010), as discussed further in Chap. 7, Laboratory Medicine and Biorepositories.

Table 6.1 lists the 18 identifiers specified by HIPAA as PHI and whether they can be included in a de-identified or a limited dataset. The traditional method

Table 6.1 HIPAA identifiers and whether they can be included in a de-identified or limited dataset

Identifier	Included in de-identified dataset	Included in limited dataset
1. Names	No	No
2. Postal address information	The initial three digits of a zip code, if, according to the current publicly available data from the Bureau of the Census: (1) The geographic unit formed by combining all zip codes with the same three initial digits contains more than 20,000 people; and (2) The initial three digits of a zip code for all such geographic units containing 20,000 or fewer people is changed to 000;	Yes
3. Social security numbers	No	No
4. Account numbers	No	No
5. Telephone and fax numbers	No	No
6. Elements of dates for dates directly related to an individual, including birth date, admission date, discharge date, date of death	No – except for year	Yes
7. Medical record numbers	No	No
8. Certificate/license numbers	No	No
9. Electronic mail addresses	No	No
10. Ages over 89 and all elements of dates indicative of such age	No	No
11. Health plan beneficiary numbers	No	No
12. Vehicle identifiers and serial numbers, including license plate numbers	No	No
13. Device identifiers and serial numbers	No	No
14. Web Universal Resource Locators (URLs)	No	No
15. Internet Protocol (IP) address numbers	No	No
16. Biometric identifiers, including fingers and voice prints	No	No
17. Full face photographic images and any comparable images	No	No
18. Any other unique identifying number, characteristic or code that is derived from or related to information about the individual	No	No

of de-identification is through the HIPAA "safe harbor" standard, which states that if all of the 18 identifiers listed under the HIPAA Privacy Rule are removed and if there is an expectation that it is not possible to use the remaining data to identify a person, then the data are de-identified (the other method involves statistical de-identification and is not routinely used because of its complexity and requirements for certification) (Title 45 Code of Federal Regulations Parts 160, 162, 164: Combined Regulation Text). Investigators may use a de-identified data set in their research without IRB approval. A limited dataset is one where 16 of the 18 HIPAA identifiers are excluded, the exceptions being full dates and zip code. The use of a limited dataset requires that the investigator submit a protocol to the IRB and receive approval before conducting research using the limited data set.

In order to provide investigators with information on patient age more specific than a year (i.e. age in months, days or weeks), and to provide data on date of admission, discharge, or service, without requiring an IRB protocol for every dataset, one approach is to use "scrambled" dates. Before loading the data into the warehouse, one applies a date shift to each patient's record. For example, for a given a patient, all dates in the record are shifted backwards a random number of days between 1 and 180. The shift is consistent within a patient's record so that sequences of events are preserved, but each patient in a cohort is shifted by a different randomly chosen number. This makes the data unusable for surveillance research, but does allow for complete information about sequences of events to be released to investigators as part of a de-identified dataset.

6.3 The i2b2 Research Patient Data Warehouse

This section describes an example of a widely used research data warehousing platform and how it can be deployed within an institution. This example focuses on i2b2 (Informatics for Integrating Biology and the Bedside), an informatics framework developed by the NIH-funded i2b2 National Center for Biomedical Computing. One of the initial, motivating use cases of i2b2 was to serve as a bridge between clinical research data and the vast data banks arising from basic science research in order to better understand the genetic bases of complex diseases. i2b2 has been released as a collection of open source software tools that facilitate the collection and aggregation of patient data, both clinical and research, into a translational research repository (Kohane et al. 2012).

The basic function of the software is to allow the user to identify cohorts of patients from a large repository of observations. The core code manages projects, files, a query ontology, data and workflows. It is designed to allow developers to add functional modules, called cells, to the core code. Cells have been developed to extend the core to a multitude of activities like natural language processing, annotation of genomic and image data, data analysis, concept mapping, and linkage to other research systems.

The functional design of i2b2 is based around a central ontology of query terms consisting of standard and local terminology to describe clinical and research observations. The i2b2 ontology is fully extensible to local terminology sets and contains terms for diagnoses, procedures, demographics, laboratory and microbiology tests, visits, service providers, genomics, proteomics and imagery. Users select terms from the ontology in logical combinations (like a Venn diagram), and i2b2 generates a query against its underlying database to deliver cohorts of patients meeting the selected criteria.

i2b2 employs a data model based on the data-warehousing-standard star schema and runs on common database engines like Oracle and SQL Server. All of the patient observations are contained in a central database table. Each observation is linked to a concept which links back to the central ontology. More detail is provided in the following section.

Communication in i2b2 occurs through a middleware of web services. These services pass information between the different components, from the database layer up to the front-end application. Currently, users access i2b2 through one of two different approaches. One is by using a derivative of the open source Eclipse workbench that runs as a workstation-based client and the other is through a browser-based Ajax (Javascript and XML) client, like the CCHMC-developed i2b2 toolkit.

A major strength of networks that build upon the i2b2 warehousing platform is the existence of a very active and diverse academic users group, consisting of sites throughout the United States, as well as internationally (Kohane et al. 2012). The users build and share software modules, which interface with i2b2 and meet diverse needs of translational researchers. As a result, this open source platform is surrounded by a wide and growing variety of interfaced services (cells), which are rapidly enhancing the ability of i2b2 to support many types of translational research. A catalog of available modules can be viewed at https://community.i2b2.org/.

6.3.1 i2b2 – Data Model

The data model embedded in i2b2 uses a variation of the entity-attribute-value model, where a number of attributes, each with a corresponding value, are stored about an entity (patient). Attributes are facts, measures or observations about a patient, such as demographics, diagnoses, or laboratory results. It is also possible to define de novo attributes or create variables that are derivations or combinations of existing ones. Observations are tagged with one or more concepts, which can come from standard terminologies like SNOMED or LOINC or from custom definitions specific to a project. These codes are used when querying the warehouse. Collectively these concept codes are the items that can be used in a query. For example, when a user submits a query asking for all patients with a diagnosis of 'asthma', the i2b2 query engine will look for anyone with an observation that has the code associated with 'Asthma.' It is also possible to assign synonym codes, so that a search for 'Asthma' would return all patients with a corresponding ICD or SNOMED diagnosis.

6.3.2 i2b2 – Privacy Mechanisms

The i2b2 framework employs a number of measures to ensure the protection of patient identity and privacy (Murphy et al. 2011). Access in i2b2 is granted on a project level, with the ability to assign a different role and access level to the same user for each project. Users can be *administrators*, which gives them full control over a project, a *manager*, which allows them to see all the previous queries of the project's users, or simply a *user*, where all they can see are their own queries. In addition, i2b2 uses a number of different access levels to restrict the amount of data that a particular user can see. The lowest level of access is at the *obfuscation* level, where the user can only see the approximate results of an aggregate query. A Gaussian (of 1.7) is applied to the aggregate count and the values are returned with an accuracy of ± 3. A user's account is locked if they try to run the same query more than a certain number of times, as doing so would allow them to try and zero in on the true aggregate value. A user with the *aggregate* level can see the exact aggregate counts, but not individual patient data. The *limited* role allows the user to see individual patient data, but only the information that is available in a limited data set (i.e. de-identified plus dates of birth and dates of service). Users who have access to a project at the *limited + de-identified* level can see all of the data in the limited dataset plus (presumably) de-identified text results, assuming they are available. Finally, users can be granted *full* access, with the ability to see fully identified patient records. Custom access levels can also be created, if needed.

6.3.3 i2b2 Workbench – Self-Service Access

The text below describes a workflow for self-service access to an i2b2 warehouse. It is not the only potential workflow, but is merely intended to represent what is possible. As mentioned previously, i2b2 allows researchers to interact with the warehouse through a web-based workbench (the screenshots represent a version of the i2b2 workbench developed at CCHMC, the code of which is available at the following web address: https://bmi.cchmc.org/svn/i2b2/i2b2/public/). The most basic functionality provided to users is the ability to generate obfuscated patient counts. The workbench also includes a mechanism to request higher-level ("elevated") access that is provided in the form of data extracts or longitudinal chart review. Users can also use the i2b2 workbench to identify and request de-identified samples (if available). In cases where the requested data resides in an institution's

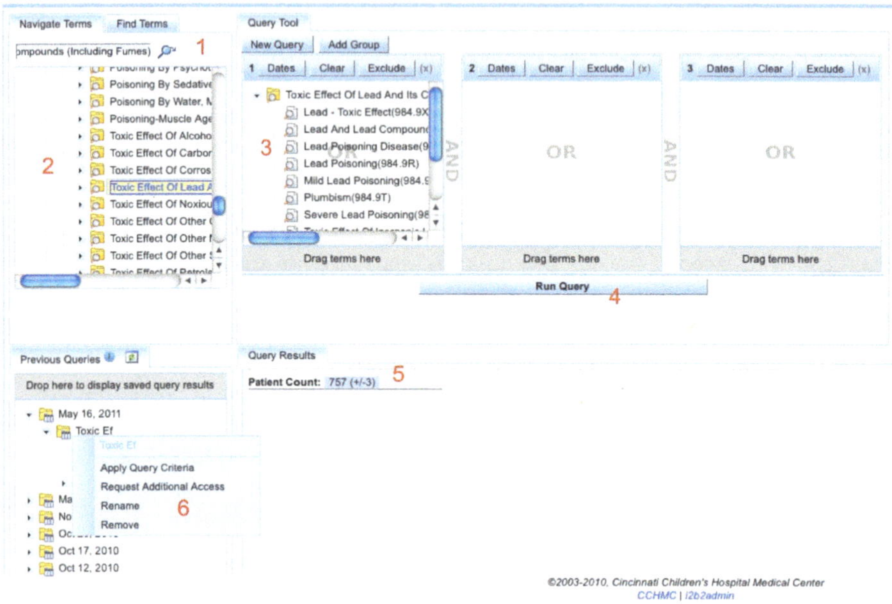

Fig. 6.2 A screenshot of the CCHMC workbench with obfuscated access. The workbench includes an autosuggest tool (*1*) that can be used to find search terms within the ontology (*2*). These terms can be dragged and dropped into a Venn diagram-like interface (*3*). Upon execution (*4*), the query will return an obfuscated count of the number of patients that meet the specified criteria (*5*). If a user wishes to gain elevated access to a query, they can request it by right-clicking on a previous query (*6*)

clinical information systems, but not the warehouse, research staff can query those systems directly to fulfill ad hoc requests for de-identified and coded datasets (i.e., mediated query). All requests for identified data or for limited datasets are reviewed by the IRB.

Users who have Obfuscated Access can use the workbench (Fig. 6.2) to execute cohort queries. As illustrated in Fig. 6.2, obfuscated access represents the default level of access and is open to any user. With this access, a user can run queries, but the results are limited to obfuscated (approximate) count data, with "<5" returned if there are fewer than 5 patients in a given cohort. The primary purpose for this level of access is to generate hypotheses, generate numbers to determine whether a suitable population exists for a grant submission, and to conduct reviews preparatory to research.

After running a query, a user may request "elevated," but still de-identified, access to the patient cohort or identified access, if patient identifiers are required. Within the i2b2 workbench, a user can right-click on a previous query and "request elevated/identified access." They can fill out a form that requests certain information: description of use/reason for access, type of access, level of access, number of patients, query criteria, query id; etc. If approved, investigators are

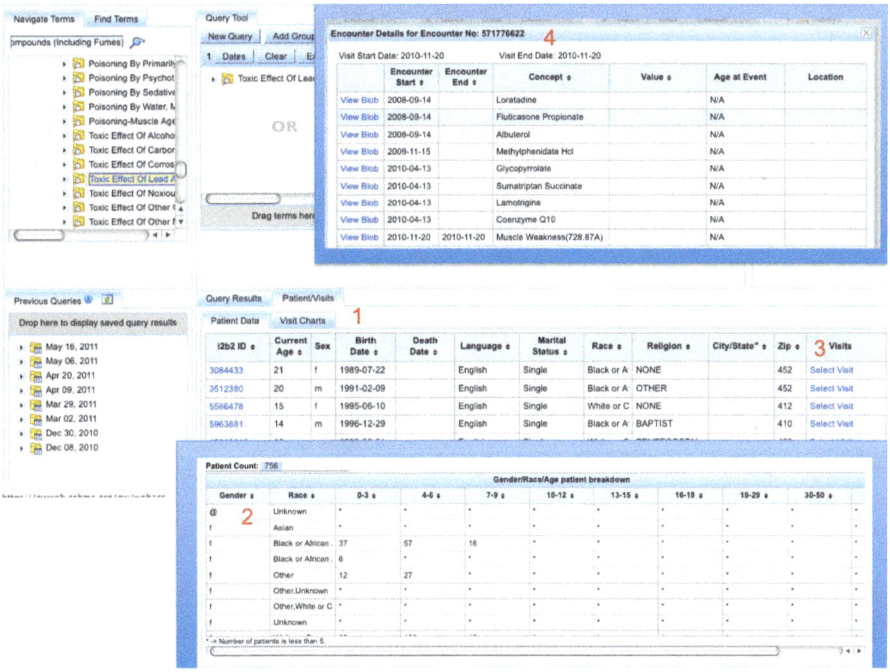

Fig. 6.3 Elevated access in the i2b2 workbench. Users can either select to view individual patients or general demographic information (*1*). In contrast to the count returned under the Obfuscated Access role, queries run with Elevated privileges return the exact number of patients (*2*). When accessing specific patients, a user can access a list of visits (*3*) and view the data collected during that encounter (*4*)

provided with a dataset or a de-identified view of the patient's record (chart review). If identified access is required, investigators must also receive IRB approval before the data can be released.

From a functionality perspective, there is no difference between elevated and identified access. The only difference is with the data in the underlying database. With elevated access, it is de-identified. With identified access, it is not. An example of elevated access is shown in Fig. 6.3. This level of access allows users to view a list of patient visits along with the data collected during that visit. Another option is to view the data in chart review mode (Fig. 6.4), which provides an EHR-like summary view of the patient's record.

Users can use a slightly modified workflow to request de-identified samples from a research biorepository. A user generates a cohort of patients as listed above, using something like "existence of a sample" as an element in the query criteria. The cohort is generated and the sample information is sent to the biobank to determine whether the request can be fulfilled. If it is, the samples are provided to the investigator, along with a data extract and/or data mart (if needed). Biorepositories and biobanking are discussed in Chap. 7, Laboratory Medicine and Biorepositories.

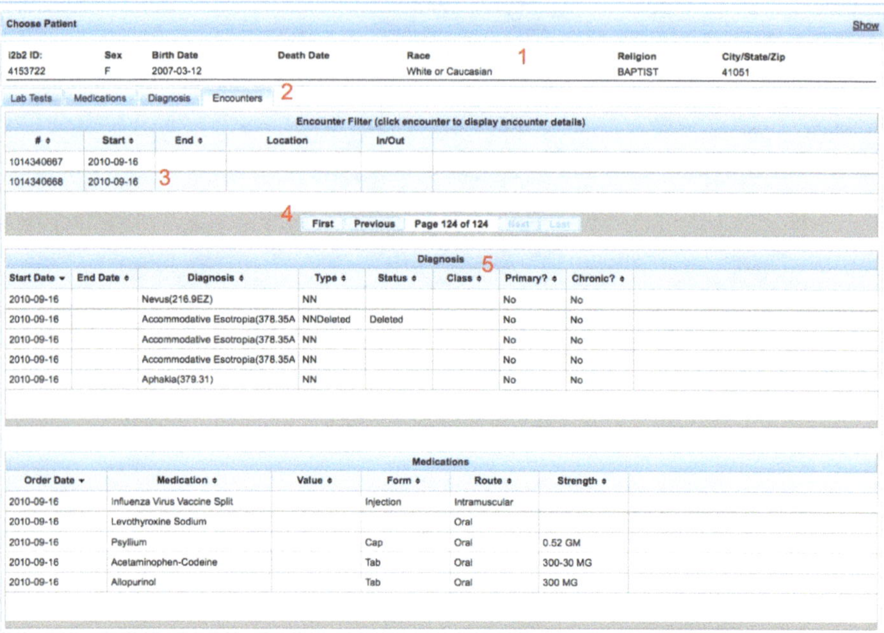

Fig. 6.4 Chart review mode in the i2b2 workbench. In this mode, a user can select a patient (*1*), and then view the different types of data available for that patient (*2*). In the "Encounter" mode, they can select a given encounter (*3*), (*4*), and view a list of data that was recorded at that particular visit. It is also possible to download this data as a report directly from the workbench, in Excel, PDF, XML or csv formats. Users would not be allowed to download these files directly to a local computer, but could place them in the virtual desktop environment

6.3.4 *Mediated Queries*

While the i2b2 workbench can be used to answer a number of research questions, it is always possible to formulate a query that is not amenable to automated tools. In such cases, it is often more efficient have a database developer manually generate and execute the query. Investigators would work directly with the data warehouse team and refine their query over several iterations.

6.4 Joint Management – Hospital and Research

As discussed in Sect. 3.8, to support research in a research oriented medical center, it is important to define clearly who is responsible for providing IT support to the different missions – patient care, business operations, and research. In the authors' institution, clinical and business operations are supported by hospital information

Table 6.2 Distribution of responsibilities for support of research between Hospital Information Services (IS) and Biomedical Informatics (BMI)

Issue	IS	BMI
Centralized data storage for research	X/J	P
Special needs data storage for research	J	X/P
Backups and archives of research data	X/J	P
Research data warehouse	J	X/P
Feeds to research warehouse from enterprise clinical systems	X/P	J
Feeds to research warehouse from research & biostatistical datasets	J	X/P
Clinical data warehouse	X/P	J
Data standards	J	X/P
Software evaluation and development for research	J	X/P
Clinical software implementation	X/P	J
Support for research applications	J	X

X primary responsibility for operational support, *P* primary responsibility to formulate policies, *J* participates in formulation of policy

services (IS), while research is supported by biomedical informatics (BMI). With regard to research patient data warehousing, IS and BMI have developed a responsibility matrix (Table 6.2). IS operates the enterprise clinical and business information systems, including the electronic medical record. IS provides data feeds to BMI, which permits BMI to extract, transform, and load data from the electronic medical record to the institution's i2b2 research patient data warehouse. BMI develops many of the software applications necessary to support research programs, including those that provide access to and analysis of data in the warehouse and other institutional data repositories.

Acknowledgements This publication was supported in part by an Institutional Clinical and Translational Science Award, NIH/NCATS Grant Number 8UL1TR000077-04. Its contents are solely the responsibility of the author and do not necessarily represent the official views of the NIH.

References

Agosta L. The essential guide to data warehousing, Essential guide series. Upper Saddle River: Prentice Hall PTR; 2000. xxvi, 499 p.

Blumenthal D. Stimulating the adoption of health information technology. N Engl J Med. 2009;360(15):1477–9.

Blumenthal D. Launching HITECH. N Engl J Med. 2010;362(5):382–5.

Dhir R, et al. A multidisciplinary approach to honest broker services for tissue banks and clinical data – a pragmatic and practical model. Cancer. 2008;113(7):1705–15.

Eggebraaten TJ, Tenner JW, Dubbels JC. A health-care data model based on the HL7 reference information model. IBM Syst J. 2007;46(1):5–18.

Gainer V, et al. Using the i2b2 hive for clinical discovery: an example. AMIA Annu Symp Proc. 2007;2007:959.

Kimball R. The data warehouse toolkit: practical techniques for building dimensional data warehouses. New York: Wiley; 1996. xxviii, 388 p.

Kimball R. The data warehouse lifecycle toolkit: expert methods for designing, developing, and deploying data warehouses. New York: Wiley; 1998. xviii, 771 p.

Kimball R, Caserta J. The data warehouse ETL toolkit: practical techniques for extracting, cleaning, conforming, and delivering data. Indianapolis: Wiley; 2004. xxxiv, 491 p.

Kohane IS, Churchill SE, Murphy SN. A translational engine at the national scale: informatics for integrating biology and the bedside. J Am Med Inform Assoc. 2012;19(2):181–5.

Lowe HJ, et al. STRIDE – an integrated standards-based translational research informatics platform. AMIA Annu Symp Proc. 2009;2009:391–5.

Lyman JA, et al. Mapping from a clinical data warehouse to the HL7 reference information model. AMIA Annu Symp Proc. 2003;2003:920.

Mackenzie SL, et al. Practices and perspectives on building integrated data repositories: results from a 2010 CTSA survey. J Am Med Inform Assoc. 2012;19(E1):e119–24.

Mendis M, et al. Integration of Hive and cell software in the i2b2 architecture. AMIA Annu Symp Proc. 2007;2007:1048.

Murphy SN, et al. Integration of clinical and genetic data in the i2b2 architecture. AMIA Annu Symp Proc. 2006;2006:1040.

Murphy SN, et al. Serving the enterprise and beyond with informatics for integrating biology and the bedside (i2b2). J Am Med Inform Assoc. 2010;17(2):124–30.

Murphy SN, et al. Strategies for maintaining patient privacy in i2b2. J Am Med Inform Assoc. 2011;18 Suppl 1:i103–8.

Musen MA, et al. The National Center for Biomedical Ontology. J Am Med Inform Assoc. 2012;19(2):190–5.

Overhage JM, et al. Validation of a common data model for active safety surveillance research. J Am Med Inform Assoc. 2012;19(1):54–60.

Research Patient Data Registry – RPDR. [cited 2012 Apr 28]. Available from: http://rc.partners.org/rpdr

SACHRP. Attachment A to January 24, 2011 SACHRP letter to the secretary – FAQs, terms and recommendations on informed consent and research use of biospecimens; 2010.

Title 45 Code of Federal Regulations; 2009 Part 46, in http://ohsr.od.nih.gov/guidelines/45cfr46.html – 2009;46.114.

Title 45 Code of Federal Regulations; 2009 Parts 160, 162, 164: Combined Regulation Text. Available from: http://www.hhs.gov/ocr/privacy/hipaa/administrative/combined/index.html

Weiner MG, Embi PJ. Toward reuse of clinical data for research and quality improvement: the end of the beginning? Ann Intern Med. 2009;151(5):359–60.

Zhang GQ, et al. VISAGE: a query interface for clinical research. AMIA Summit Transl Sci Proc. 2010;2010:76–80.

Chapter 7
Biobanking in Pediatric Research

Michael G. Barnes, John A. Lynch, Jeremy J. Corsmo, David P. Witte, and Paul E. Steele

Abstract Modern pediatric disease research requires rapid access to high quality biospecimens. Morphological studies require precise sample preservation while genetic studies often require thousands of samples to identify biologically or clinically relevant findings. To have the necessary number of samples requires specialized centers called biobanks to collect, process, store and distribute biospecimens and associated clinical data. Records of the biospecimen lifecycle must be maintained to ensure that samples are fit-for-purpose and to avoid spurious results that can

---

M.G. Barnes, Ph.D. (✉)
Division of Rheumatology, Cincinnati BioBank Core Facility, Cincinnati Children's Hospital Medical Center, 3333 Burnet Ave ML 4010, Cincinnati, OH 45229-3039, USA
e-mail: michael.barnes@cchmc.org

J.A. Lynch, Ph.D.
Department of Communication, McMicken College of Arts & Sciences, University of Cincinnati, Cincinnati, OH, USA

Center for Clinical and Translational Science and Training, University of Cincinnati College of Medicine, Cincinnati, OH, USA

J.J. Corsmo, MPH, CIP, CHRC
Research Compliance and Regulatory Affairs, Cincinnati Children's Hospital Medical Center, 3333 Burnet Ave, ML 4010, Cincinnati, OH 45229-3039, USA

D.P. Witte, M.D.
Departments of Pathology and Pediatrics, University of Cincinnati College of Medicine, Cincinnati, OH, USA

Division of Pathology and Laboratory Medicine, Cincinnati Children's Hospital Medical Center, 3333 Burnet Ave, ML 4010, Cincinnati, OH 45229-3039, USA

P.E. Steele, M.D.
Departments of Pathology and Pediatrics, College of Medicine, University of Cincinnati, Cincinnati, OH, USA

Clinical Laboratory, Cincinnati Children's Hospital Medical Center, 3333 Burnet Ave, ML 4010, Cincinnati, OH 45229-3039, USA

J.J. Hutton (ed.), *Pediatric Biomedical Informatics: Computer Applications in Pediatric Research*, Translational Bioinformatics 2, DOI 10.1007/978-94-007-5149-1_7,

cost time and money as well as possibly harming patients. Informatics solutions to sample management can vary from basic sample tracking systems to full laboratory information management systems (LIMS). These tracking systems can be 'organic' in-house software systems or commercial software packages. While attention to sample collection is imperative for biobanked samples, sample utility is generally proportional to the quality and quantity of associated clinical data. One benefit of centralized biobanking operations is economies of scale allowing development of interfaces between clinical data from a local electronic medical record (EMR) or research patient data warehouse and de-identified biobank samples. This combination allows previously un-contemplated research projects. Regulatory compliance in pediatric biospecimen collections requires consideration of many issues around informed consent, assent, and reconsent at the age of majority. Finally, issues surrounding return of incidental findings are becoming prevalent in the post-genomics era. The components and benefits of various sample management strategies will be discussed with a focus on ensuring sample quality and integration with clinical data.

7.1 Introduction

Biobanking involves the collection, processing, storage and distribution of biospecimens for clinical or research purposes. The practice of biobanking carries with it an enormous potential for discovery. Consider a recent Swedish study (Persson et al. 2011) which drew from two biobanks containing around 400,000 serum samples collected from 1968 to 2006, many of which had been obtained for clinical testing for blood-borne viral diseases. This study found a stronger than expected association of seropositivity for *Helicobacter pylori* infection (believed to be largely acquired in childhood) and the development of non-cardia stomach adenocarcinoma later in adult life.

Biospecimens can be of any sample type (e.g., blood, tissue, synovial fluid, urine, etc.) or derivative (RNA, DNA, serum, plasma, etc.) obtained from a biological source including human, animal, plant or microbe. This chapter will focus exclusively on biospecimens from human sources, especially pediatric subjects. Biobanks provide expert resources and knowledge to ensure that samples obtained reflect the true state of the donor rather than technical artifacts related to sample collection and processing methodologies. Pediatric biobanking is a specialty with unique opportunities and challenges. Cincinnati Children's Hospital has extensive experience in pediatric biobanking, identifying challenges and finding solutions that are ethical and legal, while still allowing research into human maladies with the goal of changing the outcome for the better.

Yet, access to quality biospecimens is only one piece of the overall biobanking picture. In order to provide maximal utility of samples, investigators require access to phenotype data about the donor. These data can include clinical or research data collections. The more accurately a subject is phenotyped, the more homogenous a population can be developed, supporting stronger statistical models and ultimately providing improved accuracy of results. At the same time, data overload is a common problem with researchers accessing dozens, hundreds or even thousands of data points for each individual. Refined methods for accessing phenotypic data on subjects and dealing with the mass of data are required to optimize research. In what follows, we identify practical, informatics, ethical and regulatory issues that must be considered at the start of pediatric biobanking initiatives before discussing the process of sample collection and then making banked biospecimens available for researchers along with information about the source of the specimen (patient phenotype).

7.2 Why Pediatric Biobanking

Modern society is protective of human life, especially the life and well-being of children. In the recent past, the attempt to protect children led to an avoidance of research using pediatric subjects (Grodin and Glantz 1994). Fears related to long-term damage from research procedures, many of which could not be predicted and might not be detected for many years, made the risk seemingly too high. Any damage to developing brains or reproductive systems can have consequences across the span of a person's life. Additionally, a child's physiology is rapidly changing during the transition from fetus to early childhood through puberty to the young adult. Avoidance of testing on pediatric subjects has often led to children being treated as 'little adults', as medications developed and tested in adults are prescribed to children (Kodish 2005). This inaccurate assessment could easily lead to incorrect understanding of disease and inappropriate treatment.

Biobanks specific for samples from pediatric subjects are needed to help support research in this field. Many diseases are unique to or begin in childhood. It may be impossible to study these diseases in an adult population or it may, at least, be impossible to study early stages of development of the disease. For instance, juvenile idiopathic arthritis (defined as arthritis with an onset before the 16th birthday, lasting at least 6 weeks with no known cause (Petty et al. 1998)) is a clinical entity distinct from adult onset rheumatoid arthritis (Thompson et al. 2010). Confining studies to adults who were diagnosed with this disease in childhood would make understanding mechanisms of disease difficult. Since most children are treated in a pediatric setting, the biobank must be co-located to ensure proper and prompt sample handling. Biobanks focused on pediatrics are also required in most medical centers to ensure that sufficient attention and capacity are available to address pediatric conditions and that resources are not directed exclusively toward adult research.

Research in a pediatric setting benefits from access to and active participation of family members. These individuals are more likely to provide medical information and allow biobanking of samples for research into pediatric conditions than in an adult research setting. Additionally, both parents of a child can usually be contacted, while access is more limited in adult research where family members may live in different cities or may no longer be living.

7.3 Ethical and Regulatory Issues

Before collection of samples is initiated, certain ethical and regulatory issues must be addressed. Some issues must be almost fully addressed (e.g. consent), while others can be refined as the development of the biobank proceeds.

7.3.1 Consenting for Use of Residual Clinical Samples

One of most significant and complex decisions to be made at the outset of a biobanking initiative is whether and how to obtain informed consent. This choice will have implications for a hospital's electronic health record (or similar patient tracking) system, human resources, infrastructure, and patient flow. A primary consideration in all decisions is that the consenting procedure must comply with applicable federal, state, and local laws and regulations. Many institutions begin with a firm commitment to systematically collect and store some or all residual clinical samples for future unspecified research. Residual clinical samples are a valuable source of research samples. Use of these samples is easier and safer for patients than samples obtained exclusively for research, as well as being more time and cost effective. In addition, it does not make sense to obtain a new sample, when a sample that could be used in research is being discarded by the clinical lab.

Once this decision is made, several options regarding research informed consent should be considered: (1) request a wholesale, Institutional Review Board (IRB) approved, waiver of informed consent; (2) incorporate an agreement in the standard clinical consent to treat document; (3) include a more explicit description of the project with either a verbal consent or a passive consent (i.e. an *opt out* process); (4) develop a process completed as part of the regular hospital registration process (*i.e. opt-in at registration* process); or (5) develop a process as a freestanding process completed at some point during clinical care at the institution (*i.e. a standalone opt-in process*). Each of these options presents its own unique risk/benefit profile. One of the most critical variables is whether a stored sample will be linked to the medical record or patient identifiers in the electronic medical record or sample tracking systems. Maintenance of such an identifying link can be important to ensure that stored residual clinical samples are available for additional testing/analysis, if such a need should arise. Additionally, such linkages increase sample utility by allowing association of the patient's clinical data or phenotype with the samples.

Given the complexity of the issues, it is advisable to seek input from the patient/family and local community about their perception of biobanking and the consenting process. Although both groups consulted may bring slightly different perspectives to the issues, remarkably consistent commentary was obtained in our own experience and ultimately was utilized for the local biobank. In our community, perception of the general biobanking initiative was overwhelmingly positive and viewed as an important component of the institution's overall research mission.

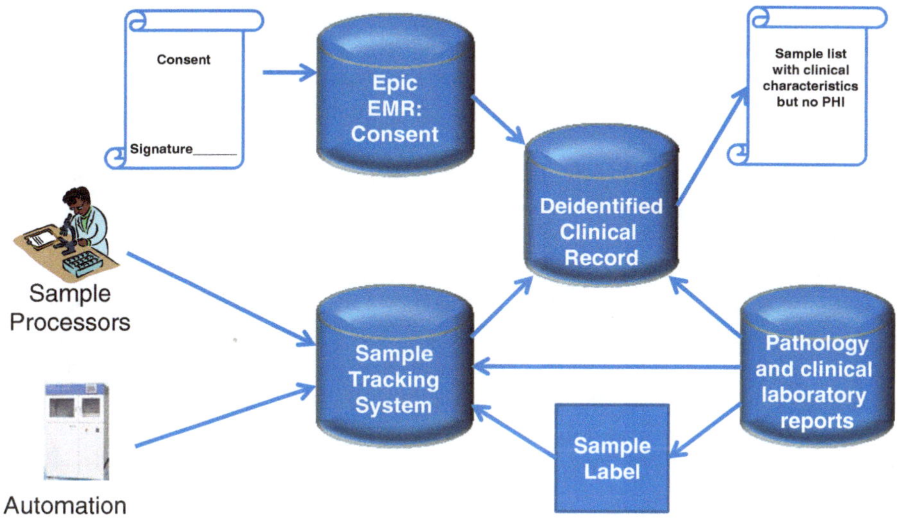

Fig. 7.1 Information flow for institutional biobanking. This diagram presents a simplified view of data flow necessary to support an institutional biobank. Consent is obtained and recorded in the electronic medical record. Data concerning samples is input to a sample tracking system from personnel or instruments in the research laboratory. Consent status and sample information are integrated using a de-identified clinical record

However, the *opt out* approach (defaulting to agreement unless the individual proactively says no) was overwhelmingly viewed as unacceptable in our community. Use of a shortened traditional consent form was much preferred.

The content of the consent document was developed through consultation with the patient and family advisory groups. The consent is presented to the patient/family as part of the routine registration process at any of our institution's numerous outpatient clinics and same day surgery facility. A variety of preferences can be selected: (1) Participation with return of incidental findings; (2) Participation without return of incidental findings; (3) Consent deferred; (4) Decline to participate; (5) Withdraw from study. The preference of the patient/family is recorded in the institution's electronic medical record system (Fig. 7.1).

Two levels of participation are tracked. One level (participation with return of incidental findings) allows the patient/family to receive information regarding individual, clinically/medically relevant information identified through research with the stored samples. A second option (participation without return of incidental findings) allows samples to be included in the biobank with the understanding that there is no mechanism for returning incidental research findings to the patient/family. (For more on this, see Return of Incidental Findings.)

From an ethical, regulatory, and informatics standpoint it is important to distinguish "consent deferred" from "decline to participate". "Consent deferred" indicates that the patient/family is unable to make a decision at the time of registration. This option is a temporary condition in which the family likely needs additional information or time to make a decision. Samples for which this option is selected are not made available for research. "Decline participation" is a definitive choice of the patient/family regarding the use of their residual clinical samples as part of the biospecimen repository. Finally, research participants must always be able to withdraw from a study. Since consent conditions are recorded in the electronic medical record, 'withdraw from study' must be maintained there to ensure no confusion when samples are considered for collection or release (see Sample Retention).

More broadly, informatics input in storing and propagating consent data is essential. At the most basic level, researchers must keep track of what study a sample and patient is consented to. This is where many sample tracking systems stop; however, as described above, there are additional details that must be recorded. It is important to retain the name and relationship of the person who signed the consent, especially with children. This information must be available for verification, especially if questions about the validity of consent are raised at a later date. As described above, modern consents often contain specific questions of consent. When selecting samples, it is important to know what type of research or research group the patient consented to allow to use their sample. With pediatric patients, a sample tracking system will often be called upon to alert investigators when a patient reaches a certain age. For instance, at18 years of age the patient must be reconsented to provide his or her own choice regarding participation in research. For an institutional biobanking effort, this is a daunting challenge and can only be accomplished using an informatics solution.

7.3.2 Return of Incidental Findings

Any discussion of biobanking in the post-genomic era would be incomplete without a discussion of return of 'incidental findings', those beyond the scope of the original research or clinical question. At the close of the twentieth century, bioethicists believed return of research results should occur rarely, if at all (Research involving human biological materials: ethical issues and policy guidance). Yet, this position was never universally accepted, and bioethicists appear to be moving toward an ethical obligation to return incidental findings to research participants, grounded in the Belmont Report's principles of respect for persons and beneficence (Wolf et al. 2008; Clayton 2005; Kohane et al. 2007; Fernandez et al. 2003). While some guidelines address return of results for adults, guidance for returning results in pediatric settings is sparse, and little research on returning results in pediatric settings has been conducted, outside of work in pediatric oncology (Avard et al. 2011; Fernandez et al. 2005).

Implementing a strategy to meet the ethical obligation to return incidental findings can be challenging. The first step toward resolution of this quandary was provided during our local community consultation. As described above, during the consenting procedure, the patient/family is asked if they would like incidental findings returned or not. This selection is recorded along with the consent and is the primary driver regarding returning results. During development of the Better Outcomes for Children (BOfC) project that proposes to collect residual blood samples from all patients of Cincinnati Children's Hospital, who consent, for preparation of DNA for genomic studies, several additional issues were considered, sometimes without complete resolution: What results should be returned; who should return them and who should receive these results; how the results should be returned; and what additional obligations researchers and clinicians have to ensure participants have sufficient resources to act on the report of incidental findings (Wolf et al. 2008; Fernandez et al. 2003). All of these issues are further complicated in pediatric settings by the tripartite relationship among clinician, child, and parent and the evolving intellectual capacity of pediatric, especially adolescent, patients.

7.3.3 What Results Should Be Returned?

Some bioethicists argue that results that have utility should be returned to patients, but the practical conception of "utility" varies: Some argue for a strict definition of *clinical utility*, while others consider *personal utility* in life planning or in *reaffirming one's identity* (Kohane et al. 2007; Fernandez et al. 2003; Bookman et al. 2006; Stack et al. 2011; Clayton et al. 2010). However "utility" is defined, the definition should include results that might not have an immediate impact on clinical care, but could play a role in life planning (e.g. carrier status, potential conditions that require monitoring, etc.). While this standard seems intuitive, it is complicated by two factors. Research results are not always obtained with an FDA approved test, and their significance is not always clear (Bookman et al. 2006; Miller et al. 2008; Shalowitz and Miller 2005). In order to be FDA approved in the United States, a test undergoes extensive testing and validation to ensure that the test is robust and results are both specific and sensitive. Tests that are not approved are of less certain validity. Also, many research laboratories are not CLIA certified to perform clinical testing (Wolf et al. 2008; Clayton 2005; Bookman et al. 2006). In the United States, only tests performed in a CLIA approved laboratory may be returned to patients and those from non-approved laboratories should not be returned even when an investigator believes that a result has medical implication for the patient.

7.3.4 Who Should Return Results and Who Should Receive Them?

It is not clear who should return incidental findings from research. While investigators are most familiar with the results, they might not have the appropriate clinical or counseling experience to effectively explain results to participants (Wolf et al. 2008; Avard et al. 2011; Bookman et al. 2006; Sharp 2011). There is also concern about investigators contacting research subjects with whom they have never had contact (i.e. as in research performed on stored samples) (Wolf et al. 2008). On the other hand, primary healthcare providers will have a rapport with their patient and his or her parents or guardians, but the primary healthcare provider might not have the expertise to interpret research results appropriately for the patient. Finally, there is the challenge of providing results to providers and patients in an easily accessed format. The electronic health record has been identified as an ideal means of returning information (Kohane et al. 2007), however, the EHR cannot be used when research results come from non-CLIA approved laboratories. In research-intensive institutions, a research patient data warehouse, research registries, and special software applications to interface researchers with patients may need to be developed (See Chap. 6, Research Patient Data Warehousing). These processes and informatics solutions require careful review by the Institutional Review Board to be certain they comply with applicable laws and regulations.

There is also some question about who should receive results. It is fairly well established that participants in a study should have the opportunity either to receive results or to refuse recontact and the return of information (Wolf et al. 2008; Clayton 2005; Fernandez et al. 2003; Bookman et al. 2006). Yet, others, like immediate family members, might need to be contacted, if a serious health risk is identified (Avard et al. 2011; Black and McClellan 2011). Unless those family members are also patients in the same clinical system, returning results will be difficult and might require an informatics system independent of the EHR. Pediatric research raises additional complications since the wishes of the parents and the child, especially adolescents near the age of majority, may clash (Wolf et al. 2008; Avard et al. 2011). Patients are consented for enrollment in the biobank during registration at their initial visit to our hospital. At the current time (April, 2012), over 15,000 families of patients have been asked to provide consent for utilization of residual clinical samples as part of our project, Better Outcomes for Children (BOfC). Eighty-six percent have agreed to participate and 83 % have asked to be informed of findings that, in the opinion of the researcher, the patient/family should know.

7.3.5 What Is Owed to the Research Subject?

A final question to be considered is what is owed to the subject. If subjects agree to enroll in a biobank, it is impossible to identify all the possible research that might occur with those samples. Neither research subjects, the researchers, nor biobank staff can anticipate what information will be produced. Therefore, a plan must be developed to help participants understand and process information about incidental findings. Yet, the extent of that help is hard to define. Grants that support research rarely (if ever) provide resources to counsel participants, and researchers do not have external financial resources to provide it. Specific guidelines about additional support are still being developed, although a consensus seems to be developing that the minimal requirements include referral to additional sources of relevant expertise or services such as genetic counseling (Wolf et al. 2008; Bookman et al. 2006).

Because of all the considerations discussed above, and the lack of a general consensus regarding return of research results in a pediatric setting, Cincinnati Children's Hospital has deliberately left this issue open-ended. Investigators, in coordination with the biobank staff and the biobank advisory committee, might identify potential results to return that would be unique to each study. If such results are found, the biobank contacts the IRB with a proposal to re-identify samples to return specific results. The IRB has the final authority to approve or disapprove return of various results.

7.4 Sample Retention

Patient and sample selection for traditional research studies using small numbers of samples is relatively straightforward. When a researcher needs samples, a clinical research coordinator (CRC) approaches the patient, obtains written informed consent and obtains the sample. Sometimes the sample will be collected at a later date, but it is the responsibility of the CRC to track the patient and ensure that appropriate samples are collected at appropriate times. As more patients are involved in a study, more time must be devoted to these activities. At a certain critical mass of patients, such as an institutional sample collection, manual tracking of consent and sample becomes an overwhelming burden.

One major challenge to the use of clinical samples for research is selecting the appropriate samples for biobank storage out of all materials processed by a clinical laboratory or pathology service. Current efforts to obtain genetic samples for research in our institution utilize samples collected for complete blood counts (CBCs), which represents about 85,000 tests for about 35,000 unique individuals annually. Determining which of these samples can be used for research is beyond the capability of manual determination. Commercially available software exists that can be loaded with IRB-approved patient lists and query clinical information systems to identify eligible specimens as they come into the clinical laboratory. Alternatively, a custom solution can be implemented that takes advantage of the existent informatics infrastructure including the electronic health record, the research patient data warehouse (See the discussion of i2b2 in Chap. 6, Research Patient Data Warehousing), and the sample tracking systems. A custom informatics system was chosen at our institution. This secure system requires user authentication for access. The clinical accession number, a unique identifier provided by the clinical laboratory, is scanned into a website interface. In cases where scanning is not possible, the barcode can be typed into a text field or searches can be performed on a medical record number (MRN) or name. These values are all part of the label on the tube containing the clinical specimen, typically whole blood, and require no additional access to patient information. The system presents the operator with either a red light (patient has not consented to use of sample) or a green light (patient has consented), hence the system is called 'red light/green light'. In all cases, the assumption is that samples may not be used until consent is proven. Therefore, lack of any response or a deferred response is interpreted as a sample being ineligible for use.

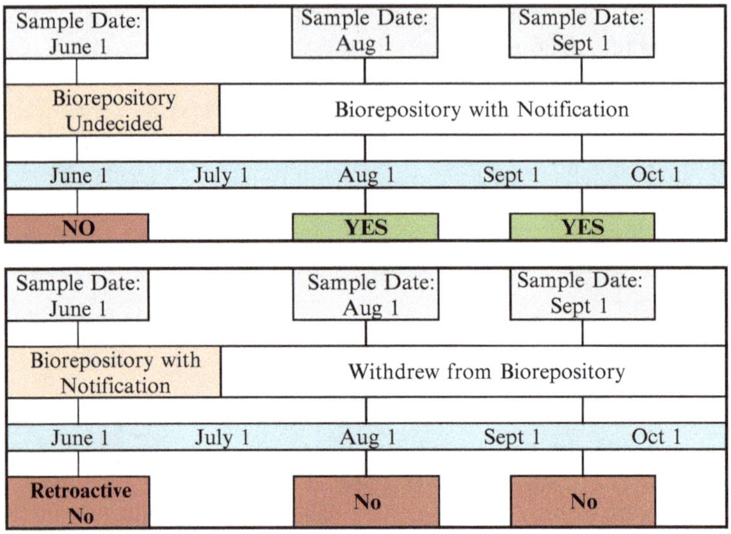

Fig. 7.2 Sample retention policy related to consent response. Samples collected for biobanking must be consented prior to use. This figure explains the logic used to determine whether a sample can be used for research. (*Top*) A patient is approached in June, but is unable to decide whether s/he wants to participate. Samples collected at this time are not kept for research. When the patient returns to the hospital in August, s/he decides to participate and future samples can be kept for research. (*Bottom*) A patient agrees to participate in BOfC at a visit in June. Subsequently, the patient decides to 'withdraw' from the study. Withdrawal overrides all previous agreements and all samples must be discarded

The system is built upon a database containing a list of all patients with BOfC responses and the response date. This information is tied to all samples from the patient allowing a determination of whether a specific sample from a patient was collected on, before or after the date of consent. After determining whether the specific sample has appropriate consent, a stop light figure with either a red or green light is displayed. The red light indicates that the sample does not have consent for research and must be discarded, while samples with green lights can be retained and used for future research purposes. Not surprisingly, more complicated cases arise requiring advanced logic. For instance, a patient may first be unsure whether they want to participate. At a later visit, they may decide to allow research use of residual samples. In this case, samples obtained before the "yes" consent may not be used for research; however, samples obtained later are available for research (Fig. 7.2). In research, there is always the option to 'withdraw' from a project. If a patient chooses to withdraw, all samples, even those collected during a time when the consent was 'yes' must be discarded. The red light/green light system will return a red light for all future samples and laboratory personnel must go through all records to identify any previous samples and discard them. In all of our logic, the default response is that samples should be discarded. Only those that are 'proven' to have appropriate consent are retained. In the case of ambiguity, samples are discarded. To further ensure that only correct samples are utilized, the red light/green light system also returns the name, date of birth and MRN of the patient. The lab worker then validates this information. After all samples have been scanned, it is possible to download a report for upload into the sample tracking system.

One unique extension of this logic related to pediatric populations is that the patient must be asked to reconfirm their consent when they turn 18 years old. The solution selected at Cincinnati Children's Hospital is that on the first visit after the child's 18th birthday, they are asked to give their own consent for research use of future samples.

A future extension of this system, the yellow light, will indicate whether DNA already exists for the patient. Many of our patients have more than one test ordered for clinical reasons, each of which generates a laboratory specimen. It is not uncommon for a single patient to have more than a dozen blood tubes. It is generally unnecessary to process multiple samples from the same patient. In a future version, the program will display a yellow light indicates to those processing samples that the sample already exists in the repository and lets the technician determine whether or not more sample is needed.

7.5 Clinical Sample Collections

Once systems are in place for consent and for identifying residual clinical samples for inclusion into the biobank, collection of those samples for research purposes can begin. These residual samples include biofluids analyzed in clinical laboratories and solid tissues assessed by the Pathology Departments. General documentation and reporting issues, as well as protocol concerns, exist for both types of samples, in addition to unique considerations for biofluids and solid tissues.

7.5.1 Sample Collection

Collection of biospecimens requires consideration of a vast array of variables including patient preparation, specimen procurement, transportation, processing, storage, and retrieval. These issues often extend to very specific guidelines, e.g. conditions of centrifugation such as *g* force, time, and temperature; time and temperature of transport to the lab; in-lab conditions prior to storage; sample sterility; documentation of deviations from standard procedures; etc. (Brisson et al. 2012; Schrohl et al. 2008; Skogstrand et al. 2008; Chaigneau et al. 2007; Bernini et al. 2011; Elliott and Peakman 2008). Additional considerations must be entertained when the protocol calls for the use of residual samples from biofluids that are obtained for routine, clinical use.

Documentation of conditions during sample collection in a clinical laboratory must be entered into the laboratory information system. This is necessary for

Table 7.1 Variables related to clinical samples

Routinely collected	Rarely/never collected	Causes of variation
Time of collection	Patient preparation (diet, posture)	Specimen types
Date of collection	Use of tourniquet	Workload
Type of tube	Collection type (arterial, capillary, etc.)	Staffing changes
Time of receipt in laboratory	Syringe or needle used	Need for repeated analysis
Time of analysis	Type of surgery	
Patient demographics	Warm and cold ischemia time	
	Temperature during transit	
	Temperature during processing and analysis (typically room temperature)	

determining the suitability of a sample for a given assay. A broad range of information on sample collection can be pertinent for biobanks. Some of that information is routinely collected by a clinical laboratory, while other types of information are rarely, if ever, recorded (see Table 7.1). For cerebrospinal fluid, for example, the average time spent prior to storage varied from less than 1 h to over 3 h in one study (Teunissen et al. 2009). Development of a system that utilizes numerous in-house informatics solutions is necessary to obtain the most accurate data providing a robust understanding of sample treatment (Fig. 7.1).

7.5.2 Standard Versus Multiple Protocols

Typically, biospecimens must be collected according to standard operating protocols developed with an eye to eventual research use (a concept called "fit-for-purpose") in order to obtain consistent results across a study. Yet, consolidation to a single protocol for all samples may mean that samples are not appropriate for additional analyses. A number of recent examples concerning RNA demonstrate this concept. In the early days of global gene expression studies, it was not understood how critical it was to rapidly process samples to ensure that observations reflected biology of the donor rather than technical artifacts due to processing delays (Barnes et al. 2010; Baechler et al. 2004). In response to this concern, technologies were developed to immediately preserve the RNA in blood samples; however, these technologies produced their own set of technical artifacts dubbed the 'globin effect' (An analysis of blood processing methods to prepare samples for GeneChip expression profiles (Technical Note) 2003). Recent advances, including new labeling technologies and globin reduction protocols, have helped overcome this issue. Additionally, it has become apparent that micro RNA, in addition to mRNA, can have significant influence on biology. Some previous sample preparation methods limit the quality or quantity of micro RNA recovered. Having samples collected using multiple procedures increases the likelihood that there will be samples available for multiple future study designs.

Regardless of whether all samples in a biobank are collected following the same protocol, consistent processing of all samples used for a specific experiment is necessary. Since modern investigations often require thousands of samples consolidated from multiple institutions and often utilize retrospective sample collections, pre-analytical variables (those which occur after the sample is obtained and before analysis) must be recorded and shared. Although a full discussion is outside the scope of this chapter, an international consortium has developed the 'Standard Preanalytical Coding' for biospecimens (SPREC) (Betsou et al. 2010). These variables include: sample type, collection container, pre-centrifugation delay, centrifugation speed and duration, second centrifugation, post-centrifugation delay, and storage condition. This 7-element code uses standard nomenclature to record major pre analytical variables in a consistent manner that can be shared electronically among biobanks.

7.6 Sample Storage

After processing, some samples may be analyzed immediately, but those for biobanking are stored prior to use. Although sometimes debated, it is generally considered best to store samples at the coldest temperature possible to prevent degradation of analytes, although new technologies are being developed to allow storage of some biomolecules at room temperature (Table 7.2).

Even with the best storage protocols, it must be kept in mind that each analyte has different storage characteristics. One clever approach to evaluation of specimen integrity is to spike isotope labeled peptides into samples prior to storage. Evaluation of their levels at the time of thawing can be used to monitor stability (Rezeli et al. 2010). While an improvement on most protocols which do not monitor for degradation, this approach does not comprehensively control for selective, extensive degradation of certain proteins such as vascular endothelial growth factor (VEGF), that occurs even in ultra-low freezers (Kisand et al. 2011). Further complicating this, the intended future assay is often unknown; therefore, so are the environmental conditions required to maintain analyte stability.

Table 7.2 Typical storage conditions for select sample types

	Storage condition				
Sample type	Room temperature	4°C	−20°C	−80°C	Liquid nitrogen
Viable cells					x
Serum				x	
Plasma				x	
DNA	x		x	x	
RNA	x			x	
Tissue	x			x	x

7.7 Sample Tracking

Methods of tracking samples come in varying sizes and complexities. Biobank staff can easily store data in written lists in laboratory notebooks or using basic spreadsheets and database programs. Alternatively, more complex systems designed specifically for sample management are available. Each level of system has benefits and deficits, although all must store unique identifiers related to samples (Fig. 7.3). Selecting sample management techniques is one of the most critical decisions a laboratory or biobank makes.

7.7.1 Laboratory Notebooks

Traditionally, data were stored on paper in laboratory notebooks. Reliance on handwritten notes can be problematic and cause loss of data when information is not copied accurately or is illegible. Laboratory notebooks are not usually maintained in duplicate. A notebook can be destroyed by water or biocontaminants (i.e., samples found in a biorepository) and all data lost. In the current information age it is especially telling that it is difficult to share hand written information, especially with those not in the same physical location. Finally, since paper-based data are not fully indexed, locating a single sample in an inventory from the past may take significant time or be impossible.

Fig. 7.3 Samples stored in a biobank. Each sample in a biobank requires a unique identifier. Shown in this picture are barcoded histocasettes containing solid tissue. Samples are stored inside biobags inside the histocasettes to protect them from dessication

7.7.2 Spreadsheets

Modern methods of data storage are great improvements over handwritten notebooks. The simplest alternative is to keep information in an electronic spreadsheet. This type of software has the benefit that most laboratory workers are familiar with spreadsheets; therefore, development, deployment, training and maintenance costs are minimal. Another benefit of electronic spreadsheets over paper is that data can be backed-up, lessening the risk of data loss. This benefit is often not fully realized because spreadsheet files are often stored on internal hard drives of personal desktop computers or external hard drives situated next to the computer. This leaves the data vulnerable to a single catastrophe such as a burst pipe.

7.7.3 Simple Databases

A higher level of complexity and security is a simple database program. Many investigators are familiar with this type of software, although they require more expertise than a spreadsheet. Design of a simple database can be done by the user, allowing flexibility to adapt to constantly and rapidly changing needs of research. In addition, uploads of data can be performed quickly and easily by the user. This allows typical laboratory workflow, where data are often recorded in a spreadsheet and then uploaded into the database, especially in the case of work done with multiple cooperating centers. One of the most significant benefits of a user controlled database system is access to the data. If a user cannot access his or her data as needed, it might as well not exist. Compared to spreadsheets, databases generally allow data access in a manner that can answer more sophisticated questions.

There are significant limitations to use of simple database systems. One major issue in today's regulatory environment is that, while files can technically be password protected in simple systems, in many cases spreadsheet and databases designed by non-informatics professionals do not require a user to login. In addition, it is unusual for a user designed database system to automatically log what changes were made to the data, why, when, and by whom. These functions are generally good laboratory practices and most importantly are required by laws and regulations such as HIPAA and/or FDA, when human subjects are involved in most types of clinical research.

Additional drawbacks are mainly related to lack of experience. When using simple databases, data may be collected in a manner that is not appropriate for future needs. For instance, if 'keys' are not set correctly, future queries may be difficult or impossible. Additionally, data uploads, if performed incorrectly can over-write previous data causing data corruption and loss. Finally, simple databases cannot generally be accessed by more than one user at a time, making use in a group difficult.

7.7.4 Dedicated Sample Tracking Software

Professionally designed sample tracking systems can support many more functions and provide much higher levels of security than simple databases. These sample tracking systems generally require significant support by informatics professionals and come in two flavors, custom or commercial. Custom software can be designed to meet the specific needs of an institution. This software is generally built with significant user input and can support many different types of research projects. The burden in direct and indirect costs to design, implement, maintain, and upgrade such a system is significant. While designing custom software provides more flexibility than using commercial software, groups often find that, over time, the costs outweigh the benefits. A commercial enterprise often has multiple staff who understand the software, while it is not unheard of for a custom piece of software to have the majority of information residing in the head of a single person. When that person leaves, all knowledge of how the software was designed and implemented may be lost.

A solution to some of these problems is to choose a commercial sample tracking software package. Commercial software is generally much more generic than custom designed software because it must meet the needs of diverse researchers in diverse fields. In fact, the company developing the software does not know the specifics of a single investigator's research when the software is designed. To provide utility for many types of users, commercial software usually has more fields available for data collection than a particular user might need. This is similar to commercial spreadsheets such as Excel, where there are many more options in dropdowns than most investigators need. The vendors build the commercial software to serve the needs of a vast array of different users. Users of these products are usually adept at selecting the options they need and ignoring those that are unnecessary for their work.

Regardless of whether sample tracking systems are custom or commercial, the advantages and disadvantages are usually similar. Sample tracking systems are generally designed with the specific goal of efficient and accurate sample management. Many of the variables discussed for clinical laboratory systems, as well as those required by the SPREC (Sect. 7.5) are recorded. These include times, temperatures, and collection vessels. Although not universal, many sample tracking systems also store some amount of demographic and/or phenotypic data to help with sample selection. A major capability of most sample tracking system, however, relates to sample relationships and storage. It is necessary, for instance, to know which serum samples came from which blood tubes. This allows patient identification related to the samples. Locating samples in storage is also a major issue because many banks have tens to hundreds of thousands of samples. Being able to define in which freezer, shelf, rack, box and slot a sample resides, is imperative for successful operation of the bank.

Large sample tracking systems generally develop with the input of many users, so that software developers and programmers can provide a robust software solution.

There is also more likely to be a core of experts to support the system than is the case with software designed and implemented by an investigator or single group. These experts evaluate the design and development, ensure system functionality, and ensure that the system complies with legal and regulatory requirements.

Large systems usually have the benefit of residing on a server system in a data center. Security of these institutional servers is maintained by a group of experts. Additionally, the physical machines are more likely to be secure from theft or damage than a typical investigator's desktop computer. Even worse, an investigator might house a database on a laptop computer, which can easily be stolen or lost, causing significant liability for the individual and the institution. For this reason, most institutions have policies prohibiting storage of Protected Health Information, such as that found in sample inventories, on laptops. Best practices in cybersecurity are discussed in Chap. 4, Institutional Cybersecurity in a Clinical Research Setting.

From the users perspective, there are disadvantages inherent in sample tracking systems. A large system takes significant time and resources to modify. This can cause significant problems for users as laboratory needs change, as regulations are modified, and when bugs are identified in the software. On the other hand, it is expected that by having a software group support the product, e.g. a vendor, the occurrence of such problems will be minimized. Local installation and maintenance of a sample tracking system is generally provided by the institution's IT group. This provides a core of experts to support the software; however, their first priority is security of the data and up-time of the software. While this is important, it may conflict with a user's priority, which is to access the data. Databases in a large software system are likely to be complex with multiple look-up tables, separate databases for Protected Health Information, and use of system generated identifiers rather than actual identifiers. Normalization is a useful database technique for storing data, but significant effort may be expended by users to denormalize the database into a format they can use.

7.8 Selection of Samples for Analysis

In the past, researchers usually examined paper-based clinical charts to identify patients for their studies. This was possible when a limited number of patients were needed and a subset of an institution's clinical records could be identified for examination. Current versions of research patient data warehouses can facilitate this task by providing investigators with the ability to search de-identified clinical records to assemble cohorts of patients with the desired phenotypes (Chap. 6, Research Patient Data Warehousing). With current genomic studies, researchers are often looking for many hundreds or even thousands of subjects (Harley 2007). In fact, in order to identify small effects of genes, over 10,000 patients and 10,000 controls may be needed. Such large numbers of patient records cannot be reviewed by hand. The informatics' solution can be the operational data warehouse of an electronic medical record (EMR) or more likely a research patient data warehouse that contains data extracted from the EMR and de-identified. Search of data in a research warehouse permits the investigator to identify patients who have a particular diagnosis, are being treated with a specific drug, or have a specific phenotype or any combination thereof. This allows for better specification of "the study cohort" and "the healthy control cohort."

There are two drawbacks to selection through the data warehouse of an EMR. First, it might be difficult to identify other conditions a patient might have in addition to the primary condition. For instance, if a researcher plans to study asthma in children, this patient population is relatively straightforward to identify. However, if the researcher wishes to exclude children with asthma who also have another autoimmune disease, this is difficult when using an EMR. Research patient data warehouses (Chap. 6) can overcome this difficulty by permitting use of multiple inclusion and exclusion criteria through automated queries (e.g. the researcher can include asthma, but exclude children who have other autoimmune diseases). Second, there are characteristics of value for some studies that simply will not appear in the medical record and an interview by study members would be required (e.g. what was the distance from an asthma patient's home to a highway and how long have they lived in that location?)

7.9 Best Practices in Biobanking

Many groups are working to develop best practices for biobanks. One way to standardize activities across diverse biobanking groups and ensure the availability of quality biospecimens is to provide generalized guidance for operations. For biobanking, several sets of best practices are currently available (International Society for Biological and Environmental Repositories (ISBER) 2008; National Cancer Institute best practices for biospecimen resources; Common Minimum Technical Standards and Protocols for Biological Resource Centres Dedicated to Cancer Research; OECD best practice guidelines for Biological Resource Centres). While each set focuses on different aspects of the biobanking enterprise, taken together they provide guidance for all stages of biobanking activity. Topics covered include staffing and management, physical facilities, data management, informed consent, conflicts of interest, and wide array of topics related to samples (acquisition, quality assurance, quality control, custodianship, and access).

In the United States, the majority of federal funding for biomedical research comes from the National Institutes of Health (NIH). Following a major disappointment during initiation of a major project, The Cancer Genome Atlas (TCGA), the National Cancer Institute (NCI) acknowledged a deficit in understanding biospecimen preservation and use. Since that time, the NCI has been at the forefront of improving biospecimen science. One step toward this goal was development of the Office of Biorepositories and Biospecimen Research (OBBR). Through OBBR and other initiatives, NCI has lead the way for NIH to develop best practices for sample collection, processing, storage and distribution. The most recent edition of the guidelines was published in 2011 (National Cancer Institute best practices for biospecimen resources).

Other groups are also working to ensure that biobanks operate in the best possible manner. One professional group, International Society of Biological and Environmental Repositories (ISBER), published a set of best practices in 2008 (ISBER 2008). ISBER also provides a self-assessment tool (SAT) for biobanks to determine how they measure up to these guidelines. The International Agency for Research on Cancer (IARC) of the World Health Organization (WHO) published standards and protocols for biobanks (Common Minimum Technical Standards and Protocols for Biological Resource Centres Dedicated to Cancer Research). This document builds on the previous general topics and includes many more details, including sample processing protocols. Finally, the Organisation for Economic Co-Operation and Development (OECD), a forum for 30 democratic governments to discuss wide ranging topics including those related to biotechnology, published "OECD Best Practice Guidelines for Biological Resource Centres" in 2007 (OECD best practice guidelines for Biological Resource Centres). This set of best practices includes biosecurity and microorganisms in addition to human samples.

Acknowledgements This publication was supported in part by an Institutional Clinical and Translational Science Award, NIH/NCATS Grant Number 8UL1TR000077-04. Its contents are solely the responsibility of the author and do not necessarily represent the official views of the NIH.

References

Affymetrix Technical Note, An analysis of blood processing methods to prepare samples for GeneChip expression profiles. 2003.

Avard D, Sénécal K, Madadi P, Sinnett D. Pediatric research and the return of individual research results. J Law Med Ethics. 2011;39(4):593–604.

Baechler EC, Batliwalla FM, Karypis G, Gaffney PM, Moser K, Ortmann WA, Espe KJ, Balasubramanian S, Hughes KM, Chan JP, et al. Expression levels for many genes in human peripheral blood cells are highly sensitive to ex vivo incubation. Genes Immun. 2004;5(5):347–53.

Barnes MG, Grom AA, Griffin TA, Colbert RA, Thompson SD. Gene expression profiles from peripheral blood mononuclear cells are sensitive to short processing delays. Biopreserv Biobank. 2010;8(3):153–62.

Bernini P, Bertini I, Luchinat C, Nincheri P, Staderini S, Turano P. Standard operating procedures for pre-analytical handling of blood and urine for metabolomic studies and biobanks. J Biomol NMR. 2011;49(3–4):231–43.

Betsou F, Lehmann S, Ashton G, Barnes M, Benson EE, Coppola D, DeSouza Y, Eliason J, Glazer B, Guadagni F, et al. Standard preanalytical coding for biospecimens: defining the sample PREanalytical code. Cancer Epidemiol Biomarkers Prev. 2010;19(4):1004–11.

Black L, McClellan KA. Familial communication of research results: a need to know? J Law Med Ethics. 2011;39(4):605–13.

Bookman EB, Langehorne AA, Eckfeldt JH, Glass KC, Jarvik GP, Klag M, Koski G, Motulsky A, Wilfond B, Manolio TA, et al. Reporting genetic results in research studies: summary and recommendations of an NHLBI working group. Am J Med Genet A. 2006;140A:1033–40.

Brisson AR, Matsui D, Rieder MJ, Fraser DD. Translational research in pediatrics: tissue sampling and biobanking. Pediatrics. 2012;129(1):153–62.
Chaigneau C, Cabioch T, Beaumont K, Betsou F. Serum biobank certification and the establishment of quality controls for biological fluids: examples of serum biomarker stability after temperature variation. Clin Chem Lab Med. 2007;45(10):1390–5.
Clayton EW. Informed consent and biobanks. J Law Med Ethics. 2005;33:15–21.
Clayton EW, Smith M, Fullerton SM, Burke W, McCarty CA, Koenig BA, McGuire AL, Beskow LM, Dressler L, Lemke AA, et al. Confronting real time ethical, legal, and social issues in the Electronic Medical Records and Genomics (eMERGE) Consortium. Genet Med. 2010;12(10):616–20.
Common Minimum Technical Standards and Protocols for Biological Resource Centres Dedicated to Cancer Research. http://www.iarc.fr/en/publications/pdfs-online/wrk/wrk2/Standards_ProtocolsBRC.pdf
Elliott P, Peakman TC. The UK Biobank sample handling and storage protocol for the collection, processing and archiving of human blood and urine. Int J Epidemiol. 2008;37(2):234–44.
Fernandez CV, Kodish ED, Weijer C. Informing study participants of research results: an ethical imperative. IRB. 2003;25(3):12–9.
Fernandez CV, Taweel S, Kodish ED, Weijer C. Disclosure of research results to research participants: a pilot study of the needs and attitudes of adolescents and parents. Paediatr Child Health. 2005;10(6):332–4.
Grodin M, Glantz L. Children as research subjects: science, ethics, and law. New York: Oxford University Press; 1994.
Harley JB. IL-7Ralpha and multiple sclerosis risk. Nat Genet. 2007;39(9):1053–4.
International Society for Biological and Environmental Repositories (ISBER). 2008 best practices for repositories; collection, storage, retrieval and distribution of biological materials for research. Cell Preserv Technol. 2008;6(1):3–58.
Kisand K, Kerna I, Kumm J, Jonsson H, Tamm A. Impact of cryopreservation on serum concentration of matrix metalloproteinases (MMP)-7, TIMP-1, vascular growth factors (VEGF) and VEGF-R2 in biobank samples. Clin Chem Lab Med. 2011;49(2):229–35.
Kodish E. Ethics and research with children: a case-based approach. New York: Oxford University Press; 2005.
Kohane IS, Mandl KD, Taylor PL, Holm IA, Nigrin DJ, Kunkel LM. Reestablishing the researcher-patient compact. Science. 2007;316:836–7.
Miller FA, Christensen R, Giacomini M, Robert JS. Duty to disclose what? Querying the putative obligation to return research results to participants. J Med Ethics. 2008;34:210–13.
National Bioethics Advisory Commission. Research involving human biological materials: ethical issues and policy guidance. Volume. 1. Report and recommendation of the National Bioethics Commission, 1999. Available from: http://bioethics.georgetown.edu/nbac/hbm.pdf. Accessed 9 Sept 2012.
National Cancer Institute. National Cancer Institute best practices for biospecimen resources. 2011 revision. Available from: http://biospecimens.cancer.gov/practices/2011bp.asp. Accessed 9 Sept 2012.
Organization for Economic Co-operation and Development. OECD best practice guidelines for Biological Resource Centres. 2007. Available from: http://www.oecd.org/sti/biotechnologypolicies/38777417.pdf. Accessed 9 Sept 2012.
Persson C, Jia Y, Pettersson H, Dillner J, Nyren O, Ye W. *H. pylori* seropositivity before age 40 and subsequent risk of stomach cancer: a glimpse of the true relationship? PLoS One. 2011;6(3):e17404.
Petty RE, Southwood TR, Baum J, Bhettay E, Glass DN, Manners P, Maldonado-Cocco J, Suarez-Almazor M, Orozco-Alcala J, Prieur AM. Revision of the proposed classification criteria for juvenile idiopathic arthritis: Durban, 1997. J Rheumatol. 1998;25(10):1991–4.
Research involving human biological materials: ethical issues and policy guidance, vol. 1. Report and Recommendations of the National Bioethics Commission. 1999. http://bioethics.georgetown.edu/nbac/hbm.pdf

Rezeli M, Vegvari A, Marko-Varga G, Laurell T. Isotope labeled internal standards (ILIS) as a basis for quality control in clinical studies using plasma samples. J Proteomics. 2010;73(6):1219–29.

Schrohl AS, Wurtz S, Kohn E, Banks RE, Nielsen HJ, Sweep FC, Brunner N. Banking of biological fluids for studies of disease-associated protein biomarkers. Mol Cell Proteomics. 2008;7(10):2061–6.

Shalowitz DI, Miller FG. Disclosing individual results of clinical research. J Am Med Assoc. 2005;294(6):737–40.

Sharp RR. Downsizing genomic medicine: approaching the ethical complexity of whole-genome sequencing by starting small. Genet Med. 2011;13(3):191–4.

Skogstrand K, Ekelund CK, Thorsen P, Vogel I, Jacobsson B, Norgaard-Pedersen B, Hougaard DM. Effects of blood sample handling procedures on measurable inflammatory markers in plasma, serum and dried blood spot samples. J Immunol Methods. 2008;336(1):78–84.

Stack CB, Gharani N, Gordon ES, Schmidlen T, Christman MF, Keller MA. Genetic risk estimation in the Coriell Personalized Medicine Collaborative. Genet Med. 2011;13(2):131–9.

Teunissen CE, Petzold A, Bennett JL, Berven FS, Brundin L, Comabella M, Franciotta D, Frederiksen JL, Fleming JO, Furlan R, et al. A consensus protocol for the standardization of cerebrospinal fluid collection and biobanking. Neurology. 2009;73(22):1914–22.

Thompson SD, Barnes MG, Griffin TA, Grom AA, Glass DN. Heterogeneity in juvenile idiopathic arthritis: impact of molecular profiling based on DNA polymorphism and gene expression patterns. Arthritis Rheum. 2010;62(9):2611–15.

Wolf SM, Lawrentz FP, Nelson CA, Kahn JP, Cho MK, Clayton EW, Fletcher JG, Georgieff MK, Hammerschmidt D, Hudson K, et al. Managing incidental findings in human subjects research: analysis and recommendations. J Law Med Ethics. 2008;36(2):219–48.

World Health Organization International Agency for Research on Cancer. Common Minimum Technical Standards and Protocols for Biological Resource Centres Dedicated to Cancer Research. 2007. Available from: http://www.iarc.fr/en/publications/pdfs-online/wrk/wrk2/StandardsProtocolsBRC.pdf.

Part II
Applications

Chapter 8
EHR-Linked Registries for Studies of Populations

Keith Marsolo

Abstract Registries are a type of database routinely used to collect a standardized set of information on a defined cohort of patients. They are used in many different areas, from monitoring drug and device safety to comparative effectiveness research to quality improvement efforts looking to standardize care practices for patients with chronic disease. For many different types of research, particularly comparative effectiveness research or research involving children, no single institution has a large enough patient population to perform a proper study. This, coupled with the growing digitization of medical records, has led to an increased effort to create distributed research networks. The widespread adoption of electronic health records (EHRs) has resulted in a strong push to use them as the primary method of collection for registry data, capturing the necessary elements as part of routine clinical care. This would achieve the long-hoped for goal of "data in once," moving clinical and translational research away from the common practice of populating registries through double data entry. This is illustrated by a discussion of the registry infrastructure utilized by the ImproveCareNow network, a collaborative of clinicians focused on improving the care of children with Inflammatory Bowel Disease.

8.1 Introduction

Broadly defined, a registry is a database of standardized clinical information that has been collected on a cohort of patients, as defined by disease state, or some other set of inclusion and exclusion criteria. They are one of the better established types of clinical information systems, having been first implemented in the late 1980s

---

K. Marsolo, Ph.D. (✉)
Department of Pediatrics, University of Cincinnati College of Medicine, Cincinnati, OH, USA
e-mail: keith.marsolo@cchmc.org

J.J. Hutton (ed.), *Pediatric Biomedical Informatics: Computer Applications in Pediatric Research*, Translational Bioinformatics 2, DOI 10.1007/978-94-007-5149-1_8,

(Metzger 2004). Registries are population-level tools that are essential to support of many types of clinical and translational research, such as studies of the comparative effectiveness of alternative interventions to prevent or treat disease, as well as the identification of genes related to disease susceptibility. They are also commonly employed in the management of chronic disease, the monitoring of product and drug safety, studies of the natural history of disease, and quality improvement (Gliklich and Dreyer 2010). Although they have been in existence for several decades, many suffer from a number of technical limitations that prevent them from being easily integrated into clinical workflows.

It is challenging to feed data directly from an electronic health record (EHR) into a registry. This is due to both the technical challenges of extracting data from the EHR and the social challenges of ensuring that clinicians enter the necessary data into the EHR, as part of the clinical workflow. While the use of the EHR as the primary method of data collection removes the need for double data entry – the need to enter data once in the EHR to support patient care and a second time in the registry to support research or population management, users of registries populated with data from EHRs frequently forget that data cannot be extracted, if it were not entered into the EHR during the care of the patient in the first place. As a result, workflows in the clinic may need to be modified if data required by a research project are to be captured in the EHR.

Another limitation of many registries is their limited reporting capabilities. Feedback is generally provided in the form of monthly quality and population reports. There is limited ability to explore underlying data. Registries that support research are frequently populated with data captured at various different geographic sites of health care delivery within a network of participants. The registry that supports a project involving a network is usually "centralized," meaning data from all participants are sent to one site. If a participating site withdraws from the network, it can be difficult for that site to withdraw all of its data and ensure that it is no longer included in the overall registry. If the central site withdraws from the network, the entire registry can collapse. While it would be appealing to use an EHR itself as the registry platform, most EHRs are patient-focused by design and do not natively support the type of functionality required by a population-focused registry. Performing population-level queries on an EHR operational data warehouse, for instance, is difficult, if not impossible. Additionally, vendor-developed EHRs are proprietary and usually are not open to modification by their customers. This makes it difficult to add new data sources and data types, integrate data from outside systems or institutions, or associate the registry data with sample information stored in a biorepository. Lastly, unless institutions are within the same health system, there is little interoperability among EHRs that support care of patients by different providers because of semantic and syntactic differences among their systems.

8.2 Population-Based Research in Pediatrics

Most pediatric chronic conditions meet the NIH definition for a rare disease (NIH Office of Rare Diseases), and no single care center has a sufficient number of patients to produce generalizable knowledge, a barrier that, unless addressed by networks or multi-center studies, slows the pace of knowledge acquisition and outcomes improvement. The American Board of Pediatrics (ABP) has supported the establishment of sub-specialty improvement and research networks in all 13 pediatric sub-specialties. This effort extends the successes of pediatric clinical networks that use data for research and (increasingly) improvement. Examples include: (1) the *Vermont Oxford Network (VON)* (est. 1988) that is dedicated to improving the quality of neonatal ICU care (Horbar et al. 2001) and (2) the *Children's Oncology Group (COG)* (est. 1998) that focuses on clinical trials of new therapies, as well as studies of how to improve the delivery of existing therapies for pediatric cancer (Children's Oncology Group). The dramatic improvement in cancer survival rates from <10% in 1962 to over 80% today is a reflection of the benefits of an infrastructure that can systematically standardize care and enroll patients in research studies (Simone et al. 1998). The common themes drawn from collaborative pediatric research networks that have demonstrated marked improvement in the outcomes of children with chronic disease are: (a) an unrelenting commitment to collecting high quality data, (b) continuously evaluating and proving their value to clinicians making in-the-trenches decisions, and (c) the long-term engagement of the participants and their institutions to sustaining the network.

Sub-specialty research and improvement networks offer advantages that are foundational for research "laboratories" (Margolis et al. 2009). Creation of total population registries at each site provides large and diverse study samples. By standardizing practice, they reduce variations in outcomes that are caused by variations in the way care is delivered, thereby increasing statistical power. By linking research to care delivery and engaging clinicians directly, networks provide a forum for user-led comparative effectiveness research (CER) (Von Hippel 2005), a core attribute of the learning healthcare systems (Olsen et al. 2007). Despite the pressing need to scale these research networks, expansion is hampered by the lack of an informatics infrastructure capable of supporting needed expansions to include a large number of participating clinical sites, while reducing the costs of conducting research. Sub-specialty research networks typically collect data manually, using specially designed disease-specific forms. As a result, they do not capture data directly in electronic health records or incorporate the data into federated data warehouses. Furthermore, they do not make the data available to other researchers, allowing the data to be used to address a wide variety of secondary questions. These manual methods are costly, typically paper based, ad hoc (i.e., may depend on grant funding), and slow the pace of discovery by creating silos of data that are difficult or impossible to integrate.

8.3 Data Sharing Within Distributed Research Networks

Distributed research networks are gradually replacing the traditional central data coordinating center model because of advances in Internet and data sharing technologies. A small, but growing, number of distributed research networks have been implemented across a spectrum of clinical conditions and practice settings (caBIG Strategic Planning Workspace 2007; Loeffler and Winter 2007; Maro et al. 2009; Pace et al. 2009). Syntactic interoperability (the ability to package and exchange data among EHRs) remains a challenge. Open-source solutions for health information exchange such as CONNECT, which uses Nationwide Health Information Network (NHIN) standards and governance, are in the nascent stage (U.S. Department of Health and Human Services). Even if (and when) the syntactic barriers to EHR-based communication were resolved, semantic interoperability (the ability to understand the meaning of the data through a shared vocabulary among EHRs) would remain an outstanding issue (Gliklich and Dreyer 2010). There is little consensus on how to describe and label EHR data. Efforts by CDISC (Clinical Data Interchange Standards Consortium), with projects like BRIDG (Fridsma et al. 2008) and SHARE (Clinical Data Interchange Standards Consortium), aim to solve some of these challenges, but it remains to be seen whether they will achieve meaningful adoption in clinical operations. Relying on a distributed network removes the need for semantic interoperability among EHRs, because translation happens outside each individual EHR.

In a federated network, each individual site retains administrative control of its data, but sites trust one another for certain types of data sharing. They expose the information needed for agreed upon analyses. One example is the Shared Health Research Information NEtwork (SHRINE) (Weber et al. 2009), which is based on tools developed for the Shared Pathology Informatics Network (SPIN) (Namini et al. 2004; Drake et al. 2007). In its current implementation, a SHRINE network is composed of a single Query Aggregator and a set of SHRINE adaptors, one for each repository in the network. Authorized researchers can query the network for the total number of patients at participating institutions who meet a given set of criteria—a combination of demographics, diagnoses, medications and laboratory tests. Because counts are aggregate, patient privacy is protected. The Query Aggregator consists of a user interface and a set of web services that broadcast queries to each SHRINE adaptor. A SHRINE adaptor translates each SHRINE query into the corresponding local terminology and queries the attached data repository. The local results are translated into the SHRINE terminology and sent back to the Query Aggregator, which compiles the results as a set of aggregated counts. While SHRINE adaptors already exist for the i2b2 research patient data warehouse (Sect. 6.3, Murphy et al. 2006, 2010), they can be created for any repository, provided that the queries have been converted to run locally and a mapping has been created between the network terminology and the local codes. An added benefit to the SHRINE infrastructure is that there is no "central" site, whose demise would threaten the entire system, allowing sites to leave the network if they wish. As a real-world example, SHRINE has recently been used to examine the co-morbidities of children with Autism Spectrum Disorder (ASD) across several hospitals, all without the need to share patient-level data (Kohane et al. 2012).

An alternative to the SHRINE model is the one employed by PopMedNet, which was created to facilitate comparative effectiveness research (PopMedNet 2011). The major differences between an approach like SHRINE and one like PopMedNet are: (a) PopMedNet queries are not executed automatically, but require manual mediation and (b) PopMedNet users have the ability to request the results of an analysis that is run on a dataset, not just counts of patients.

On a large scale, as part of its learning healthcare initiative (Olsen et al. 2007), the Institute of Medicine in the United States stimulated development of Electronic Health Record Innovation Collaboratives, one of which is the Pediatric EHR Data Sharing Network (PEDSNet). PEDSNet was launched in late 2009 as a cooperative effort of 18 children's hospitals. PEDSNet was created to provide an enabling platform for open-science pediatric clinical and health services research. Semantic interoperability is an especially difficult challenge in pediatrics because ontologies and standard vocabularies are less developed for children than for adults. An early focus of PEDSNet is expansion and improvement of pediatric terminologies.

8.4 Workflow and Data Collection

The typical workflow for many chronic disease registries requires double data entry. This process is not efficient and is prone to error. First, data are captured in the EHR during the process of clinical care, in many instances as a clinical note. Then, data in the EHR are loaded into the registry. Generally, this requires that the data be abstracted onto a case report form and manually re-entered into a data collection system. At the end of each month, reports are generated and provided back to the participating centers. To be most efficient and to reduce errors, patient data should be collected once during routine clinical care and then transferred directly from the EHR to the registry by ETL without human intervention. In order for data to be easily extractable from the EHR, however, it is best to have it collected in a discrete field, not as part of a free-text (or semi-structured) note. The data elements utilized in many chronic disease registries (particularly those in pediatrics) are not standard in most EHRs. This means that forms for data capture in the EHR frequently need to be custom designed at significant cost for both development and maintenance.

Ambulatory clinic visits, where data would be collected in the EHR, are typically very short. Because of the time pressure, data collection must be well planned so that a minimum number of essential elements must be entered. In order to determine which elements to collect, the best approach is to start with the desired end-result, then work backwards to determine minimum, essential data to capture and data flow, an approach advocated by James (2003) and others (Horbar et al. 2001). This structured approach collects a parsimonious set of data that include only those elements that lead to actionable reports. Otherwise, one can be left with a lengthy

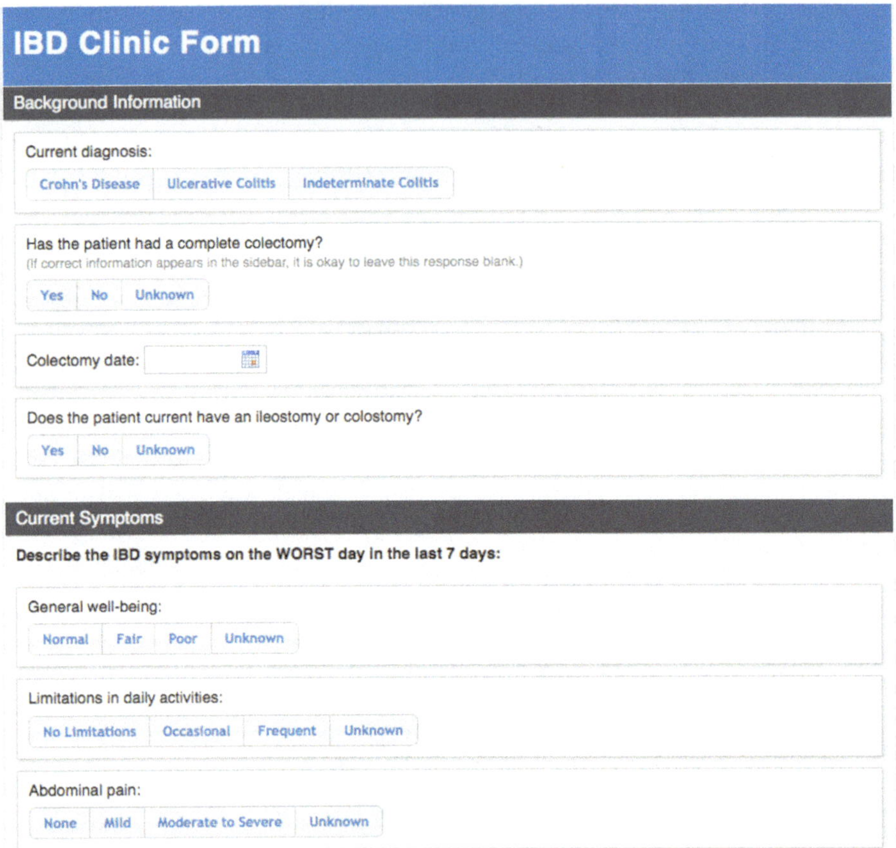

Fig. 8.1 A portion of a form that could be embedded within an EHR to enable clinicians to capture registry data as part of the clinical workflow. The look-and-feel of the form may differ by vendor, but the basic functionality is replicable. This form represents the data that would be collected by a gastroenterologist for patients with Inflammatory Bowel Disease

data collection form with fields that are rarely completed in a clinical setting because of lack of time and commitment. Figure 8.1 shows an example of a form used to collect data for a pediatric Inflammatory Bowel Disease (IBD) registry as part of routine clinical care. The data collected are valuable both to the physician providing care and to the researcher, thereby assuring more conscientious data capture.

8.5 Design of an EHR-Linked Registry

To avoid double data entry and associated errors, it is highly desirable to develop registries that are linked to the EHR (i.e., the EHR serves as the primary method of collection of data that will populate the registry). These registries could be

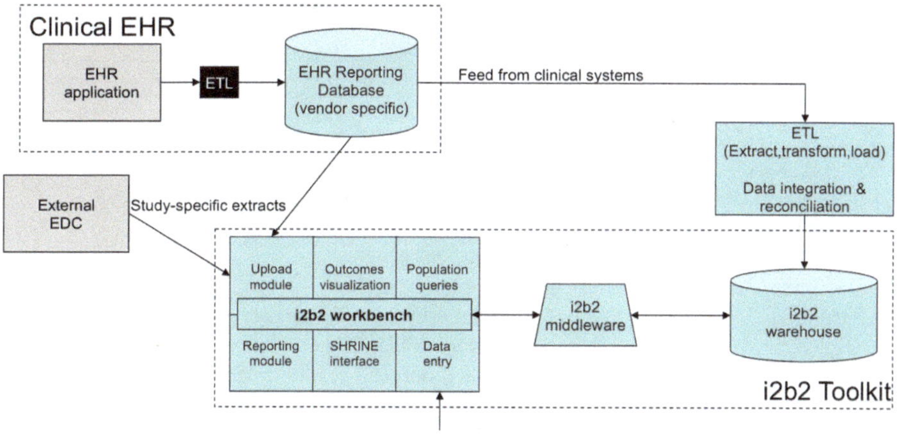

Fig. 8.2 An illustration of the i2b2 Toolkit and extensions to the i2b2 workbench to support multiple methods of transferring data into an EHR-linked registry. The supported transfer methods include direct database connections, file upload and direct data entry

stand-alone or federated (linked) into a multi-center network. From an informatics perspective, the architecture of a network of federated EHR-linked registries would include the following components: (a) linkage of a local registry to the EHR at each member site; (b) an application layer that allows the development of functionality to support research; and, (c) software to support federated queries among a network of local registries, allowing for the creation of a larger "virtual" registry.

One approach to instantiating this registry architecture is to combine elements of the i2b2 informatics framework, the SHRINE federated query architecture, and application modules, in the form of an i2b2 toolkit (https://community.i2b2.org/wiki/display/CCHMC/Home). This is illustrated in Fig. 8.2. Table 8.1 lists the different needs of the registry users and how the EHR-linked registry architecture detailed here satisfies those needs.

Data are extracted from the EHR, transformed, and loaded into an i2b2 database, as represented in the ETL process of Fig. 8.2 (*component a*). The i2b2 toolkit provides the necessary functionality in a modular fashion, allowing users to choose the capabilities that are appropriate for their research project (*component b*). The SHRINE federated query architecture can be used to link local registries into a larger network, allowing for queries on a larger, more diverse patient set (*component c*). Such a design gives registries the ability to capture data directly from the EHR and other sources, and integrate it into a single repository. The modular construction allows the registries to support the diverse needs of various types of research projects.

The EHR-linked registry architecture addresses two major limitations of current registries: (1) data will be captured once during the routine clinical care of the patient so that manual double data entry is not necessary, and (2) data will be housed locally in federated data warehouses, rather than in a single central registry.

Table 8.1 Functionality required of an EHR-linked registry and how the proposed architecture satisfies those needs

Functionality	Options supported by the EHR-linked registry
Registry population definition	1. Specified by user
	2. Derived from a larger population (i.e., entire EHR) based on inclusion/exclusion criteria
	3. Specific sub-populations of existing cohort
Data input	1. File upload
	2. Direct data entry
	3. Direct feed from EHR
Validation	1. Field-level validation (during data entry and upload)
	2. Cohort-level validation during file upload
	3. Validation scripts can be user-defined or selected from a library of existing procedures
Reporting and analysis	1. Pre-defined or ad-hoc reports
	2. Supports population and patient management reports, along with risk stratification
	3. Ability to generate data extracts for statistical processing
Visualization	1. Display data in run charts and dashboards
	2. Allow users to drill down into data
Variable definition	1. Ability to share measures across instances/projects/sites
	2. Ability to define data using multiple terminologies

Information is exchanged using distributed queries (creating a "virtual" repository), as shown in Fig. 8.3. Interoperable modules permit the assembly of specialized registries with little cost and technical effort. If data standardization is in place at the point of care, the creation of a second disease registry requires only the creation of a new data dictionary. In addition, concepts and variables can be shared, promoting reuse and allowing for cross-registry queries (where permissible). If a participating site chooses not to participate as a node on the network, it can send its data to a central node, as in the traditional model. In this manner, the technical architecture supports distributed, partially distributed and centralized models.

8.5.1 Importing Data into a Registry

In many cases, a large network will contain sites with EHRs in various stages of implementation and maturity. Some may still be paper based. To permit networks to function under these circumstances, the registries and networks are designed so that data can enter the system in multiple ways (Fig. 8.3). Sites can feed data directly from their EHR, they can upload files or reports extracted from an operational data store (ODS), and sites without EHRs can enter data directly through a web browser.

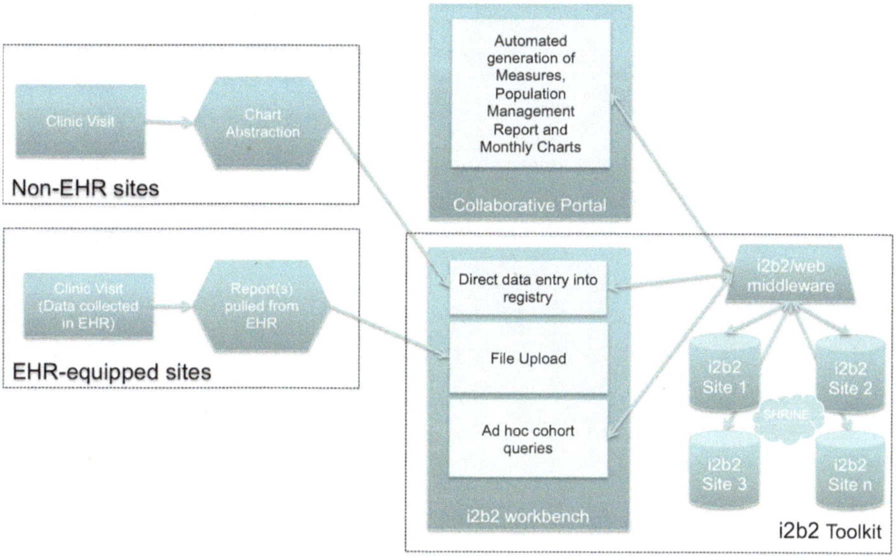

Fig. 8.3 A variation of the EHR-linked registry architecture. Sites with EHRs can upload data directly into the registry. Sites without EHRs can utilize direct data entry. The data for each site is stored in their site-specific database and these databases are federated using SHRINE. These queries are used in the generation of population management and pre-visit planning reports, which can be part of the i2b2 workbench or a separate web portal

To implement a direct link between an EHR database (or ODS) and the registry, the data in the EHR undergoes an ETL process and then is imported. ETL programming requires database developers to create a mapping between fields in the EHR and those in the registry. If an institution has multiple clinical systems, then overlapping fields must be reconciled. The process of refreshing or updating the data also has to be designed and monitored to make sure that there are no exceptions or failures. Such a process is complex, resource-intensive, and therefore a significant barrier to adoption.

An alternative to ETL through a direct database connection is to use a web-based upload module. This permits users to upload data stored in common file formats (Excel, comma separated value, etc.). Once uploaded, the data are processed by a validation module and matched to a registry terminology (i.e., data dictionary). For many medical institutions, file uploads are cost effective and readily available because data extracts can be generated from their ODS by an enterprise reporting team using the same processes that are used to support the institution's routine business and clinical operations.

Even with the above approaches, direct data entry into the registry should still remain an option. Not every participating center will have an EHR. In addition, for research projects, it may not be possible to collect every registry data element within the EHR. The data may be something not captured during the clinical workflow, may

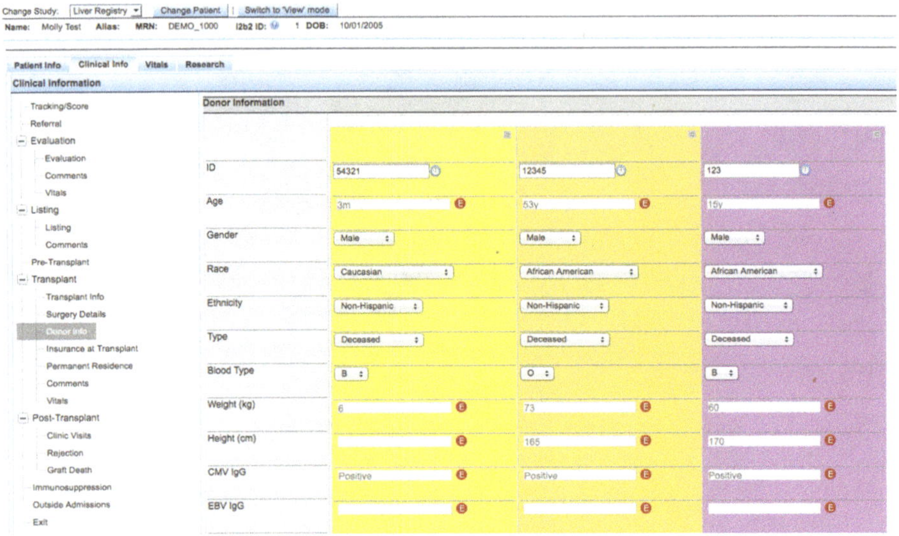

Fig. 8.4 An example registry form for Liver Transplant. The steps in the transplant workflow are listed on the *left side* (Evaluation, Listing, Pre-Transplant, Transplant, etc.). Fields where the primary source of data collection is the EHR are denoted with an "E." The patient in this example had three previous transplants, each listed in a separate column

not be directly "mappable" between the EHR and the registry (data derived from a medication order, for instance), or may need to be collected within the registry before the appropriate fields can be built in the EHR. Figure 8.4 shows an example of a form from a Liver Transplant registry, where portions of the data are collected in the EHR (fields denoted by an "E") and the rest are populated manually.

8.5.2 Data Mapping

Data mapping is one of the first steps required to populate a registry with data extracted from an EHR. EHRs and registries have very different data models. EHRs are transactional in nature and may store hundreds of observations about patients spread across multiple visits. Registries, particularly chronic disease registries, tend to view the patient record as a series of standard observations collected at specific time periods. Depending on the clinical workflow, registry data may come from multiple locations within the EHR, or be entered across multiple visits. There may even be data model differences between the different types of registries. A chronic disease registry, where the same data are collected on the patient during every visit, is very different from one focused on patients with a liver transplant, where a single "transplant episode" may span multiple years. Figure 8.5 shows an example workflow for a Liver Transplant registry, with the different "steps" that a patient

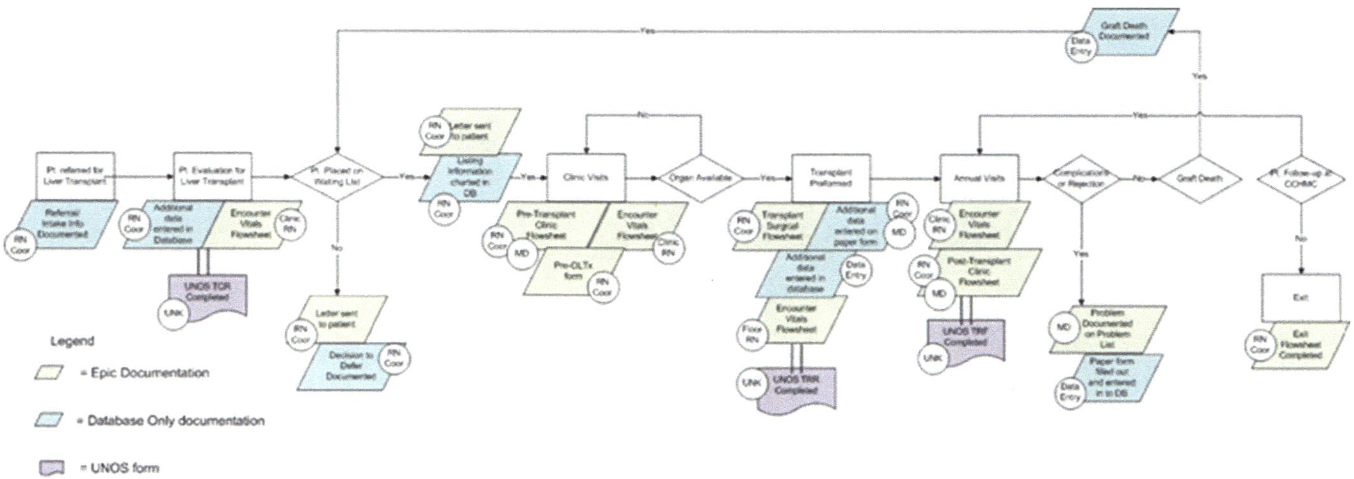

Fig. 8.5 A diagram of the data collection process used in a Liver Transplant clinic as the patient moves through the transplant workflow. Certain elements are collected in the EHR, others are entered directly into the registry. With this particular clinic, data is abstracted out of their registry so it can be submitted to a second registry, this one run by the United Network for Organ Sharing (*UNOS*), a private, non-profit organization that has a government contract to manage organ transplants in the United States

follows through the transplant process and the breakdown of data collected directly in the EHR versus the data that are captured manually. Maintenance of the EHR to registry mapping process is also a challenge. For example, vendors frequently update versions of the EHR with changes in data fields, a new laboratory procedure may be introduced which changes the normal value and range of a test result (discrete element), or new procedures may be introduced. If the underlying data elements change within the EHR, then the mapping between the EHR and the registry must be updated.

8.5.3 Data Quality and Validation

A key element to ensuring high data quality within a registry is the ability to run procedures that will clean and validate the incoming data. While the procedures will vary depending on the context, they should provide feedback so the end user can correct any discrepancies in the data, identify sources of error, and monitor quality over time. In the United States, the EHR is the patient's legal medical record. Federal and state laws govern who is permitted to enter data and/or view the record. Once an encounter is closed and signed by the clinician, all of the data in a patient's EHR are deemed to be "correct" and cannot be changed except by an authorized health care provider. If changes are made in an EHR, there must be an audit trail listing what was changed and by whom. This makes it difficult to correct errors in the data source, when errors are found during the cleaning of the data that are being loaded from the EHR into the registry. Policies for dealing with such discrepancies must be developed by each institution that is implementing an EHR-linked registry. An alternative is to give users the ability to augment and annotate the EHR data in the registry, i.e., overwriting uploaded EHR data in the registry through a data entry form. This workflow can also be used to add missing data to the registry. This approach, however, will not allow users to change or correct the EHR from within the registry so that discrepancies between the EHR as data source and the registry will persist.

8.6 Use Case – Registry Populated by Federated Network in Support of Comparative Effectiveness Research

As stated previously, most pediatric chronic conditions meet the NIH definition for a rare disease (NIH Office of Rare Diseases) and no single care center has a sufficient number of patients to produce generalizable knowledge. One example is pediatric Crohn's disease, which is a type of inflammatory bowel disease (IBD) that is associated with significant medical, psychosocial, and economic impact. Despite the availability of new drugs, there has been little or no improvement in outcomes over the past four decades (Wolters et al. 2004). Lack of evidence on how to best treat the condition is highlighted by widespread practice variation (Kappelman et al. 2007). Pediatric gastroenterologists have formed a network of providers and families at approximately 30 different geographic sites called ImproveCareNow (ICN) to study and improve the care of children with inflammatory bowel disease through data sharing (Crandall et al. 2011). Data entry in the network previously occurred via web-based data capture into a central electronic FDA-validated registry. Members of the network were unable to capture data directly in electronic health records or incorporate the data into federated data warehouses, but relied on double data entry. These methods were costly and slowed the pace of discovery by siloing datasets.

The Agency for Health Care Research and Quality (AHRQ) has supported a research project to develop a registry that permits capture of research-grade data directly from EHRs at multiple sites and its incorporation into a federated registry platform ("Building Modular Pediatric Chronic Disease Registries for QI and CE Research," R01HS020024). The technology has been tested in collaboration with ICN. The informatics infrastructure is versatile, built with open-source tools, and generalizable, which makes it attractive to other investigators. The primary goal of the ICN registry is to collect the data necessary to compare the effectiveness of alternative treatment strategies for pediatric Crohn's Disease, with a special focus on timing of biologic agents. To achieve sufficient statistical power, the ICN network aims to enroll several hundred new patients per year.

The implementation of this EHR-linked "enhanced" registry will permit: (1) data to be captured once during the routine clinical care of the patient so that manual double data entry is not necessary, (2) data to be housed locally in federated registry databases rather than in a single central registry and (3) on-demand generation of patient and population management reports. The technical architecture of this registry is shown in Fig. 8.3. While it is a work in progress, it serves as an example of how registries can be built with contemporary informatics tools to support research on rare diseases.

8.7 Biomedical Informatics and Data Management Core

Registries are a type of database routinely used to collect a standardized set of information on a defined cohort of patients. In a research setting, registries are populated with data elements needed to support one or more clinical research projects. The need for data management resources (people and tools) is pervasive across research disciplines (especially clinical, but also basic). User-friendly access to data management services, tools, and expertise remain a constant challenge. To facilitate cooperation and optimize the quality of data-driven clinical research, our institution has created a Data Management Core (DMC) to complement Biomedical Informatics and Biostatistics. Faculty and staff of Biomedical Informatics and the DMC work closely together to support data management/quality as a core service to research programs. It is important to have a clear understanding of the distribution

Table 8.2 Matrix of Responsibilities – Biomedical Informatics and Data Management Core Biomedical Informatics and the Clinical Data Management professionals work closely together to support data management/quality, as illustrated for our institution

Responsibility	Comment	BMI	DMC
Liaison with Biostatistics	For many tasks, DMC staff will consult with biostatisticians		X
Study consultation and design			X
Design of enterprise case report forms	Design a set of common, reusable forms that leverage standard terminologies such as CDASH, CDISC	J	J
Auditing, cleaning, querying, tracking and resolution process			X
Locking and archiving clinical research databases for dissemination			X
Enterprise data dictionary	Developing a common metadata dictionary for the institution	J	J
Software selection for data capture, tracking, cleaning, auditing, clinical studies	A variety of software applications are needed to support data management	J	J
Software development, licensing vendor products, maintenance	Development of custom software, where necessary	X	
Hosting and administration of databases		X	
Portals, data storage, permissions management, authentication, security		X	
Hosting of web sites and stand- alone applications		X	

Responsibilities are assigned: Biomedical Informatics (*BMI*); Data Management Core (*DMC*)
X one unit has the responsibility, *J* joint responsibility

of responsibilities for data management, some of which are outlined in Table 8.2. DMC responsibilities differ from those of Biomedical Informatics and include:

- Development and review of protocols
- Staffing, budgeting and resourcing of the data management component of clinical studies
- Design and construction of case report forms (CRF)
- Database design and testing
- Data entry and cleaning
- Data management training
- Document development (study-specific and enterprise-wide)
- Report generation
- Database lock and archiving.

Studies that are supported range from small projects of clinical trainees to large multi-site studies. Included are registries, multi-site observational studies, and clinical trials While not all institutions will create a Data Management Core to complement and collaborate with Biomedical Informatics, conducting many types of translational research of high quality does require access to an appropriate infrastructure for data management. Defining the services that are needed to accomplish this should be a high priority for an academic health center.

Acknowledgements This work was supported in part by an award from the Agency for Healthcare Research and Quality, grant number R01HS020024, Building Modular Pediatric Chronic Disease Registries for QI and CE Research.

References

caBIG Strategic Planning Workspace. The Cancer Biomedical Informatics Grid (caBIG): infrastructure and applications for a worldwide research community. Stud Health Technol Inform. 2007;129(Pt 1):330–4.

Children's Oncology Group. 2011 [cited 2011 July 20]. Available from: http://www.childrensoncologygroup.org/

Clinical Data Interchange Standards Consortium. CDISC shared health and clinical research electronic library. [cited 2012 June 4]. Available from: http://www.cdisc.org/cdisc-share

Clinical Data Interchange Standards Consortium. CDISC. 2012 [cited 2012 June 4]. Available from: http://www.cdisc.org

Crandall W, et al. ImproveCareNow: the development of a pediatric inflammatory bowel disease improvement network. Inflamm Bowel Dis. 2011;17(1):450–7.

Drake TA, et al. A system for sharing routine surgical pathology specimens across institutions: the Shared Pathology Informatics Network. Hum Pathol. 2007;38(8):1212–25.

Fridsma DB, et al. The BRIDG project: a technical report. J Am Med Inform Assoc. 2008;15(2):130–7.

Gliklich RE, Dreyer NA, editors. Registries for evaluating patient outcomes: a user's guide. Rockville: U.S. Department of Health and Human Services, Public Health Service, Agency for Healthcare Research and Quality; 2010.

Horbar J, et al. Collaborative quality improvement for neonatal intensive care. NIC/Q Project Investigators of the Vermont Oxford Network. Pediatrics. 2001;107(1):14–22.

James B. Information system concepts for quality measurement. Med Care. 2003;41(1 Suppl):171–9.

Kappelman MD, et al. Intercenter variation in initial management of children with Crohn's disease. Inflamm Bowel Dis. 2007;13(7):890–5.

Kohane IS, et al. The co-morbidity burden of children and young adults with autism spectrum disorders. PLoS One. 2012;7(4):e33224.

Loeffler M, Winter A. Editorial: networking clinical research. Methods Inf Med. 2007;46(5):572–3.

Margolis P, et al. Quality improvement, clinical research and quality improvement research: opportunities for integration. Pediatr Clin N Am. 2009;56(4):831–41.

Maro JC, et al. Design of a national distributed health data network. Ann Intern Med. 2009;151(5):341–4.

Metzger J. Using computerized registries in chronic disease care. Oakland: California Healthcare Foundation; 2004.

Murphy SN, et al. Integration of clinical and genetic data in the i2b2 architecture. AMIA Annu Symp Proc. 2006;2006:1040.
Murphy SN, et al. Serving the enterprise and beyond with informatics for integrating biology and the bedside (i2b2). J Am Med Inform Assoc. 2010;17(2):124–30.
Namini AH, et al. A submission model for use in the indexing, searching, and retrieval of distributed pathology case and tissue specimens. Stud Health Technol Inform. 2004;107(Pt 2):1264–7.
NIH Office of Rare Diseases. Rare diseases and related terms. 2010 [cited 2010 Mar 27]. Available from: http://rarediseases.info.nih.gov/RareDiseaseList.aspx
Olsen L, Aisner D, McGinnis JM. The learning healthcare system. Washington, DC: National Academy Press; 2007.
Pace WD, et al. An electronic practice-based network for observational comparative effectiveness research. Ann Intern Med. 2009;151(5):338–40.
PopMedNet. PopMedNet. 2011 [cited 2011 Sep 8]. Available from: http://www.popmednet.org/Home.aspx
Simone J, Lyons J. Superior cancer survival in children compared to adults: a superior system of cancer care? Background paper, National Cancer Policy Board 1998. [cited 2010 Jan 18]; Available from: http://www.simoneconsulting.com/PDF/children_survival.pdf
U.S. Department of Health and Human Services, Office of the National Coordinator for Health Information Technology, CONNECT Community Portal. About connect. 2009 [cited 2010 Mar 26]. Available from: http://www.connectopensource.org/about/what-is-CONNECT
Von Hippel E. Democratizing innovation. Cambridge, MA: MIT Press; 2005.
Weber GM, et al. The Shared Health Research Information Network (SHRINE): a prototype federated query tool for clinical data repositories. J Am Med Inform Assoc. 2009;16(5):624–30.
Wolters FL, Russel MGVM, Stockbrugger RW. Systematic review: has disease outcome in Crohn's disease changed during the last four decades? Aliment Pharmacol Ther. 2004;20(5):483–96.

Chapter 9
Natural Language Processing – The Basics

John P. Pestian, Louise Deleger, Guergana K. Savova, Judith W. Dexheimer, and Imre Solti

Abstract Natural language processing (NLP) emerged in the 1900s to support the wartime efforts. It's dubious performance, however, slowed research initiatives until the 1960s when advances in machine learning provided novel approaches to text analysis. Increased processing speed and widespread availability of digital text accelerated this trend in the late 1990s. At the present time, there are extensive efforts to use NLP on clinical text and to incorporate this technology into software applications that support clinical care. In this chapter, the first of two about NLP, we will present: basic principles of NLP, the lexical resources required to produce high quality output from clinical text, the process (called annotation) of creating and NLP gold standard, the statistical methods used to evaluate and the role of shared tasks for evaluating and facilitating standardization in the field. Subsequent chapters will discuss ongoing research dedicated improving the quality and utility of NLP in the clinical setting.

J.P. Pestian, Ph.D. (✉) • I. Solti, M.D., Ph.D., M.A.
Department of Pediatrics, University of Cincinnati College of Medicine, Cincinnati, OH, USA

Division of Biomedical Informatics, Cincinnati Children's Hospital Medical Center, 3333 Burnet Avenue, ML-7024, Cincinnati, OH 45229-3039, USA
e-mail: john.pestian@cchmc.org

L. Deleger, Ph.D.
Division of Biomedical Informatics, Cincinnati Children's Hospital Medical Center, 3333 Burnet Avenue, ML-7024, Cincinnati, OH 45229-3039, USA

G.K. Savova, Ph.D.
Boston Children's Hospital and Harvard Medical School, Harvard University, 300 Longwood Avenue, Enders 138, Boston, MA 02115, USA

J.W. Dexheimer, Ph.D.
Department of Pediatrics, University of Cincinnati College of Medicine, Cincinnati, OH, USA

Divisions of Emergency Medicine and Biomedical Informatics, Cincinnati Children's Hospital Medical Center, 3333 Burnet Avenue, ML-7024, Cincinnati, OH 45229-3039, USA

J.J. Hutton (ed.), *Pediatric Biomedical Informatics: Computer Applications in Pediatric Research*, Translational Bioinformatics 2, DOI 10.1007/978-94-007-5149-1_9,

9.1 Introduction

9.1.1 Definition

Definition: "Natural Language Processing is a theoretically motivated range of computational techniques for analyzing and representing naturally occurring written or oral texts at one or more levels of linguistic analysis for the purpose of achieving human-like language processing for a range of tasks or applications" (Liddy 2003).

9.1.2 NLP as an Enabling Tool

The Electronic Health Record (EHR) is a vast repository of information about a patient, much of which is in free text written by various health professionals during the course of care. The importance of information included only in the narrative clinical text of the EHR is gaining increasing recognition as a critical component of computerized decision support, quality improvement, and patient safety (Meystre et al. 2008). Natural Language Processing (NLP) is an enabling tool for accessing the vast information residing in the EHR notes (Jha 2011). For example, NLP could extract information from clinical free-text to populate decision rules or represent clinical knowledge and procedures in a standardized format (Demner-Fushman et al. 2009). Approximately 30–50% of data elements useful for quality improvement are available only in the text of the EHR (Hicks 2003). In addition, patient safety and clinical research could also benefit from information stored in text that is not available in either structured EHR entries or administrative data (Melton and Hripcsak 2005; Murff et al. 2011; Zhan and Miller 2003). Three examples of the use of NLP to support of translational research will be provided here to help the reader understand the power of the tool and to set the stage to learn more about the basic principles of NLP (Denny et al. 2010; Kho et al. 2011; Kohane 2011):

- ***Deidentification of Text in EHRs***: NLP permits the automated removal of personal health information from vast numbers of records in the electronic patient data warehouse now maintained by many medical institutions to support clinical care delivery. Deidentification permits an investigator to review the EHRs for large numbers of patients prior to receiving institutional review board approval of specific projects requiring consent of patients to review personal health information.
- ***Phenotyping in Support of Genomics Research***: Discrete data in EHRs typically include coding of diagnoses by ICD-9 and of procedures by CPT in the format of claims data. Assembling cohorts of patients with a specific diagnosis, e.g. diabetes, appendicitis, for genomic studies requires highly reliable phenotyping. Coding in medical records in the United States is generally for billing purposes and not for research. For this reason claims data are not sufficiently reliable to

support genomics research. Searches of clinical notes in the EHR using NLP to confirm diagnoses greatly increase the accuracy of phenotypes and facilitate genetic and genomic studies (Denny et al. 2010; Kho et al. 2011; Kohane 2011). The successes of eMERGE and PGRN (McCarty 2011; Giacomini 2007) networks clearly demonstrated the essential role of NLP. This is further discussed in Chap. 10.

- ***Pharmacovigilance and Novel Discoveries***: Detection of unanticipated, post marketing, adverse drug reactions (Brownstein et al. 2007) and of early phases of epidemics of infectious diseases can utilize NLP of text of electronic health records and social media (Brownstein et al. 2009; Health Map). Through an automated process, updating 24/7/365, HealthMap monitors, organizes, integrates, filters, visualizes and disseminates online information about emerging diseases in nine languages, facilitating early detection of global public health threats.

9.2 History

In our brief overview of the history of NLP, we provide a summary of Sparck Jones' paper and categorization of the evolution of general NLP and merge her summary with Jurafsky's and Martin's brief overview of NLP history (Jones 2001; Jurafsky and Martin 2000). There are five phases of the development of general NLP that correspond to six decades of NLP research. For more details we refer the reader to Jones' article and book chapters and Jurafsky's and Martin's book (Jones 2001; Jurafsky and Martin 2000; Bright 1992; Zampolli et al. 1994).

- ***Phase One***: The first phase started in the late 1940s and continued until the late 1960s. Initially, this phase was dominated by optimism and great expectations for impending success in Machine Translation (MT). Automatic MT focused on Russian to English, driven by the needs of the military and intelligence gathering communities. Translation efforts were based on word-for-word processing and dictionary lookups. Syntactic and semantic ambiguities were handled by analyzing local context and utilizing elaborate hand crafted rules. The ALPAC Report in 1966 described the MT efforts in critical terms: "… *after 8 years of work, the Georgetown University MT project tried to produce useful output, in 1962 they had to resort to postediting. The postedited translation took slightly longer to do and was more expensive than conventional human translation.... Unedited machine output … makes slow and painful reading*" (National Research Council 1966). However, the report also provided recommendations to spend more on "*computational linguistics as a part of linguistics–studies of parsing, sentence generation, structure, semantics, statistics, and quantitative linguistic matters, including experiments in translation*" (National Research Council 1966). All these components became major lines of research within the computational linguistics community, which led to breakthroughs in computational text processing. During the first decade of that phase Chomsky's work led to formal language theory (including the context-free grammar) and Shannon's ideas of noisy channel and entropy led to probabilistic algorithms for speech and language processing.

- ***Phase Two***: The second phase of NLP history occurred between the late 1960s and 1970s. Artificial Intelligence (AI) researchers mostly dominated this phase and question-answering systems focusing on narrowly defined topics were developed. Partly because of the lessons learnt during the previous decades and because transformational grammar (the dominant linguistic theory of the time) was considered ill-suited for computational tasks, the potential contribution of the linguistics field was mostly ignored. A number of methods were explored in the areas of syntactic parsing, mainly incorporating rule-based approaches.
- ***Phase Three***: The third phase of NLP research (between the late 1970s and 1980s) was dominated by developments in computational grammar theory (e.g. functional, categorical, and generalized phrase structure grammars) and linking them with artificial intelligence logics for meaning and knowledge representation. DARPA (Defense Advanced Research Projects Agency) funded conferences on speech recognition and message understanding and for the first time promoted rigorous performance evaluation of NLP.
- ***Phase Four***: The DARPA initiated focus on evaluation was crucial for the successes of the fourth phase (late 1990s–2000s). Statistical (also called stochastic or probabilistic) language processing became the dominant approach in the general NLP field. Large-scale practical tasks were successfully attacked and corresponding robust systems were developed (e.g. Internet search engines). In addition, government funded evaluation efforts (e.g. for robustness and portability in public NLP challenges), and development of linguistic resources (e.g. British National Corpus (The British National Corpus 2007), WordNet (WordNet: a lexical database for English), Penn Treebank (Penn Treebank Project), PropBank (PropBank Project), etc.) and tools (e.g. part of speech taggers, chunkers) dominated this period.
- ***Phase Five***: The fifth phase (late 2000s–2012) demonstrated further progress. One of the highlights of the most recent NLP developments is the well-publicized and spectacular success of IBM's Watson on Jeopardy. The success of Watson is no doubt an important milestone in the evolution of general NLP. Many other applications have become ubiquitous, such as search engines (Google, Bing). Chapter 10 discusses these in greater detail. The main theme within this phase has been working with big data (for example the text data on the entire internet) and computationally powerful numerical methods for representing domain knowledge tackling each discrete NLP task – syntactic parsing, predicate-argument structure, entity mention discovery, relation identification, temporality, overall text cohesiveness, text generation, summarization, machine translation. Each of these tasks has grown into a major line of computational linguistics research, the output of which has matured to a level where it can be successfully incorporated into applications (see Chap. 10 for more details on applications).

The Text Retrieval Conference (TREC), the Text Analysis Conferences (TAC), the Computational Natural Language Learning Conferences (CoNLL (ACL and SIGNLL 2010)), and Senseval (SENSEVAL) have provided independent competitive evaluation of many of the core NLP components. In the clinical domain NLP has been an integral part of large-scale applications (eMERGE, PGRN, etc.), which are applied to translational research. These are further described in Chap. 10.

9.3 State-Of-The-Art NLP

Dr. Robert C. Moore, who was formerly a Senior Researcher at Microsoft Research, Director of the Research Institute for Advanced Computer Science (RIACS) at NASA Ames Research Center, and the Director of SRI's Natural-Language Research Program in Menlo Park, California, stated in his 2003 Lecture Notes in Computer Science presentation, one of the most significant splits in NLP concerns how knowledge is acquired by intelligent systems. He defined the two paradigms as knowledge engineering and data-driven learning. He described knowledge engineering as a process where human experts enter large amounts of knowledge, in NLP this is usually in the form of manually crafted rules. In the case of data-driven learning, a statistical model is developed and the model is trained on large amounts of data in a supervised or unsupervised manner. He observed that during the 1988–2003 period, *"NLP has gone from being almost completely in the knowledge engineering camp to predominantly focused on data-driven-learning-based approaches"* (Moore et al. 2003). This conversion is completed in the general NLP domain and is happening in the biomedical NLP domain with a 10–14 year lag.

Figure 9.1 shows the trends for machine learning (ML) and knowledge engineering (KE) based approaches in the general and biomedical NLP domains. As we described in the history section, ML has dominated the general, non-biomedical, NLP domain, since the late 1990s (Moore Slides (n.d.)). In 1988 only 6% of submitted papers at the annual conference of the Association of Computational Linguistics (ACL) included ML methods. In 1993 32 and in 1998 approximately 52% of the papers utilized ML methods. By 2003 the ML was almost 90% and in 2011 (based on the classification of full length ACL papers), fully 90% of the submissions utilized ML. Practically, only papers that described a new annotation tool or theory based grammar did not use ML.

The trend is similar for biomedical NLP papers submitted to the American Medical Informatics Association's (AMIA) annual conferences (AMIA Proceedings Library (n.d.)). Six percent of the 1997 AMIA papers were ML-based. Fourteen years later, in 2011, 49% of the biomedical NLP submissions to AMIA utilized ML.

The two trend lines indicate that the biomedical NLP domain lags 10–14 years behind the general NLP field in utilizing ML over KE. Reading the history of the non-biomedical NLP domain, it is hard to avoid the conclusion that the lack of publicly available biomedical resources (e.g. large-scale Penn Treebank-like

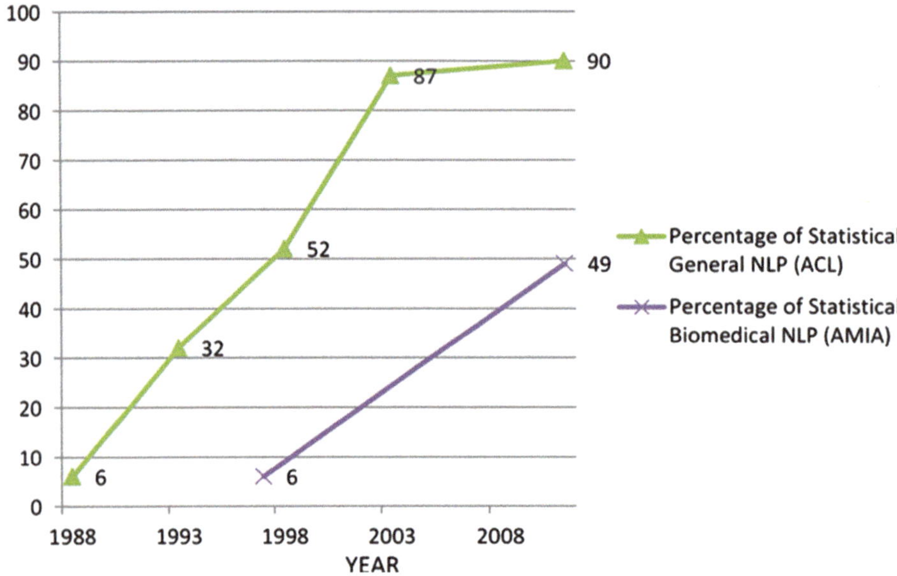

Fig. 9.1 Percentage of statistical/machine learning based NLP submissions at annual meetings of the Association of Computational Linguistics (*ACL*) and the American Medical Informatics Association's (*AMIA*)

annotated corpora) that are necessary for ML/statistical/corpus-based NLP are the reason for the lag. Recognizing this trend, the National Institute of Health funded two major initiatives to develop annotated clinical corpora (SHARPn Main Page (n.d.); U.S. Department of Health and Human Services (n.d.)). There are two other clinical corpora under development, for question answering (MiPAC) (Allbright et al. 2012), and a text corpus based on clinical trial announcements (Deleger et al. 2012).

An example of a supervised machine learning (data-driven) versus a rule-based approach is presented in an Acute Lung Injury (ALI) chest X-ray classification paper (Solti et al. 2009). In the rule-based (knowledge-engineering) approach, the classification keywords are predefined by domain experts (Pulmonary Medicine physicians). In the supervised ML (data-driven) approach, the algorithm "learns" the important keywords from the annotated corpus and assigns weights of importance to the identified keywords on its own. For a highly accurate supervised ML algorithm, a well-annotated large training corpus is vital. In the ALI example, the annotated corpus consists a set of chest X-ray reports that were classified by physicians as consistent or not with the diagnosis of ALI. When a sufficiently large corpus is annotated with proper annotation methodology, the supervised automated learning is very robust and usually provides better results than the rule-based system.

Independently from its underlying core (ML, KE or a hybrid of the two), NLP applications must accomplish major tasks. A non-comprehensive list of these tasks includes:

- ***Tokenization***: Separating punctuation from words. It includes differentiating end of sentence punctuation from part-of-word punctuation.
- ***Part-Of-Speech (POS) Tagging***: Assigning POS tags (e.g. noun or verb) to words.
- ***Constituency Parsing***: Assigning group of words to constituents that behave as a single unit (e.g. noun phrase).
- ***Dependency Parsing***: Determining the dependency relations between words and phrases.
- ***Semantic Analysis***: Linking the linguistic input to its potential meaning in the world and the human knowledge of it.

9.4 Lexical Resources

A variety of textual documents written in natural language exist in the medical domain. Accessing and using the information contained in those documents is the goal of medical NLP systems. To evaluate the robustness of such systems, high quality gold standards are required. That is, sets of manually labeled instances that are relevant to the specific NLP tasks must be created. For information extraction tasks, creating such gold standards involves looking at text corpora and labeling named entities (relevant text phrases) with the proper category (e.g. *disease*, *symptom*, *medication*, etc.) and/or the relations between them (e.g. *locationof* relation). Such a process is called manual annotation. Annotation is essential to evaluate the performance of NLP systems, but also to train them when they are based on machine-learning methods.

9.4.1 Developing Gold Standard Annotations

High quality gold standards are a necessary resource, when building and evaluating NLP systems. That is, sets of manually labeled instances that are relevant to the specific NLP tasks must be created. This is essential to: (1) evaluate the performance of NLP systems, which means comparing the system output with the reference manually created by humans, and (2) train machine-learning based NLP systems, which need reference examples to learn from. The depth of the current NLP systems is proportional to the availability of linguistic and domain-specific gold standard annotations because machine learning systems can only learn what is given to them. To achieve the complexities of language understanding that state-of-the-art NLP systems aspire to, well-thought out, theoretically motivated, interoperable and layered annotations in high volumes are needed. Therefore, the demand for such annotated data is unprecedented.

In the general domain, the creation of the Penn Treebank (PTB, Marcus et al. 1993) and the word sense-annotated SEMCOR (Fellbaum et al. 1998; Miller 1995) showed how even limited amounts of annotated data can result in major improvements in complex natural language understanding systems. Since then, there have been many other successful efforts including the Automatic Content Extraction (ACE) annotations (named entity tags, nominal entity tags, co-reference, semantic relations and events); semantic annotations, such as more coarse grained sense tags (Palmer et al. 2007); semantic role labels as in PropBank (Palmer et al. 2005), NomBank (Meyers et al. 2004), and FrameNet (Baker et al. 1998); and pragmatic annotations, such as coreference (Poesio 2004; Poesio and Vieira 1998), temporal relations as in TimeBank (Pustejovsky et al. 2003; Pustejovsky and Stubbs 2012), the Opinion corpus (Wiebe et al. 2005), and the Penn Discourse Treebank (Miltsakaki et al. 2004), to name just a few. Many of the core NLP components listed in Sect. 9.5 are trained and evaluated on these annotations.

In the following sections, we will summarize annotation efforts relevant to the research using the clinical narrative. For details on general annotations, consult the excellent exposition in Palmer et al. (2010) and Pustejovsky and Stubbs (2012).

9.4.2 Lexical Resources in the Clinical Domain

Several studies have built corpora of clinical notes annotated with medical information to train and evaluate the performance of NLP systems. They are often focused on a specific entity type or on a specific topic. Ogren et al. (2008) built a gold standard corpus of clinical notes annotated for disorders in order to measure the performance of a NLP system that processes a repository of clinical notes and identifies mentions of disorders, as well as the context (e.g. "current", "history of", etc.) and status (e.g. "confirmed", "negated", etc.) of those mentions. By constructing such a gold standard, they are able to automatically compare the output of the system with the manual annotation and compute performance scores. Uzuner et al. (2010) evaluated the performance of several medication extraction systems against a corpus manually annotated for medication names and related information such as dosage and frequency. This corpus provided a benchmark for accurately comparing the various systems. South et al. (2009) developed a manually annotated corpus for a use case of identifying phenotypic information for inflammatory bowel disease.

A few studies reported annotation efforts on a larger scale, including multiple entities as well as relations among them, although they are often limited in size and/or to specific types of clinical documents. The annotated corpus of the Clinical E-Science Framework (the CLEF corpus) is one of the richest semantically annotated resources (Roberts et al. 2009). It includes a variety of named entities, relationships, co-reference, and temporal information, and gathers data from the whole patient record, although only a couple hundred documents were annotated. The authors demonstrated its use in developing and evaluating an information extraction system that identifies medical entities such as conditions, drugs, devices and interventions, as well as a variety of relationships between those entities (e.g. has_finding, has_location, etc.). The corpus from the 2010 i2b2 (Informatics for Integrating Biology and the Bedside (i2b2: Informatics for Integrating Biology and the Bedside 2012)) NLP shared task (Uzuner et al. 2011) is also annotated for several entities (treatments, problems and tests), modifiers (assertion such as negation and uncertainty) and relationships, but is limited to discharge summaries. It has been used as a benchmark to evaluate and compare a number of NLP systems extracting medical problems, treatments and tests along with their modifiers and relationships. Chapman et al. (Chapman and Dowling 2006; Chapman et al. 2008) developed an extensive annotation schema to capture medical conditions and focused their annotation effort on emergency department reports. Their goal was to provide standard guidelines from annually annotating variables relevant for the indexing of clinical conditions (such as symptoms, diagnoses, radiological findings, etc.) in order to create strong reference corpora to evaluate automated NLP-based indexing applications. Automatically indexing clinical conditions from emergency reports can be useful for many purposes, including identifying patients eligible for clinical trials.

Because of privacy issues, these annotated resources are usually not publicly available to be used by other teams for research purposes. The NLP shared tasks from i2b2 currently are the most important effort to provide benchmarks to evaluate NLP systems on the same annotated corpora. Corpora from those challenges are manually de-identified and can be accessed with user agreements. They are annotated for various types of information, to permit the evaluation of the performance NLP systems on various use cases, such as automatic evaluation of the smoking status of patients, which is important for many health studies (such as studies on asthma) (Uzuner et al. 2008); automatic extraction of information about obesity and co-morbidities which can be used for indexing, classifying, and summarizing obesity-related facts found in discharge summaries; and automatic extraction of medication information which can be useful for instance for pharmacovigilance, for comparative effectiveness studies, and for detection of adverse events (Uzuner et al. 2008, 2010).

Apart from annotation of medical information, clinical texts have also been annotated for non-medical entities. For instance, Pakhomov et al. (2006) have annotated a corpus of clinical notes for part-of-speech information in order to adapt part-of-speech taggers to the medical domain, as part-of-speech tagging is a pre-processing task essential to many information extraction systems. De-identification of clinical narrative text is also an example of a useful preliminary task. In order to re-use and distribute more easily clinical text for research purposes (e.g. for translational research), de-identification, i.e. removal of elements of Protected Health Information (PHI) (e.g. names, dates, phone numbers) is required. Automated NLP-based de-identification methods have been developed and a number of corpora have been built for measuring the performance of those systems (Meystre et al. 2010). They often consist of a few document types (e.g. discharge summaries (Neamatullah et al. 2008; Uzuner et al. 2007) and pathology reports (Beckwith et al. 2006; Gupta et al. 2004; Thomas et al. 2002)) and do not always include annotation of all required elements (e.g. geographical locations, institutions and contact information are ignored (Thomas et al. 2002; Gardner and Xiong 2008)). None of those corpora are publicly available, with the exception of the corpus from the 2006 i2b2 shared task on de-identification, which is available with a Data Use Agreement (Uzuner et al. 2007).

Most of the annotation effort in the clinical domain has been focused on notes from the EHR. Other types of medical documents, such as clinical trial announcements or FDA drug labels, have been much less frequently annotated. Clinical trial announcements are starting to draw interest and have been annotated on a few occasions: for a preliminary study on using Amazon Mechanical Turk (Yetisgen-Yildiz et al. 2010) and for detecting temporal constraints in a sample of 100 eligibility criteria sentences (Luo et al. 2011). On-going efforts at CCHMC are investigating large-scale annotation of both clinical trial announcements and FDA drug labels. These corpora are being annotated for a variety of medical concepts, including medications, symptoms and diagnoses. Intended applications are the building of automated NLP systems to identify children eligible for clinical trials and to mine adverse drug events.

A very recent trend in the clinical NLP is to create sharable, large-scale, multi-layered, extendable annotations similar in style to those available in the general domain as outlined in the very beginning of Sect. 9.4. Recently, several complementary initiatives have begun to leverage large de-identified clinical corpora and establish schemas and guidelines for their annotations at the syntactic, semantic, and discourse levels. The goal of these efforts is to provide the research community with shared annotated corpora to allow for the development and testing of novel clinical NLP methods and tools to meet the aspirations of sophisticated natural language understanding systems that can successfully be used in translational and healthcare research. Special care has been taken to ensure that the initiatives are complementary and aligned, so that the resulting resources can be merged transparently whenever possible. The general principles are to:

- Ensure that the annotated resources are available to the research community through Data Use Agreements with the contributing institutions
- Establish smooth mechanisms of delivery and maintenance of the annotation
- Create a set of annotations which are for general NLP purposes, yet specific enough to enable the development of meaningful clinical applications
- Adhere to standards, where such exist
- Rely on and normalize to clinical terminologies and ontological knowledge, when annotating the clinical core concepts in the corpus (e.g., SNOMED-CT, RxNORM)
- Ensure compatibility with existing general-domain annotated resources by utilizing community-adopted conventions whenever possible (e.g., the PTB part-of-speech tagset)
- Rely on existing schemas and guidelines in the general domain and in the clinical domain whenever possible (e.g., syntactic chunking and full dependency parses)
- Design schemas with modular extensions for the clinical-specific annotations, which act as additional layers on top of existing, established general-domain schema

In the *Shared Annotated Clinical Resource project* (Shared Annotated Resource for the Clinical Domain (ShaRe)), 500,000 tokens of clinical narratives from

MIMIC corpus are being annotated with linguistic information (POS tags and phrasal chunks) and full semantic parses of disorder mentions (e.g., core clinical concept normalized to terminology, along with modifiers such as anatomical location, temporality, severity, etc.). The gold standard annotations developed under the Multi-source Integrated Platform for Answering Clinical Questions (*MiPACQ*) (Allbright et al. 2012) amount to approximately 200,000 tokens of clinical narrative, Medpedia content (Medpedia), and clinical questions from Ely et al. (2005). In addition to constituent syntactic parses and semantic roles, the corpus includes UMLS entities, co-reference, UMLS relations and answer types. In the Temporal Histories of Your Medical Events (THYME), an additional layer is being added – that of temporal relations between clinical events – based on the ISO TimeML specifications (TimeML specifications) to annotate a corpus of 200,000 tokens of clinical narrative. Under the *Strategic Health Advanced Research Project Area 4* (SHARPn Main Page (n.d.)) 500,000 tokens of clinical narrative spread across specialties, patients, types of notes, and two geographic sites are being annotated. Layers of annotations comprise constituent parses, dependency parses, semantic role labels, UMLS entities, co-reference, UMLS relations and five templates based on the Clinical Element Models (CEMs).

9.4.3 Annotation Tools

Annotation of text corpora is typically performed using special tools that provide graphical interface to visualize a text, highlight relevant instances and export the resulting annotations in various formats (text, XML). There exist several available open-source tools for manually annotating text corpora. In the medical domain, Knowtator (Ogren 2006) has been most often used (Ogren et al. 2008; South et al. 2009; Roberts et al. 2009). Knowtator is a plug-in of the Protégé (Grosso et al. 1999) knowledge representation system and is specifically designed for text annotation. Its features include definition of complex annotation schemas, visualization of annotations with colors, export of annotations to an XML format, computation of inter-annotator agreement measures and a "consensus mode" to visualize disagreements between annotators. Figure 9.2 shows the Knowtator user interface for a de-identification annotation task (i.e. annotation of Protected Health Information elements such names and dates). Typical steps after manually annotating a corpus using Knowtator (or another software) include exporting the annotations in text or XML format, and using them to (1) train a machine-learning based NLP system, (2) evaluate the output of the system, i.e. compare the output of the system to the manual annotations and compute the performance score of the system. In the case of a de-identification task, this means using part of the annotated corpus as example data to train a system to automatically learn how to recognize Protected Health Information elements in text, and then using a separate portion of the corpus to measure how well the trained system is able to identify PHI compared to what was manually annotated.

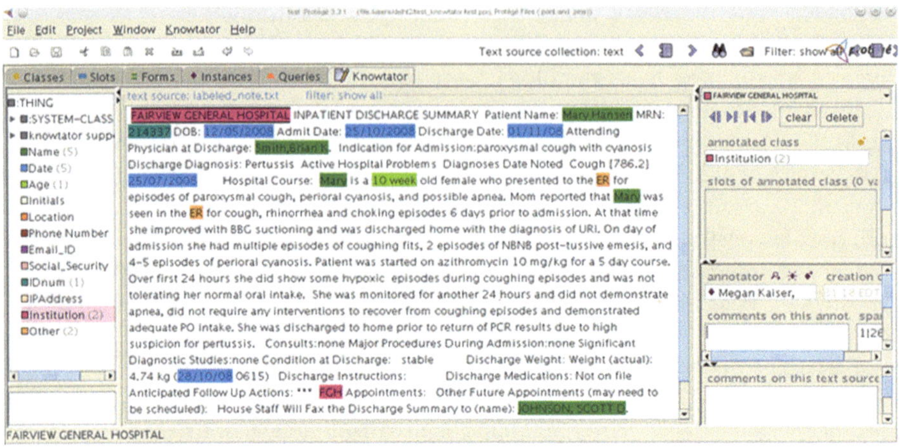

Fig. 9.2 Knowtator user interface: example of the annotation of PHI elements (NB: no real PHI element has been used in this example, all elements are synthetized PHI)

9.4.4 Human Annotator

Manually annotating medical documents can require different levels of skill, depending on the complexity of the annotation task. When annotating non-medical entities, annotators are not required to have a medical background. For instance, annotating elements of Protected Health Information such names and dates for evaluating a de-identification system can be performed perfectly well by non-clinicians. Medical knowledge is most useful when annotating medical entities and/or their relations (e.g. disorders, medications, location of relation, etc.). However, with proper training, non-clinicians can accomplish high quality annotation for this type of task. Chapman et al. (2008) used both clinician and non-clinician annotators to annotate medical conditions in emergency department reports and found that the annotation by both was high quality, but non-clinicians needed more extensive training.

9.4.5 Annotation Process

A typical annotation process involves using pairs of annotators so that it will be possible to measure inter-annotator agreement to evaluate the reliability of the gold standard. A pair of annotators annotates the same documents working independently, and disagreements between them are later resolved, usually by adjudication from a third party (e.g. an experienced annotator, medical informatician or expert clinician). Double-annotation of documents reduces cases of mislabeling and builds stronger gold standards. Annotators rely on guidelines to perform annotation. Guidelines define the annotation schema, i.e. the specific items to be annotated, and provide specific examples to guide the annotation. Ideally, guidelines should be non-ambiguous and in a stable form when given to the annotators. In actuality, guidelines are often revised or extended with more examples during the annotation process, when previously unforeseen items arise. It is also good practice to include a pilot phase in an annotation project (Palmer et al. 2010). This phase is essential to familiarize annotators with the guidelines and avoid misinterpretations.

9.4.6 Alleviating the Cost of Annotation

Because annotation is time-consuming and expensive, methods have been investigated to alleviate the cost of annotation, such as automatic pre-annotation and active learning. Pre-annotation involves automatically labeling the texts to be given to the human annotators, using an existing NLP system. The hope is that it will speed up the annotation process by providing preliminary annotations that will be corrected by the annotators. The potential issues with this method are that it could introduce a bias in the annotation and that it could actually slow down the process, if the pre-annotation is of low quality (Ogren et al. 2008). Active learning is a machine learning technique that allows reducing the amount of necessary manual annotations, by selecting only the most "informative" instances to be annotated rather than sampling them randomly (Settles 2009). The algorithm starts learning from a small set of labeled examples, then selects one or more informative instances from a large set of unlabeled examples to be labeled by a human annotator and then adds the newly labeled instance(s) to the training set, and repeats. Chen et al. (2012) successfully used active learning for annotating and classifying assertion of concepts in clinical notes. Miller et al. (2012) applied active learning to the co-reference task.

9.5 Pipeline

The process of moving from text to a desired endpoint can be called a pipeline. To develop a pipeline, start with the end in mind: given a starting document, what do the users want to know, what is the format and content of the report they want to receive or what is the task they want carried out, e.g. removal of protected health information from clinical text so that it can be freely shared. There is an alluring fallacy that one can develop a system that is all things to all people. It sings like a siren that leads to failure. Rather, a well-understood approach to success is to start with the end in mind (Covey 2004). Begin with a prototype report. Then, circulate it to the end-users, all of the end-users. Request comments. Revise the prototype. Circulate it again. Request comments again. Revise it again. Do this over and over until the comments have been exhausted or your colleagues ignore you.

Next, each element on the report must be operationalized. Operationalization is defining concepts so they can be measured. For example, say you want to measure the date when a patient's epilepsy was deemed medically intractable and the patient became a candidate for surgery. First create the functional equation (Table 9.1). Now that it is clear what needs to be accomplished, one can apply these requirements to the traditional NLP pipeline, or a process for analyzing natural language. The Tools Section describes some open-source software for each step.

9.5.1 Creating a Pipeline – Core NLP Components

The first step in creating a pipeline is to identify the data or text to which NLP is to be applied. What data are needed to carry out a specific analysis and where are the data? There are many potential sources, among the most common are clinical notes within the electronic health record. Once sources have been defined, the process of text analysis must be defined and core NLP components selected. These components are shown in Table 9.2. They include tokenization, sentence segmentation, part-of-speech tagging, named entity extraction, chunking, parsing, and co-reference resolutions. Each step is described in Table 9.2.

There are a large number of open-source and commercial tools available for each core component. These can be found by using various Internet search tools and evaluations (Sparck Jones and Galliers 1995).

9.6 Evaluation

While several NLP systems have been developed to process medical documents, their evaluation has varied in scope and quality (Meystre et al. 2008). A number of pre-requisites are necessary for proper evaluation of NLP systems, among which are strong gold standards and appropriate methods of evaluation (Friedman and Hripcsak 1998).

9.6.1 Different Types of Evaluation

Several distinctions can be made in the way NLP systems are evaluated. First, *formative* evaluation should be distinguished from *summative* evaluation (Resnik and Lin 2010). Formative evaluation is performed (often iteratively) during the development of a system to measure progress, while summative evaluation assesses the performance of the system once it is complete. Second, *intrinsic* and *extrinsic* evaluations can be performed (Resnik and Lin 2010). Intrinsic evaluation assesses the quality of the output that the system is specifically designed to produce. Extrinsic

Table 9.1 Operationalization is defining concepts so they can be measured. For example, say you want to measure the date the epilepsy is considered medically intractable and the patient becomes a candidate for surgery. First, create the functional equation as shown in this table

Intractable Epilepsy = *f* (*number of failed medications, side effects, medication impact*)

					Data Sources			
ARGUMENTS of the indicator	Definition of the ARGUMENT	NLP needed?		Notes	DB	Field/Table	*VALUES* of the indicator	*Expected value range*
The date of diagnosis of intractable epilepsy.	The date is after the third drug has failed. If switching a drug is due to side effects or reason other than no effect then this is not counted as a failed drug.	Yes	Training data is needed	Intractable epilepsy diagnosis is not always written in clinical notes.	EPIC	NOTE_TEXT/ TBL_NEURO_ AMB_NOTE_ LINES		

Table 9.2 Core NLP components

Some core NLP tasks/components	Description	Examples
Tokenization	A token is a group of characters that are grouped together to provide information, often as a worked. The process of dividing the input text into words, numbers or a punctuation mark.	Sally became drowsy when she was taking carbmazepine. /Sally/ /became/ /drowsy/ /when/ /she/ /was/ /taking/ /carbamazepine/
Sentence segmentation	Dividing written text in meaningful units.	//Sally became drowsy when she was taking carbmazepine// // This occurs in 11.8% of patients//.
Part-of-speech tagging	Adding parts of speech to segmented text	NNP Sally VBD became NN drowsy WRB when PRP she VBD was VBG taking NN carbmazepine. NNP = proper singular noun NN = Singular noun VBD = Verb, past tense VBG = Verb, gerund/present participle WRB = wh-Adverb IN = Preposition DT = Determiner
Named entity extraction	Classifying text into predetermined categories like: people, expression of time, location and so forth	Sally was admitted to Cincinnati Children's Hospital Medical Center in Cincinnati OH on 4/20/2012 Cincinnati Children's Hospital Medical Center Cincinnati OH 4/20/2012
Chunking	Sometimes called shallow or light parsing. Chunking *up* refers to increasing abstraction. Chunking *down* refers to increasing details	Chunking up: A discharge summary is found in the medical record. Chunking down: What data fields describe a patient demographics?
Parsing	Analyzing text to determine grammatical structure.	nsubj(went-2, Sally-1) root(ROOT-0, went-2) det(hospital-5, the-4) prep_to(went-2, hospital-5) partial output from Stanford parser. de Marneffe and Manning (2008)
Co-reference resolution	Resolving when multiple words mean the same thing	Sally had severe side effects. She was admitted to the hospital. *Sally* had severe side effects. *She* was admitted to the hospital.

evaluation measures the impact of the system on a task external to the system itself, i.e. its contribution to a larger application. Finally, another important distinction is made between *component* based evaluation and *end-to-end* evaluations (Resnik and Lin 2010). Because NLP systems are often built from distinct components, performance can be measured focusing on each component individually (component evaluation) or on all components at once (end-to-end evaluation). Both evaluations are useful, as component evaluation allows researchers to understand the impact of each component, while end-to-end evaluation measures the effectiveness of the system in a real world setting.

9.6.2 Gold Standards

Performance of automatic NLP systems is usually assessed by comparing the systems' output to manually created reference gold standard. Special attention should be given to the creation of strong reference standards. Section 9.4 described in detail the task of manually annotating text corpora to serve as reference standard and emphasized the benefit of using at least two human annotators and measuring inter-annotator agreement to ensure reliability of the created gold standard. Gold standards can be used both for developing and evaluating a system, in which case the set used for evaluating the performance (i.e. the *test set*) should be held out from the data used for developing the system and remain unseen.

9.6.3 Metrics for Evaluating the Reliability of Manually Annotated Gold Standards

Reliability of manually created gold standards is usually measured through inter-annotator agreement (IAA). Good IAA is important because if agreement is low then it is likely that the task is too difficult or is not well defined (Resnik and Lin 2010). Moreover it is generally considered as an upper limit on our ability to measure automated performance (Resnik and Lin 2010; Gale et al. 1992).

Cohen's Kappa and F-measure are two commonly used measures of IAA.

Cohen's kappa coefficient (κ) (Cohen 1960) is defined using observed agreement (Ao) and agreement expected by chance (Ae):

$$\kappa = \frac{Ao - Ae}{1 - Ae}$$

The observed agreement Ao is the percentage of instances the annotators agree on. The expected agreement Ae is based on the probability of the two annotators agreeing on any given category, which is the product of the chance of each annotator assigning an instance to that category. Ae is the sum of this product over all categories. The chance of each annotator assigning a given category is estimated by looking at the observed distribution.

Inter-annotator agreement can also be measured using standard performance measures in information retrieval and NLP (Brants 2000; Hripcsak and Rothschild 2005): precision, recall, and more particularly the F-measure. Precision is equivalent to positive predictive value and recall is equivalent to sensitivity. F-measure is the harmonic mean of precision and recall and is written as (with β usually equal to 1):

$$F = \frac{\left(1 + \beta^2\right) Precision * Recall}{\beta^2 * Precision + Recall}$$

When computing IAA, the annotations of one annotator are considered as the reference, while the annotations of the other annotator are considered as the predictions of a system. Switching the annotators' roles gives the same F-measure value so it does not matter who plays the role of the reference (Brants 2000; Hripcsak and Rothschild 2005).

Cohen's kappa is often the standard measure of inter-annotator agreement used for classification tasks. However, kappa is not the most appropriate measure for annotation of named entities in textual documents (Hripcsak and Rothschild 2005), because it requires the number of negative cases, which is unknown for named entities. Named entities are sequences of words, and there is no pre-existing fixed number of items to consider when annotating a text. Proposed solutions include considering individual tokens as the items to be marked to compute a "token-level" kappa (Tomanek and Hahn 2009), considering only noun phrases, considering all possible n-grams in a text (sequence of n tokens) or considering only items annotated by one or two of the annotators (Grouin et al. 2011). However none of those options is completely accurate (Grouin et al. 2011), which is why the F-measure, which does not require the number of negative cases, is usually used to measure IAA for this type of annotation tasks. An F-measure of 0.8 or above is generally considered sufficient inter-annotator agreement.

9.6.4 Metrics for Evaluating the Performance of NLP Systems

Standard metrics to evaluate the performance of (medical and general-language) NLP systems are Recall (*R*), Precision (*P*) and F-measure (*F*). Recall (i.e. sensitivity) is the number of instances correctly predicted by a system divided by the total number of instances in the gold standard. Precision (i.e. positive predictive value) is the number of correct instances divided by the total number of instances predicted by a system. They can be written in terms of true positives (*tp*), false positives (*fp*) and false negatives (*fn*) as follows:

$$R = \frac{tp}{tp + fn}$$

$$P = \frac{tp}{tp + fp}$$

F-measure is the weighted harmonic mean of precision and recall (as detailed in Sect. 9.6.3).

9.6.5 Cross-Validation

A common method for evaluating the performance of statistical systems is to use cross-validation (Geisser 1975; Kohavi 1995). The classical type of cross-validation is k-fold cross validation. In k-fold cross validation, the annotated data are randomly partitioned into k subsets. K rounds of validation are then performed, each time using a different subset as the validation data for testing the system and the remaining k-1 subsets for training the system. The k performance results can then be averaged to obtain a single performance measure. The advantage of this procedure is that it uses all data instances to validate a system while keeping the test sets independent (thereby preserving the integrity of the evaluation), and each instance is used only once for validation. In machine-learning-based NLP, tenfold cross-validation is most commonly used.

9.6.6 Statistical Significance Testing

When comparing the performance of two systems (or of two versions of the same system), it is good practice to perform statistical significance testing to rule out the possibility that the performance difference is due to chance. The probability that the difference between two results is due to chance can be estimated by statistical significance tests. Statistical significance tests start from a formulated null hypothesis (e.g. "there is no difference between the scores of system A and system B") and test it by estimating the probability of observing data as extreme as that observed (i.e. as extreme as the observed difference in scores between the two systems) if the null hypothesis were true. If this probability (referred to as the p-value) is below a certain threshold, then the hypothesis can be rejected and the difference between the systems is accepted to be real (Cinchor 1992). A threshold of 0.05 is often chosen.

There exists a variety of significance tests. Classic statistical significance tests such as the t-test require knowledge of the distribution of the data. If the distribution of the data is unknown, as in many NLP applications, computationally-intensive methods can be used to simulate the distribution (Cinchor 1992; Noreen 1989). A method of significance testing commonly used in NLP is approximate randomization (Noreen 1989), which involves shuffling of the data to generate a distribution.

9.6.7 Shared Tasks for Evaluating Medical NLP Systems

There have been several efforts to provide benchmarks to evaluate NLP systems. Competitive shared tasks were designed to encourage development of medical NLP applications and to evaluate them in a standard framework. A 2007 NLP challenge

(Computational Medicine Center (CMC) 2007) focused on automatic mapping of ICD-9-CM codes to radiology free-text reports (Pestian et al. 2007). The shared tasks organized by i2b2 (i2b2: Informatics for Integrating Biology and the Bedside 2012) are the largest initiative to evaluate NLP systems. Challenges have been held each year for several years on a variety of NLP tasks: on de-identification of clinical text (Uzuner et al. 2007), on identification of patient smoking status (Uzuner et al. 2008), on recognition of obesity and co-morbidities (Uzuner 2009), on medication information extraction (Uzuner et al. 2010), on identification of concepts, assertions and relations (Uzuner et al. 2011) and on co-reference resolution (Uzuner et al. 2012). The 2011 i2b2 challenge was held in conjunction with the Computational Medicine Center, which proposed a second shared task on classification of emotions in suicide notes (Computational Medicine Center (CMC)). Additional challenges are planned (i2b2: Informatics for Integrating Biology and the Bedside 2012). Finally, The Text Retrieval Conference (TREC), which organizes annual evaluations on a variety of information retrieval tasks, recently created a Medical Records Track (Guidelines for the 2011) that focuses on the retrieval of patients who meet specific criteria (i.e. specific diseases and treatments).

Acknowledgements Dr. Savova's work was supported in part by NIH grants U54LM008748 and 1U01HG006828. Drs. Deleger's and Solti's work was supported in part by NIH grant 5R00LM010227.

References

Allbright D, et al. Towards comprehensive syntactic and semantic annotations. Submitted for publication, is under review (2012)
AMIA Proceedings Library. (n.d.) [cited 2012 Apr 5]. Available from: http://proceedings.amia.org/
Association of Computational Linguistics (ACL), Special Interest Group on Natural Language Learning (SIGNLL). CoNLL: the conference of SIGNLL. December 9, 2010 [cited 2012 June 1]. Available from: http://ifarm.nl/signll/conll/
Baker CF, Fillmore CJ, Lowe JB. The Berkeley FrameNet project. In: Proceedings of the 17th international conference on computational linguistics, vol. 1. Montreal: Association for Computational Linguistics; 1998. p. 86–90.
Beckwith BA, et al. Development and evaluation of an open source software tool for deidentification of pathology reports. BMC Med Inform Decis Mak. 2006;6:12.
Brants T. Inter-annotator agreement for a German newspaper corpus. In: Proceedings of the second international conference on language resources and evaluation (LREC 2000), Athens, Greece. Paris: European Language Association; 2000.
Bright W. International encyclopedia of linguistics. New York: Oxford University Press; 1992.
Brownstein JS, et al. The tell-tale heart: population-based surveillance reveals an association of rofecoxib and celecoxib with myocardial infarction. PLoS One. 2007;2(9):e840.
Brownstein JS, Freifeld CC, Madoff LC. Influenza A (H1N1) virus, 2009–online monitoring. N Engl J Med. 2009;360(21):2156.
Chapman WW, Dowling JN. Inductive creation of an annotation schema for manually indexing clinical conditions from emergency department reports. J Biomed Inform. 2006;39(2): 196–208.

Chapman WW, Dowling JN, Hripcsak G. Evaluation of training with an annotation schema for manual annotation of clinical conditions from emergency department reports. Int J Med Inform. 2008;77(2):107–13.

Chen Y, Mani S, Xu H. Applying active learning to assertion classification of concepts in clinical text. J Biomed Inform. 2012;45(2):265–72.

Cinchor N. The statistical significance of MUC4 results. In: MUC4'92 proceedings of the 4th conference on message understanding. San Mateo: Morgan Kaufmann; 1992.

Cohen J. A coefficient of agreement for nominal scales. Educ Psychol Meas. 1960;20:37–46.

Computational Medicine Center (CMC). 2007 international challenge: classifying clinical free text using natural language processing. [cited 2012 June 1]. Available from: http://computationalmedicine.org/challenge/previous

Computational Medicine Center (CMC). Suicide notes. [cited 2012 June 12]. Available from: http://computationalmedicine.org/home-0

Covey SR. The 7 habits of highly effective people: powerful lessons in personal change. Rev. ed. New York: Free Press; 2004. p. 372.

de Marneffe MC, Manning CD. The Stanford typed dependencies representation. In: Coling 2008: proceedings of the workshop on cross-framework and cross-domain parser evaluation. Manchester: Coling 2008 Organizing Committee; 2008. p. 1–8.

Deleger L, et al. Building Gold Standard Corpora for Medical Natural Language Processing Tasks. In: American Medical Informatics Annual Symposium Proceedings. Chicago, November 1–6, 2012

Demner-Fushman D, Chapman WW, McDonald CJ. What can natural language processing do for clinical decision support? J Biomed Inform. 2009;42(5):760–72.

Denny JC, et al. PheWAS: demonstrating the feasibility of a phenome-wide scan to discover gene-disease associations. Bioinformatics. 2010;26(9):1205–10.

Ely JW, et al. Answering physicians' clinical questions: obstacles and potential solutions. J Am Med Inform Assoc. 2005;12(2):217–24.

Fellbaum C, Grabowski J, Landes S. Performance and confidence in a semantic annotation task. In: Fellbaum C, editor. WordNet: an electronic lexical database. Cambridge, MA: MIT Press; 1998.

Friedman C, Hripcsak G. Evaluating natural language processors in the clinical domain. Methods Inf Med. 1998;37(4–5):334–44.

Gale W, Church KW, Yarowsky D. Estimating upper and lower bounds on the performance of word-sense disambiguation programs. In: Proceedings of the 30th annual meeting on Association for Computational Linguistics. Newark/Delaware: Association for Computational Linguistics; 1992. p. 249–56.

Gardner J, Xiong L. HIDE: an integrated system for health information DE-identification. In: Proceedings of the 21st IEEE international symposium on computer-based medical systems. Los Alamitos: IEEE Computer Society; 2008. p. 254–9.

Geisser S. Predictive sample reuse method with applications. J Am Stat Assoc. 1975;70(350): 320–8.

Giacomini KM, Brett CM, et al. The pharmacogenetics research network: from SNP discovery to clinical drug response. Clin Pharmacol Ther. 2007;81(3):328–45.

Grosso WE, et al. Knowledge modeling at the millennium (The design and evolution of Protégé-2000). In: Twelfth Banff workshop on knowledge acquisition, modeling, and management, Alberta; 1999.

Grouin C, et al. Proposal for an extension of traditional named entities: from guidelines to evaluation, an overview. In: Proceedings of the 5th linguistic annotation workshop (LAW V'11). Portland: Association for Computational Linguistics; 2011. p. 92–100.

Guidelines for the 2011 TREC Medical Records Track. [cited 2012 June 1]. Available from: http://www-nlpir.nist.gov/projects/trecmed/2011/tm2011.html

Gupta D, Saul M, Gilbertson J. Evaluation of a deidentification (De-Id) software engine to share pathology reports and clinical documents for research. Am J Clin Pathol. 2004;121(2):176–86.

Health Map. [cited 2012 June 1]. Available from: http://www.healthmap.org/en/
Hicks J. The potential of claims data to support the measurement of health care quality. Santa Monica: RAND Corporation; 2003.
Hripcsak G, Rothschild AS. Agreement, the f-measure, and reliability in information retrieval. J Am Med Inform Assoc. 2005;12(3):296–8.
i2b2: Informatics for Integrating Biology and the Bedside. 2012 NLP shared task: shared-tasks and workshop on challenges in natural language processing for clinical data. [cited 2012 June 1]. Available from: https://www.i2b2.org/NLP/TemporalRelations/Call.php
i2b2: Informatics for Integrating Biology and the Bedside. Datasets. [cited 2012 June 1]. Available from: https://www.i2b2.org/NLP/DataSets/Main.php
Jha AK. The promise of electronic records: around the corner or down the road? JAMA. 2011;306(8):880–1.
Jones K. Natural language processing: a historical review (Paper). Current issues in Computational Linguistics: in honour of Don Walker. 2001 [cited 2012 June 1]. Available from: http://www.cl.cam.ac.uk/archive/ksj21/histdw4.pdf
Jurafsky D, Martin JH. Speech and language processing: an introduction to natural language processing, computational linguistics, and speech recognition, Prentice Hall series in artificial intelligence. Upper Saddle River: Prentice Hall; 2000. xxvi, 934 p.
Kho AN, et al. Electronic medical records for genetic research: results of the eMERGE consortium. Sci Transl Med. 2011;3(79):79re1.
Kohane IS. Using electronic health records to drive discovery in disease genomics. Nat Rev Genet. 2011;12(6):417–28.
Kohavi R. A study of cross-validation and bootstrap for accuracy estimation and model selection. In: Proceedings of the fourteenth international joint conference on artificial intelligence. San Mateo: Morgan Kaufmann; 1995. p. 1137–43.
Liddy ED. Natural language processing. In: Drake MA, editor. Encyclopedia of library and information science. New York: Marcel Dekker; 2003. p. 2126–36.
Luo Z, et al. Extracting temporal constraints from clinical research eligibility criteria using conditional random fields. AMIA Annu Symp Proc. 2011;2011:843–52.
Marcus MP, Marcinkiewicz MA, Santorini B. Building a large annotated corpus of English: the penn treebank. Comput Linguist. 1993;19(2):313–30.
McCarty CA, Chisholm RL, et al. The eMERGE Network: a consortium of biorepositories linked to electronic medical records data for conducting genomic studies. BMC Med Genomics. 2011;4:13.
Medpedia: an open platform connecting people and information to advance medicine. [cited 2012 June 1]. Available from: http://www.medpedia.com/
Melton GB, Hripcsak G. Automated detection of adverse events using natural language processing of discharge summaries. J Am Med Inform Assoc. 2005;12(4):448–57.
Meyers A, et al. The NomBank project: an interim report. In: HLT-NAACL 2004 Workshop: frontiers in corpus annotation. Boston: Association for Computational Linguistics; 2004.
Meystre SM, et al. Extracting information from textual documents in the electronic health record: a review of recent research. Yearb Med Inform. 2008;2008:128–44.
Meystre SM, et al. Automatic de-identification of textual documents in the electronic health record: a review of recent research. BMC Med Res Methodol. 2010;10:70.
Miller GA. WordNet: a lexical database for English. Commun ACM. 1995;38(11):39–41.
Miller T, Dligach D, Savova GK. Active learning for coreference resolution. In: BioNLP workshop at the Conference of the North American Association of Computational Linguistics (NACCL), Montreal; 2012.
Miltsakaki E, et al. The Penn Discourse Treebank. In: Proceedings of the LREC 2004 fourth international conference on language resources and evaluation, Lisbon, Portugal; 2004. pp. 2237–2240.
Moore Slides. (n.d.) [cited 2012 Apr 5]. Available from: http://old-site.clsp.jhu.edu/ws04/calendar/School/Moore_slides.ppt

Moore RC. Two paradigms for natural-language processing. In: Cappelli A, Turini F, editors. AI*IA 2003: advances in artificial intelligence: 8th Congress of the Italian Association for Artificial Intelligence, Pisa, Italy, September, 2003: proceedings (Associazione italiana per l'intelligenza artificiale. Congress (8th: 2003: Pisa Italy)). Berlin/New York: Springer; 2003. p. 548.

Murff HJ, et al. Automated identification of postoperative complications within an electronic medical record using natural language processing. JAMA. 2011;306(8):848–55.

National Institutes of Health and National Institute of General Medical Sciences. The NIH Pharmacogenomics Research Network (PGRN). [cited 2012 June 1]. Available from: http://www.nigms.nih.gov/Research/FeaturedPrograms/PGRN/

National Research Council. "Recommendations." Language and machines: computers in translation and linguistics. 1966 [cited 2012 Apr 5]. Available from: http://www.nap.edu/openbook.php?isbn=ARC000005.

Neamatullah I, et al. Automated de-identification of free-text medical records. BMC Med Inform Decis Mak. 2008;8:32.

Noreen EW. Computer-intensive methods for testing hypotheses: an introduction. New York: Wiley; 1989.

Ogren P. Knowtator: a protégé plug-in for annotated corpus construction. In: Proceedings of the North American chapter of the association for computational linguistics on human language technology. East Stroudsburg: Association for Computational Linguistics; 2006.

Ogren PV, Savova GK, Chute C. Constructing evaluation corpora for automated clinical named entity recognition. In: Proceedings of the sixth international conference on language resources and evaluation (LREC'08), Marrakech, Morocco; 2008.

Pakhomov SV, Coden A, Chute CG. Developing a corpus of clinical notes manually annotated for part-of-speech. Int J Med Inform. 2006;75(6):418–29.

Palmer M, Gildea D, Kingsbury P. The proposition bank: an annotated corpus of semantic roles. Comput Linguist. 2005;31(1):71–106.

Palmer M, Dang HT, Fellbaum C. Making fine-grained and coarse-grained sense distinctions, both manually and automatically. Nat Lang Eng. 2007;13(02):137–63.

Palmer M, Xue N. Linguistic annotation. In: Clark A, Fox C, Lappin S, editors. The handbook of computational linguistics and natural language processing. Chichester/Malden: Wiley-Blackwell; 2010. p. 13 and 21.

Penn Treebank Project. [cited 2012 June 1]. Available from: http://www.cis.upenn.edu/treebank/

Pestian JP, et al. A shared task involving multi-label classification of clinical free text. In: Proceedings of the workshop on BioNLP 2007: biological, translational, and clinical language processing. Prague: Association for Computational Linguistics; 2007. p. 97–104.

Poesio M. Discourse annotation and semantic annotation in the GNOME corpus. In: Proceedings of the 2004 ACL workshop on discourse annotation. Barcelona: Association for Computational Linguistics; 2004. p. 72–9.

Poesio M, Vieira R. A corpus-based investigation of definite description use. Comput Linguist. 1998;24(2):183–216.

PropBank Project. [cited 2012 June 1]. Available from: http://verbs.colorado.edu/~mpalmer/projects/ace.html

Pustejovsky J, Stubbs A. Natural language annotation for machine learning. Sebastopol: O'Reilly Media; 2012.

Pustejovsky J, et al. The TIMEBANK Corpus. In: Proceedings of the Corpus Linguistics, Lancaster, UK; 2003. pp. 647–656.

Resnik P, Lin J. Evaluation of NLP systems. In: Clark A, Fox C, Lappin S, editors. The handbook of computational linguistics and natural language processing. Chichester/Malden: Wiley-Blackwell; 2010. p. 271–96.

Roberts A, et al. Building a semantically annotated corpus of clinical texts. J Biomed Inform. 2009;42(5):950–66.

SENSEVAL: evaluation exercises for the semantic analysis of text. [cited 2012 June 1]. Available from: http://www.senseval.org/

Settles B. Active learning literature survey, in computer sciences technical report 1648. Madison: University of Wisconsin-Madison; 2009.

Shared Annotated Resource for the Clinical Domain (ShaRe). Clinical NLP annotation. [cited 2012 June 1]. Available from: https://www.clinicalnlpannotation.org/index.php/Main_Page

SHARPn Main Page. Strategic Health IT Advanced Research Projects (SHARP): Research Focus Area 4. (n.d.) [cited 2012 Apr 5]. Available from: http://informatics.mayo.edu/sharp/index.php/Main_Page

Solti I, et al. Automated classification of radiology reports for acute lung injury: comparison of keyword and machine learning based natural language processing approaches. Proceedings (IEEE Int Conf Bioinformatics Biomed). 2009;2009:314–19.

South BR, et al. Developing a manually annotated clinical document corpus to identify phenotypic information for inflammatory bowel disease. BMC Bioinformatics. 2009;10 Suppl 9:S12.

Sparck Jones K, Galliers JR. Evaluating natural language processing systems: an analysis and review, Lecture notes in computer science. Berlin/New York: Springer; 1995. xv, 228 p.

Text Analysis Conference (TAC). [cited 2012 June 1]. Available from: http://www.nist.gov/tac/

Text REtrieval Conference (TREC). [cited 2012 June 1]. Available from: http://trec.nist.gov/

The British National Corpus, version 3 (BNC XML Edition). Distributed by Oxford University Computing Services on behalf of the BNC Consortium; 2007. URL: http://www.natcorp.ox.ac.uk/

Thomas SM, et al. A successful technique for removing names in pathology reports using an augmented search and replace method. Proc AMIA Symp. 2002;2002:777–81.

THYME: Temporal history of your medical events. [cited 2012 June 1]. Available from: http://clear.colorado.edu/compsem/index.php?page=endendsystems&sub=temporal

TimeML specifications. [cited 2012 June 1]. Available from: http://www.timeml.org/site/publications/specs.html

Tomanek K, Hahn U. Timed annotations – enhancing MUC7 metadata by the time it takes to annotate named entities. In: Proceedings of the linguistic annotation workshop. Singapore: Association for Computational Linguistics; 2009.

U.S. Department of Health and Human Services. Project information 5R01GM090187-03. (n.d.) [cited 2012 Apr 5]. Available from: http://projectreporter.nih.gov/project_info_description.cfm?aid=8133360

Uzuner O. Recognizing obesity and comorbidities in sparse data. J Am Med Inform Assoc. 2009;16(4):561–70.

Uzuner O, Luo Y, Szolovits P. Evaluating the state-of-the-art in automatic de-identification. J Am Med Inform Assoc. 2007;14(5):550–63.

Uzuner O, et al. Identifying patient smoking status from medical discharge records. J Am Med Inform Assoc. 2008;15(1):14–24.

Uzuner O, et al. Community annotation experiment for ground truth generation for the i2b2 medication challenge. J Am Med Inform Assoc. 2010;17(5):519–23.

Uzuner O, et al. 2010 i2b2/VA challenge on concepts, assertions, and relations in clinical text. J Am Med Inform Assoc. 2011;18(5):552–6.

Uzuner O, et al. Evaluating the state of the art in coreference resolution for electronic medical records. J Am Med Inform Assoc. 2012;19(5):786–91.

Wiebe J, Wilson T, Cardie C. Annotating expressions of opinions and emotions in language. Lang Resour Eval. 2005;39(2):165–210.

WordNet: a lexical database for English. Princeton: Princeton University. Available from: http://wordnet.princeton.edu/

Yetisgen-Yildiz M, Solti I, Xia F. Using Amazon's Mechanical Turk for annotating medical named entities. AMIA Annu Symp Proc. 2010;2010:1316.

Zampolli A, et al. Current issues in computational linguistics : in honour of Don Walker. Linguistica computazionale. Pisa/Norwell: Giardini ; Distributed in the U.S.A. and Canada by Kluwer Academic Publishers; 1994. xxv, 595 p.

Zhan C, Miller MR. Administrative data based patient safety research: a critical review. Qual Saf Health Care. 2003;12 Suppl 2:ii58–63.

Chapter 10
Natural Language Processing: Applications in Pediatric Research

Guergana K. Savova, Louise Deleger, Imre Solti, John Pestian, and Judith W. Dexheimer

Abstract We discuss specific biomedical Natural Language Processing-based applications that cover a wide spectrum of use cases within the field of translational and health services research. In our use cases we focus on four categories of applications: (1) Information Extraction (IE), (2) Document Classification, (3) Patient Classification, and (4) Sentiment Analysis. We show how the extracted information could be used for (a) Phenotype identification, (b) Comparative effectiveness studies, (c) Cohort identification, (d) Meaningful Use, and (e) Linking patients' phenotype and genotype. In addition, we discuss the use of Natural Language Processing components for de-identification of large collections of patient notes. We review the literature for examples of pediatric natural language processing applications and show the transferability of select adult clinical natural language processing applications to the pediatric population.

G.K. Savova, Ph.D. (✉)
Boston Children's Hospital, Harvard Medical School, Harvard University, 300 Longwood Avenue, Enders 138, Boston, MA 02115, USA
e-mail: Guergana.Savova@childrens.harvard.edu

L. Deleger, Ph.D.
Divisions of Biomedical Informatics, Cincinnati Children's Hospital Medical Center, 3333 Burnet Avenue, ML-7024, Cincinnati, OH 45229-3039, USA

I. Solti, M.D., Ph.D., M.A. • J. Pestian, Ph.D.
Department of Pediatrics, University of Cincinnati College of Medicine, Cincinnati, OH, USA

Divisions of Biomedical Informatics, Cincinnati Children's Hospital Medical Center, 3333 Burnet Avenue, ML-7024, Cincinnati, OH 45229-3039, USA

J.W. Dexheimer, Ph.D.
Department of Pediatrics, University of Cincinnati College of Medicine, Cincinnati, OH, USA

Divisions of Emergency Medicine and Biomedical Informatics, Cincinnati Children's Hospital Medical Center, 3333 Burnet Avenue, ML-7024, Cincinnati, OH 45229-3039, USA

J.J. Hutton (ed.), *Pediatric Biomedical Informatics: Computer Applications in Pediatric Research*, Translational Bioinformatics 2, DOI 10.1007/978-94-007-5149-1_10,

10.1 NLP Applications in Pediatric Research

As Chap. 9 on the basic principles of Natural Language Processing (NLP) pointed out, NLP focuses on developing technologies for our most ubiquitous product: human language (Coursera.org). The electronic version of that product has exploded because of pervasive computing, and interest in and applications of NLP have been unprecedented. In fall of 2011, Stanford University offered the first free world-wide online class in NLP taught by renowned NLP experts Profs. Jurafsky and Manning. The reaction of the student community has been nothing short of amazing – the class drew an enrollment in the tens of thousands of students, unofficial numbers are 60,000 students (Online Colleges.net 2012).

In this section we outline some of the technologies enabling translational science and health care research. Two of the most well-known applications of NLP technologies are Information Retrieval (IR) and Question Answering (QA). IR systems focus on developing methods and algorithms for retrieving documents from a given repository (Jurafsky and Martin 2000; Manning et al. 2000). IR has become mainstream through search engines like *Google, Bing* and *Yahoo!*. An IR engine returns a set of documents ranked by an algorithm, the end user is then expected to sift through them to find the relevant information. On the other hand, QA systems aim to present results at a finer level, usually at the level of the sentence or the paragraph.

In 2011, IBM introduced to the world their Watson QA system and demonstrated that it can outsmart a highly knowledgeable human by competing in, and winning, the quiz show *Jeopardy!* (IBM – Watson n.d.). The future will tell to what extent its proprietary version for healthcare, *Dr. Watson*, will be able to address the needs of the clinical community. The Watson QA system brilliantly showcased the end product of a well-thought out and efficient software architecture weaving NLP technologies (constituency parsers, dependency parsers, entity recognizers, relation extractors, co-reference modules, semantic role labelers, expected answer classifiers), lexical resources (a variety of encyclopedias and gold standard parsed data off which machine learners were built) and purely engineering solutions (Apache Unstructured Information Management Architecture (UIMA) and fast computing hardware/databases.

Clinical QA systems are specifically tailored to the domain of medicine (for an overview of the QA for biomedicine consult (Athenikos and Han 2010)). Demner-Fushman and colleagues (Demner-Fushman and Lin 2007) developed a system of feature extractors that identify subject populations, interventions, and outcomes in medical studies. Later they extended this research to create a general-purpose QA system. Weiming et al. (2007) experimented with the Unified Medical Language System (UMLS) (Lindberg et al. 1993) semantic relationships and created a system that extracts exact answers from the output of a Lucene-based (Lucene) IR system; however, their approach requires an exact semantic match, which limits performance on more complex and non-factoid questions. Athenikos et al. (2009) propose a rule-based system based on human-developed question/answer UMLS semantic patterns; such an approach is likely to perform well on specific question types but might have difficulty generalizing to new questions. AskHermes (Yu and Cao 2008) finds answers based on keyword searching. The Multi-source Integrated Platform for Answering Clinical Questions also uses the UMLS as the backbone ontology and returns answers at the paragraph level (Cairns et al. 2011; Nielsen et al. 2010).

A more middleware application is Information Extraction (IE). IE converts the seemingly unstructured natural language (although linguistic studies show that language per se has inherent expressible structure) into explicitly structured computable data elements. For example, the text in the following sentence, "Patient was diagnosed with cancer located in the ascending colon", will be converted to a computable event data structure anchored at the event of *cancer*, where the experiencer is the *patient* and the body location is the *ascending colon*. This data structure is then used as the text metadata tags which are stored in databases to facilitate retrieval and data mining.

A pioneering clinical IE system is the Medical Language Extraction and Encoding System (MedLEE) (Friedman 1997, 2000; Hripcsak et al. 1998) developed at Columbia University. MedLEE has been applied to many use cases, however it is not open-sourced. Other examples of IE systems (focusing on English clinical text) are University of Sidney's Health Information Technologies Research Laboratory's NLP tool suite (HITRL), Health Information Text Extraction (HITEx; Zeng et al. 2006), IBM's MedTAS (Mack et al. 2004), SymText and MPLUS (Christensen et al. 2002; Haug et al. 1995), and University of Pittsburgh's Cancer Tissue Information Extraction System (caTIES; Crowley et al. 2010). Other tools developed primarily for processing biomedical scholarly articles include the National Library of Medicine MetaMap (Aronson et al. 2000), providing mappings to the UMLS Metathesaurus concepts, those from the National Center for Text Mining (NaCTeM), JULIE lab (JULIE Lab), and U-compare (U-Compare), with some applications to the clinical domain (Meystre and Haug 2005).

A recently developed NLP IE system is the Apache open source Clinical Text Analysis and Knowledge Extraction System (cTAKES) platform (cTakes). cTAKES is built within the UIMA engineering framework, which provides a solid basis for scalability, expandability and collaborative software development. The overarching principle is *process once, use many times* thus allowing pre-processing of the data, the result of which is the generation of indices to be used for retrieval purposes spanning a variety of use cases – from phenotype discovery, cohort identification, patient care to "meaningful use" of the EHR and comparative effectiveness. These topics are discussed in the subsequent subsections.

The machine learning components of cTAKES are trained on gold annotations of clinical text. cTAKES processes clinical narratives and identifies clinical named entity mentions (NEs) of types Drugs, Diseases/Disorders, Signs/Symptoms, Anatomical sites, and Procedures per UMLS semantic types and groups. Each discovered NE is assigned attributes such as text span, ontology mapping code (SNOMED CT, RxNORM), context (family history of, current, unrelated to patient), and a negation indicator. The cTAKES components include following annotators in this specific order (for details, consult Savova et al. 2010a):

- Sentence boundary detector, which is a wrapper around OpenNLP's sentence boundary detector (OpenNLP Tools 1.5.0 API: sentence boundary detector), but trained on clinical data. The module finds the end of the sentence.
- Rule-based tokenizer to separate punctuations from words.
- Normalizer, which is wrapper around LVG (Lexical Systems Group) to standardize for example morphologically different phrases with same meaning, e.g. "infection", "infecting", "infects" normalize to the same form of "infect".
- Context dependent tokenizer grouping tokens to create Date, Digits and other types of groupings.
- Part-of-speech tagger, which is a wrapper around OpenNLP's but trained on clinical data. The module assigns a part-of-speech label to each word, e.g. "infects" is a verb.
- Phrasal chunker, which is a wrapper around OpenNLP's but trained on clinical data to detect phrases such as noun phrases, verb phrases, prepositional phrases. Most concepts are expressed as noun phrases thus the noun phrases become the lookup window for the next module.
- Dictionary lookup annotator that performs a permutational lookup against a dictionary database. The permutations are computed off the components within the lookup window, which is the noun phrase.
- Context annotator based on NegEx (Chapman et al. 2001) to link information with the patient or family members.
- Negation detector based on NegEx to discover whether a concept is negated or not.
- Dependency parser (Choi and Palmer 2011a, b) to detect dependency relations between words.
- Semantic role labeler based on the dependency parse labels to parse the sentence into a predicate with its arguments (Choi and Palmer 2011b). For example, "In 2007, patient was diagnosed with autism", there is a predicate associated with "diagnosed" which has three arguments: "the patient", "autism" and "2007" each associated with a semantic role (Palmer et al. 2005).
- Drug mention annotator which is a module to populate the drug template with attributes for change status (started, discontinued, increased, decreased, no change), dosage, duration, end date, form, frequency, route, start date, strength.

There is an additional module for the identification of the patient smoking status (Sohn et al. 2010): past smoker, current smoker, non-smoker, smoker, unknown. In a typical IE application scenario, the SNOMED CT and RxNORM codes generated by cTAKES are retained as indices to the clinical narrative. Any subsequent queries directly go against these indices, obviating real time processing of the text. An NLP-based IE application thus creates metadata tags used for indexing and retrieval.

The open source and modular nature of cTAKES facilitate scenarios in which each of its NLP components can be re-purposed within the architecture of a variety of applications such as QA and patient summarization to name a few. Such repurposing is already under way to port cTAKES to other medical domains (e.g.

extracting information from clinical trial announcement text (Deleger et al. 2012)). cTAKES has been used in a number of translational science research project, which are described in more detail below (Kullo et al. 2010).

10.2 De-identification

10.2.1 NLP-Based Automated De-identification of Clinical Narrative Text

A substantial portion of clinically-relevant information for computerized decision support, patient safety, quality improvement and translational research studies is included in narrative text of the Electronic Health Record (EHR) (Jha 2011; Meystre et al. 2008). However, the HIPAA Privacy Rule (Health Insurance Portability and Accountability Act of 1996) requires that before clinical text can be used for research, either (1) all elements of protected health information (PHI) should be removed through the process of de-identification, (2) a patient's consent must be obtained, or (3) the Institutional Review Board should grant a waiver of consent. Studies have shown that requesting consent reduces participation rate, and is often infeasible when dealing with large populations (Dunlop et al. 2007; Wolf and Bennett 2006), for example, thousands of patients. Even if a waiver is granted, documents that include PHI data should be tracked to detect and prevent unauthorized disclosure. On the other hand, de-identification removes the requirements for consent, waiver and tracking, and facilitates clinical NLP research, and consequently, the utilization of information stored in narrative EHR notes.

While manual de-identification on a large scale is all but impossible, NLP can provide a scalable solution. Several studies have investigated the use of NLP to remove PHI elements from clinical narrative texts (Meystre et al. 2010). Rule-based approaches (Friedlin and McDonald 2008; Gupta et al. 2004; Neamatullah et al. 2008) make use of dictionaries and manually designed rules to match patterns of PHI elements. They often lack generalizability and require both time and skill for the creation of the rules, but perform better for rare PHI elements. Machine-learning-based approaches (Gardner and Xiong 2008; Szarvas et al. 2007; Uzuner et al. 2008; Wellner 2009) automatically learn to detect patterns of PHI elements based on a set of examples and are more generalizable, but require a large set of manually-annotated examples. Systems using a combination of both methods usually obtain better results (Meystre et al. 2010; Uzuner et al. 2007). Overall, the best systems report high recall and precision, often above 90 and sometimes as high as 99. The i2b2 2006 de-identification challenge (Uzuner et al. 2007) provided a benchmark to evaluate and compare a number of automated de-identification systems (Szarvas et al. 2007; Arakami et al. 2006; Guo et al. 2006; Hara et al. 2006). In this evaluation the top systems obtained F_1-measures around 96 for overall PHI detection and often superior to 90 for individual categories, with a few exceptions for categories such as locations (below 80) and phone numbers (below 90).

Although recent advancements in NLP carry the promise of a highly accurate, automated approach to de-identification, existing de-identification systems present some limitations. These include exclusion of some PHI elements (Uzuner et al. 2008; Fielstein et al. 2004; Ruch et al. 2000), or institution and contact information (Gardner and Xiong 2008; Taira et al. 2002), use of only a limited set of types of clinical note to build and evaluate systems (such as pathology reports (Gupta et al. 2004; Beckwith et al. 2006; Berman 2003)), discharge summaries (Szarvas et al. 2007; Uzuner et al. 2008; Arakami et al. 2006; Guo et al. 2006; Hara et al. 2006), or medical message boards (Benton et al. 2011), or use of documents with synthetic PHI elements (manually de-identified documents re-identified with fake PHI elements (Uzuner et al. 2007)). Furthermore, the effect of de-identification and potential over-scrubbing on subsequent automated information extraction tasks is rarely investigated (Meystre et al. 2010).

Recent automated de-identification efforts are moving towards overcoming those limitations. Aberdeen et al. (2010) evaluated their de-identification toolkit (the MITRE Identification Scrubber Toolkit, MIST) on four classes of clinical notes from the Vanderbilt University Medical Center (discharge summaries, laboratory reports, letters, and order summaries) and reported high performance (token-level F_1-measures ranging 93.4–99.6). They found variations between document classes (highest performance was achieved on the classes which were the richest in PHI elements, such as discharge summaries).

Experiments were also conducted at Cincinnati Children's Hospital Medical Center (CCHMC) to evaluate the MITRE Identification Scrubber Toolkit (MIST) as well as an in-house de-identification system on a larger set of clinical notes representing the variety of notes available in the EHR of a specialized children's hospital (over 20 note types were included in the study). Furthermore, the effect of de-identification on a subsequent IE task (the extraction of medications) was investigated. Evaluation of MIST and the CCHMC system showed performance that was indistinguishable from human annotators: overall precision (P) and recall (R) of the systems were 92.79 (P) and 92.8 (R) for MIST and 95.08 (P) and 91.92 (R) for the CCHMC system, while precision and recall of the human annotators against the gold standard were 97.68 (P) and 90.01 (R) for one annotator and 97.18 (P) and 98.07 (R) for the other annotator. Experiments on extracting medications from the de-identified version of clinical notes as opposed to the original version of the notes demonstrated that the automated de-identification caused no significant decrease in the performance of the medication extractor. Results also further emphasized the findings that performance varies depending on note types and their density of PHI elements (Aberdeen et al. 2010).

There are many take-home messages in these recent de-identification experiments that are likely to influence the decisions of Institutional Review Boards regarding whether to accept the output of automated de-identification systems as comparable to manual de-identification. First, no single manual de-identification is 100% accurate. Second, different note types have different density of PHI elements (and potentially different context of the same elements), and a de-identification system that is trained on a mix of note types will show varying performance on

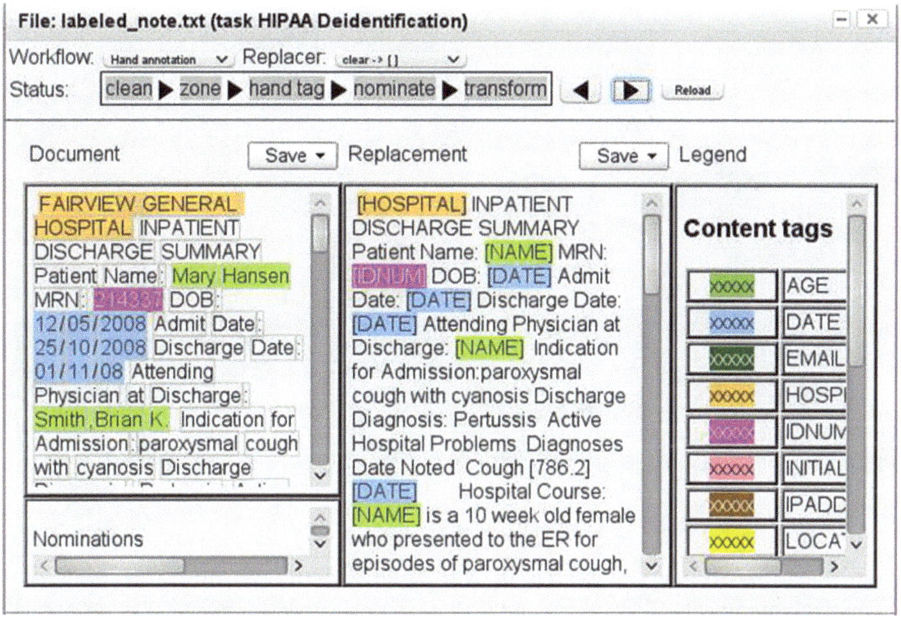

Fig. 10.1 MIST's web-based user interface (NB: this document does not contain any real PHI elements, all PHI elements were replaced)

these note types. As a result, the de-identification performance of machine learning systems will depend on the frequency of PHI types in the training data. This is not unexpected, but health services researchers and hospital management should be aware of the potential differences in performance for different note and PHI types. Finally, the impact of de-identification seems minimal on the effectiveness of subsequent IE of clinically relevant concepts.

10.2.2 A Closer Look at an Open-Source De-identification Toolkit: MIST

The most recent publicly released automatic de-identification tool is the previously mentioned MITRE Identification Scrubber Toolkit (MIST) (Aberdeen et al. 2010). MIST presents the advantage of being open-source and freely downloadable (MIST). The kit provides a full suite of tools to train and build a de-identification model, starting from the manual annotation of example clinical notes to train a machine-learning algorithm, up to the automatic replacement of PHI elements with fake identifiers. Components of the toolkit (Aberdeen et al. 2010) are:

- A web-based graphical interface to manually annotate clinical documents for PHI elements (see Fig. 10.1 for an example of the user interface).

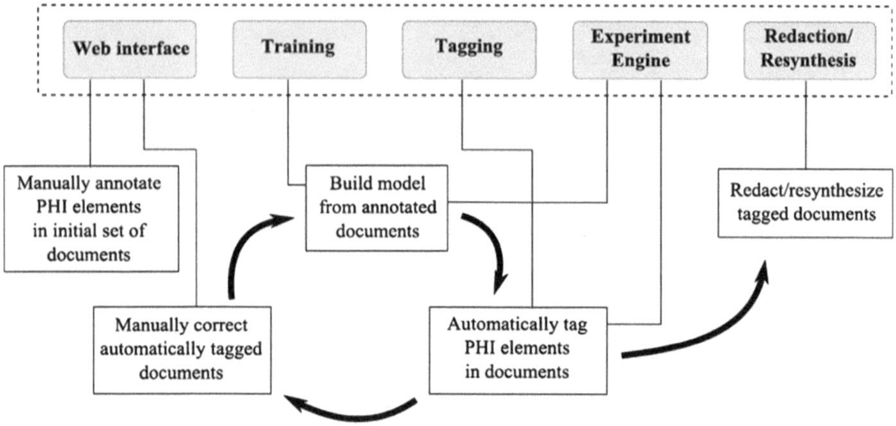

Fig. 10.2 MIST de-identification workflow

- A machine-learning component used to build a model to automatically identify PHI elements based on the manually annotated examples. This component is based on the Carafe implementation of Conditional Random Fields (CRFs) (Wellner 2009).
- A component to automatically tag PHI elements using the model previously built.
- A redaction/resynthesis component to replace tagged PHI elements with synthetic (fake) identifiers.
- An experiment engine to iteratively evaluate the performance of the models.

Figure 10.2 (as described in Aberdeen et al. 2010) depicts the MIST de-identification workflow. MIST can be used in a single iteration, i.e. annotating all documents first and then building and evaluating the model, or in an iterative fashion, i.e. starting from only a small set of annotated notes to build increasingly better models.

Setting up MIST and using it in its default configuration can be achieved easily in a few hours. The toolkit also provides some customization options, such as defining a new de-identification task (with new PHI types) and providing new types of features to the machine-learning algorithm.

Aberdeen et al. (2010) reported high performance of the system on four classes of documents (overall token-level F_1-measure of 93.4–99.4). Experiments conducted at Cincinnati Children's Hospital also showed high precision and recall, when using MIST in its default configuration (overall span-level F_1-measure of 91.5) and higher after customization (overall span-level F_1-measure of 92.8) averaged over more than 20 different document types (see Chap. 9 – NLP Basics for further information about measures of performance such as F-measure).

10.3 Phenotype Identification

In May, 2012 the pediatric eMERGE network (Electronic Medical Records and Genomics) was established. Its overarching goal is to combine the phenotype and genotype data into a cohesive EHR, benefiting the patients through appropriate levels of genetic counseling. Success would be a significant advance in the field of Electronic Health Record Driven Genomic Research (EDGR) (Kohane 2011). One aim of the informatics arm of a pediatric eMERGE project is set forth to demonstrate real-time execution of phenotypic selection across two different pediatric institutions (Boston Children's Hospital and CCHMC) with disparate EHR systems as a model for ensuring phenotypic standardization and for national scalability.

Pediatric eMERGE's software stack in this case includes the i2b2 platform (Informatics for Integrating Biology and the Bedside (Murphy et al. 2010)) for unifying phenotypic and genotypic information, cTAKES for NLP processing of the clinical narrative, and the distributed query engine called the Shared Health Informatics Research Network (SHRINE) (Weber et al. 2009). The i2b2 software suite integrates the EHR and genomics data with clinical research specifically for phenotype/genotype association studies. It has been successfully used to identify adult phenotypes such as rheumatoid arthritis and major depressive disorder by combining information extracted through NLP with codified data such as lab results and medication orders (Liao et al. 2010). SHRINE is a general purpose clinical querying protocol that can be adapted to many types of data repositories (i2b2, commercial EHR data warehouses) and has been deployed across multiple institutions (See Chap. 6, Research Patient Data Warehousing). In 2011, the Catalyst SHRINE went into production at the four main Harvard hospitals for the sharing of aggregate counts of patients with defined exclusion/inclusion criteria for labs, diagnoses, demographics, and medications. CARRAnet, the pediatric rheumatology consortium across 60 institutions, and ICN, a pediatric inflammatory bowel consortium across 40 institutions, are working to test whether i2b2 and SHRINE represent generalizable approaches to federated data sharing among institutions.

The pediatric eMERGE members are linked with the adult eMERGE network that was established in 2007 (eMERGE Network; McCarty et al. 2011). At the time of writing, we have 5 years of experience with the adult eMERGE. One of the findings is the importance of NLP in achieving the eMERGE goals. As Dan Masys, MD, former Professor and Chair of the Department of Biomedical Informatics at Vanderbilt University, the NIH funded center of the eMERGE initiative, noted in his presentation at an Institute of Medicine workshop: *"Structured data alone (ICD9 codes, labs, meds) is not enough to accurately define clinical phenotypes in EHR data."* and *"Natural language processing of clinician notes improves both sensitivity and specificity of phenotype selection logic* (Institute of Medicine (IOM) 2010)."

Numerous publications from the eMERGE research program support the importance of NLP of text of the EHR in phenotyping. Denny et al. (2010), demonstrated that scanning phenotypes in EHR and grouping patients by EHR-derived phenotypes and analyzing the genetic material of the grouped patients for known mutations is a successful approach to duplicate the findings of Genome Wide Association Studies (eMERGE Network). In the above study Denny et al. used ICD9 codes alone, but in a subsequent publication they proved the importance of NLP for EHR-based phenotyping. Kho et al., reported that "*To assess the additional benefit of NLP, we performed a comparison of the number of cases identified using structured data alone compared with that using both structured data and NLP at one site (Vanderbilt University). At this site, the use of NLP tools identified 129% more cases of QRS duration (2950 versus 1288) than did the use of structured data and string matching only, while maintaining a PPV of 97%* (Kho et al. 2011)." Within the adult eMERGE, Kullo and colleagues mine the EHR clinical narrative to discover patient cohorts of confirmed, possible and absent peripheral arterial disease (Savova et al. 2010b).

Another network of researchers focuses exclusively on studying the pharmacogenomics of treatments – the PharmacoGenomics Network (PGRN). Wilke and colleagues (2011) provide an overview of the use of the EHR within PGRN and the role NLP plays in it. EHR data including the clinical narrative processed through NLP has proven accurate for quantifying disease phenotypes and treatment outcomes such as toxicity and efficacy. The NLP-processed clinical narrative combined with medication prescription information was used to develop an algorithm to find the drug treatment patterns a breast cancer patient has undergone (Savova et al. 2011). This cohort was used to study genetic determinants of breast cancer treatments. Within the current PGRN efforts, the NLP-based cTAKES is used to discover the disease activity of patients with rheumatoid arthritis, build a timeline off it to identify responders and non-responders to inflammation treatments (Lin et al. 2012) and personal communication, Savova).

Within the pediatric domain, the team of investigators at CCHMC studied predictive models to determine whether an epilepsy patient is a candidate for neurosurgery (personal communication, Pestian). More than three million people in North America have epilepsy. Nearly 70% can be controlled with medication. The International League Against Epilepsy guidelines suggest that when seizures cannot be controlled by two or more anti-epileptic drugs, the patient can be considered "medically intractable" and may be a candidate for neurosurgery. The team used all progress notes from 125 epilepsy patients considered medically intractable who had evaluation for epilepsy surgery, all progress notes from 127 epilepsy patients who were retrospectively determined not to be intractable, and all progress notes from 51 patients whose progress notes were too incomplete to determine their status. Machine learning algorithms were trained and evaluated to achieve an accuracy of 90% in distinguishing tractable and intractable epilepsy based on the text in the progress notes. Twenty-four uninformative patients were classified as "maybe non-intractable", 25 as "non-intractable", and 2 as intractable. A subsequent chart review confirmed that two patients were indeed intractable, as classified by the machine-learning algorithm and should not have been classified as uninformative by medical experts. While this is an example of research in progress, the machine learning algorithms show promise as tools to analyze clinical records and to classify patients with epilepsy.

10.4 Document and Patient Classification

Depending on the type of the translational science and health care research question, the unit of analysis could be one document from the patient's chart or the entire patient's chart. For example, if the overall patient medical history is relevant, then approaches that permit mining the patient's entire medical record are necessary. On the other hand, if the patient's status at a particular time point is of interest, then it is more suitable to mine only specific documents. Thus, document classification is a process to separate a large corpus of diverse documents into smaller homogenous subsets based on predefined criteria (supervised classification) or criteria discovered from the corpus directly without preconceptions of the criteria (unsupervised classification). The most intuitive method for document classification is based on a so-called bag-of-words approach. The documents are converted into vectors that represent unordered collection of words in the documents. In a supervised method, class labels are manually assigned to the training and evaluation gold standard sets (e.g. indicating that one set of radiology reports are consistent with a diagnosis of Acute Lung Injury (ALI), while the second set of documents is not). The vectors are used to train machine-learning algorithms to develop models. Finally, the models are used to classify previously unseen document sets (e.g. radiology reports to assign the ALI label on a large scale automatically) (Solti et al. 2009).

Patient classification could be considered a special case of document classification. Some research questions require the aggregation of information over the patient's entire chart, such as whether the patient is a responder or non-responder to a particular treatment. In the first step of patient classification, all the documents and structured EHR entries are collected for each patient. In subsequent steps, using NLP the information is extracted from the clinical narrative documents and combined with information derived from the structured entries. An aggregate vector representing the patient's chart is assembled. Finally, the vectors are classified with a similar procedure as in document classification. In patient classification the vectors represent patients instead of documents and the vectors might include information from the structured entries of the EHR, in addition to the documents. Many of the phenotype extractions discussed in Sect. 10.3 are examples of document- and patient- level classifications (Liao et al. 2010; Savova et al. 2010b, 2011).

10.5 Cohort Identification

Many translational science and health care studies require the identification of large cohorts of patients who match a set of complex criteria. The implementation of large EHRs has enabled large-scale cohort identification for a variety of investigations – epidemiology, phenotype/genotype associations, biosurveillance. Brownstein et al. (2005), conducted an influenza epidemiology study through monitoring patient visits for respiratory illness using a real-time syndromic surveillance system. Their results added to a growing body of research in support of vaccinating preschool age children. Hansen et al. (2007) mined the EHR data in a large urban academic medical system in northeast Ohio to identify a cohort of 14,187 children and adolescents aged 3–18 to study the underdiagnosis of pediatric hypertension. The following variables were retrieved from the EHR: race/ethnicity, age, sex, weight, height, blood pressure, diagnosis codes, family history of hypertension codes, past medical history codes, problem list codes for each participant from all visits. The investigators concluded that hypertension and prehypertension were frequently underdiagnosed in the studied pediatric population.

Cohort identification from the EHR has been used in studies related to health services research. The American Academy of Neurology (AAN) acknowledges that current evidence is insufficient for robust recommendations for an accurate diagnostic assessment and medical management of children with status epilepticus (SE) (Riviello et al. 2006). SE is defined in the AAN guidelines as a seizure with >30 min duration that includes two or more sequential seizures without full recovery of consciousness between seizures (AMA 1993). Outstanding translational research questions include (a) whether routine diagnostic testing such as blood cultures, cerebrospinal fluid analysis and neuro-imaging are indicated for the evaluation of these patients and; (b) what the prevalence of acute bacterial meningitis, significant intra-cranial pathology or bacteremia among these patients is. These are key questions for emergency department physicians, given the risks of radiation exposure (Stein et al. 2008), sedation, and invasive procedures in children.

Prospective SE studies have been limited by low incidence (Singh et al. 2010) and low recruitment rates. Analysis of retrospective data has the advantage of the large number of cases potentially available. However, existing data are often limited by the inability to identify a valid cohort of children with SE. Current approaches to data extraction based on chief complaint or ICD9 codes have proven to be rather limited. Such searches normally miss a large proportion of potential cases. Kimia and colleagues (2010) studied a pediatric cohort evaluated in a pediatric emergency department for their first complex febrile seizure. To identify the cohort, the authors created a NLP text screening tool which identified 5,744 potential patients out of which the final cohort of 526 eligible patients was selected. Kimia et al. (2009), employed the same approach. They report a sensitivity of 1 and specificity of 0.957 for the NLP text screening tool. The patient cohort was used to study the diagnostic value of lumber punctures for first simple febrile seizure and to assess the rate of bacterial meningitis detected among these children.

The translational science work within eMERGE, PGRN and i2b2 project described in Sect. 10.3 is another example of cohort identification, where the cohort members match phenotypes. Within the adult eMERGE, Kullo and colleagues (2010) mine the EHR clinical narrative to discover patient cohorts of confirmed, possible and absent peripheral arterial disease. Within i2b2, Liao et al. (2010), developed an EHR algorithm to identify cohorts of patients with confirmed, possible and absent rheumatoid arthritis. Patient cohort identification is thus closely tied to phenotype discovery using specific exclusion and inclusion criteria.

One promising area of cohort identification is to automate the matching of clinical trials and eligible patients by learning eligibility criteria directly from the text of trial announcements. This will be the next step in cohort discovery because the textual eligibility criteria descriptions (e.g. on the ClinicalTrials.gov website) will be converted into a computable format, using NLP without or with only minimal human intervention. There are multiple groups working on finding solutions for this task. To identify cohorts for pediatric clinical trials, researchers at CCHMC are developing a software tool that uses the cTAKES NLP pipeline to extract: (1) the eligibility criteria compiled from the inclusion and exclusion criteria listed in trial announcements; and (2) the corresponding clinical information from patients' EHR notes and structured entries. The ultimate goal of this NIH funded study (Solti) is to highly automate the process and provide the patient a list of clinical trials for which s/he is eligible.

10.6 Sentiment Analysis

Who cares? Who doesn't? Why? It is the role of the sentiment expert to find these answers in written text and transcriptions of conversations. Texts filled with sentiment are texts filled with opinions and emotions. They present specific challenges to investigators who wish to extract relevant information from free text, as for example within an EHR. NLP tools are being developed that can support computational approaches to understanding opinions, sentiments, subjectivity, views, and emotions that are found in text. These subjective data are essential to understanding why people make certain decisions (Hu and Liu 2004).

Concepts of opinions and sentiment need to be well defined because opinion mining and sentiment analysis are very different types of tasks. Opinion mining is an identification problem. Opinions are a positive or negative sentiment, view, attitude, emotion or appraisal. Analysis of sentiment is a classification problem. Sentiments are often embedded within opinions and render the opinion positive, negative or possibly neutral. Sentiment analysis is conducted on text, although it can be done on speech as well. Text can be the electronic medical record, blogs, patient portals or short messages. The common thread is that the text must be digitized and not on paper. These texts can be semi-structured or unstructured. Semi-structured texts are ordered and can be assigned to a predefined data model. They have clearly identified headings, sub-headings, tables and figures. Human language is usually dubbed as "unstructured information" although any linguist would argue that language does have structure. The term unstructured information is used to distinguish this kind of information from information residing in well-structured databases. Unstructured information does have some level of irregularities and ambiguity, which are challenges NLP addresses. Sentiments are usually found in unstructured text.

Table 10.1 Examples of semi-structured and unstructured text

Semi-structured	Unstructured
Current medications: 50 mg of Lamictal	*Jane said that since her mother has a new boyfriend she doesn't want to leave her room.*

In sentiment analysis we want to find the structure in unstructured data (Liu et al. 2011). Then, we have the opportunity to identify patterns, because they provide knowledge (Ackoff 1989). Developing the tools to analyze sentiments is a challenge in translational research and, if successful, could play a valuable role in supporting patient care. In the example of unstructured text in Table 10.1, a clinician first asks: "Who doesn't want to leave her room, Jane or the mother?" If it is Jane, what is the chance that Jane is afraid, abused, shy or just doesn't like the new boyfriend? It is clear that computational analysis will require special algorithms. On the other hand, taking 50 mg of Lamictal is semi-structured. A simple query shows which drug and how much is dispensed. This illustrates the challenge of sentiment analysis, when applied to clinical text.

Sentiment analysis, like most NLP, uses statistical machine learning because these methods can find structures in large data sets. There are a large number of machine learning algorithms that can be used, but characteristics like sample size, variability and the complexity of the data guide the choice of algorithm (Pestian et al. 2008; Sebastiani 2002).

Sentiment analysis has great potential to support clinical decisions. One example is discovering sentiments embedded in suicide notes, a task requiring NLP technologies. An international competition among over 20 international teams was held to determine which team and algorithm were most accurate in identifying emotions in 1300 suicide notes, compared to a manually annotated gold standard. Their accuracy, as measured by the F_1-measure, ranged from 29 to 61 (Pestian et al. 2012), showing that current tools must be considerably improved to be clinically useful. Pestian et al. (2012), conducted a test of accuracy of sentiment identification by NLP by analysis of suicidal patients. Suicidal and control patients were interviewed with open ended questions, transcripts were prepared, and machine learning was used to analyze the text. Based on the analysis, the origin of the text was assigned to either the control or suicidal group. The accuracy of assignment by machine learning compared to the actual origin (F_1-measure) was 93.3 (Pestian et al. 2012). This shows that it is possible to accurately classify text of notes of suicidal and control patients, using presently available tools of NLP.

10.7 "Meaningful Use" of the Electronic Health Record and Comparative Effectiveness

"Meaningful Use" and comparative effectiveness are hot topics widely discussed in healthcare. As part of the Health Information Technology for Economic and Clinical Health Act of 2009, health care providers using a certified EHR are eligible for financial incentives if they meet "meaningful use" objectives (U.S. Department of Health and Human Services (HHS) 2010). Reporting clinical quality measures (CQM) is one such objective. CQMs aim to quantify outcomes and quality of patient care (Centers for Medicare and Medicaid Services (CMS)). Two examples of CQMs relevant to pediatrics are:

- *Measure 0001: Percentage of patients aged 5 through 40 years with a diagnosis of asthma and who have been seen for at least 2 office visits, who were evaluated during at least one office visit within 12 months for the frequency (numeric) of daytime and nocturnal asthma symptoms.*
- *Measure 0002: The percentage of children 2–18 years of age who were diagnosed with Pharyngitis, dispensed an antibiotic and received a group A streptococcus (strep) test for the episode.*

To provide the answers to the above measures, the EHR needs to be queried. Some of the information resides in already structured databases (for example, diagnosis with ICD-9 codes, number of visits, medication prescription), another part is within the clinical narrative (for example nocturnal asthma symptoms, the patient's current medications). In previous subsections, we already introduced one application of NLP, specifically IE. If the clinical narrative has been preprocessed with an NLP-based IE system (for example, cTAKES), a query can then be executed against the SNOMED CT and RxNORM codes generated by the system and used as indices for the clinical narrative. Thus, queries specific to "meaningful use" of the EHR can consume previously processed text.

Comparative effectiveness aims to measure the healthcare outcomes from one approach of disease management to another. As a result, new evidence of the effectiveness of a new intervention or health care service is generated. The Institute of Medicine (IOM) states that "*the overall goal of Comparative Effectiveness Research is the generation and synthesis of evidence that compares the benefits and harms of alternative methods to prevent, diagnose, treat and monitor a clinical condition, or to improve the delivery of care. The purpose of CER is to assist consumers, clinicians, purchasers, and policy makers to make informed decisions that will improve health care at both the individual and population levels* (Institute of Medicine (IOM) 2009)." One such study is a report commissioned by the Agency for Healthcare Research and Quality (AHRQ) on Therapies for Children with Autism Spectrum Disorders conducted by a team of investigators from Vanderbilt Evidence-based Practice Center (Warren et al. 2011). The variables and axes are complex and comprehensive; the authors used publications as their primary source of information, which were retrieved through literature searches over PubMED. This study addresses one of the IOM comparative effectiveness priority topics "Compare the effectiveness of therapeutic strategies (e.g., behavioral or pharmacologic interventions, the combination of the two) for different autism spectrum disorders (ASD) at different levels of severity and stages of intervention."

It should be obvious from these examples that one of the main goals of the meaningful use initiative is to facilitate the widespread adoption of informatics in support of ongoing health services and translational research. Another important point to make is that the methods to "query" the EHR to support health services and translation research are not standard methods, yet. The field is under ongoing research and new methods are constantly developed and existing methods refined.

Acknowledgements Dr. Savova's work was supported in part by NIH grants U54LM008748 and 1U01HG006828. Drs. Deleger's and Solti's work was supported in part by NIH grant 5R00LM010227.

References

Aberdeen J, et al. The MITRE identification scrubber toolkit: design, training, and assessment. Int J Med Inform. 2010;79(12):849–59.

Ackoff RL. From data to wisdom. J Appl Syst Anal. 1989;16(1):3–9.

AMA. Treatment of convulsive status epilepticus. Recommendations of the Epilepsy Foundation of America's Working Group on Status Epilepticus. JAMA. 1993;270(7):854–9.

Arakami E. Automatic deidentification by using sentence features and label consistency. In: Workshop on challenges in natural language I2b2 processing for clinical data, Washington, DC; 2006.

Aronson AR, et al. The NLM indexing initiative. Proc AMIA Symp. 2000;2000:17–21.

Athenikos SJ, Han H. Biomedical question answering: a survey. Comput Methods Programs Biomed. 2010;99(1):1–24.

Athenikos SJ, Han H, Brooks AD. A framework of a logic-based question-answering system for the medical domain (LOQAS-Med). In: Proceedings of the 2009 ACM symposium on applied computing. Honolulu: ACM; 2009. p. 847–51.

Beckwith BA, et al. Development and evaluation of an open source software tool for deidentification of pathology reports. BMC Med Inform Decis Mak. 2006;6:12.

Benton A, et al. A system for de-identifying medical message board text. BMC Bioinformatics. 2011;12 Suppl 3:S2.

Berman JJ. Concept-match medical data scrubbing. How pathology text can be used in research. Arch Pathol Lab Med. 2003;127(6):680–6.

Brownstein JS, Kleinman KP, Mandl KD. Identifying pediatric age groups for influenza vaccination using a real-time regional surveillance system. Am J Epidemiol. 2005;162(7):686–93.

Cairns BL, et al. The MiPACQ clinical question answering system. AMIA Annu Symp Proc. 2011;2011:171–80.

cancer Text Information Extraction System (caTIES). [cited 2012 Mar 19]. Available from: https://cabig.nci.nih.gov/community/tools/caties

Centers for Medicare and Medicaid Services (CMS). Clinical Quality Measures (CQMs). [cited 2012 Mar 19]. Available from: http://www.cms.gov/Regulations-and-Guidance/Legislation/EHRIncentivePrograms/ClinicalQualityMeasures.html

Chapman WW, et al. Evaluation of negation phrases in narrative clinical reports. Proc AMIA Symp. 2001;2001:105–9.

Choi JD, Palmer M. Getting the most out of transition-based dependency parsing. In: Proceedings of the 49th annual meeting of the Association for Computational Linguistics: Human Language Technologies. Portland: Association for Computational Linguistics; 2011a. p. 687–92.

Choi JD, Palmer M. Transition-based semantic role labeling using predicate argument clustering. In: Proceedings of the ACL 2011 workshop on relational models of semantics. Portland: Association for Computational Linguistics; 2011b. p. 37–45.

Christensen LM, Haug PJ, Fiszman M. MPLUS: a probabilistic medical language understanding system. In: Proceedings of the ACL-02 workshop on natural language processing in the biomedical domain, vol. 3. Phildadelphia: Association for Computational Linguistics; 2002. p. 29–36.

Coursera.org [Standford University]. Natural language processing. [cited 2012 June 1]; Available from: https://class.coursera.org/nlp/auth/welcome

Crowley RS, et al. caTIES: a grid based system for coding and retrieval of surgical pathology reports and tissue specimens in support of translational research. J Am Med Inform Assoc. 2010;17(3):253–64.

cTakes (Clinical Text Analysis and Knowledge Extraction System). [cited 2012 June 4]. Available from: http://ohnlp.svn.sourceforge.net/viewvc/ohnlp/trunk/cTAKES/

Deleger L, et al. Building gold standard corpora for medical natural language processing tasks. In: American medical informatics annual symposium proceedings. Chicago, 1–6 November 2012.

Demner-Fushman D, Lin J. Answering clinical questions with knowledge-based and statistical techniques. Comput Linguist. 2007;33(1):63–103.

Denny JC, et al. PheWAS: demonstrating the feasibility of a phenome-wide scan to discover gene-disease associations. Bioinformatics. 2010;26(9):1205–10.

Dunlop AL, et al. The impact of HIPAA authorization on willingness to participate in clinical research. Ann Epidemiol. 2007;17(11):899–905.

eMERGE Network: electronic medical records & genomics. A consortium of biorepositories linked to electronic medical records data for conducting genomic studies. [cited 2012 Mar 19]. Available from: http://gwas.net/

Fielstein FJ, Brown SH, Speroff T. Algorithmic De-identification of VA medical exam text for HIPAA privacy compliance: preliminary findings. In: Fiesch M, Coiera E, Li YCJ, editors. MEDINFO 2004: proceedings of the 11th World Congress on Medical Informatics. Fairfax: Ios Press; 2004. p. 1590.

Friedlin FJ, McDonald CJ. A software tool for removing patient identifying information from clinical documents. J Am Med Inform Assoc. 2008;15(5):601–10.

Friedman C. Towards a comprehensive medical language processing system: methods and issues. Proc AMIA Annu Fall Symp. 1997;1997:595–9.

Friedman C. A broad-coverage natural language processing system. Proc AMIA Symp. 2000;2000:270–4.

Gardner J, Xiong L. HIDE: an integrated system for health information DE-identification. In: Proceedings of the 21st IEEE international symposium on computer-based medical systems. Los Alamitos: IEEE Computer Society; 2008. p. 254–9.

Guo Y, et al. Identifying personal health information using support vector machines. In: I2b2 workshop on challenges in natural language processing for clinical data, Washington, DC; 2006.

Gupta D, Saul M, Gilbertson J. Evaluation of a deidentification (De-Id) software engine to share pathology reports and clinical documents for research. Am J Clin Pathol. 2004;121(2):176–86.

Hansen ML, Gunn PW, Kaelber DC. Underdiagnosis of hypertension in children and adolescents. JAMA. 2007;298(8):874–9.

Hara K. Applying a SVM based Chunker and a text classifier to the deid challenge. In: I2b2 workshop on challenges in natural language processing for clinical data, Washington, DC; 2006.

Haug PJ, et al. Experience with a mixed semantic/syntactic parser. Proc Annu Symp Comput Appl Med Care. 1995;19:284–8.

Health Information Technologies Research Laboratory (HITRL). [cited 2012 Mar 19]. Available from: http://hitrl.it.usyd.edu.au/

Health information Text Extraction (HITEx). HITEx manual v2.0. [cited 2012 Mar 19]. Available from: https://www.i2b2.org/software/projects/hitex/hitex_manual.html

Health Insurance Portability and Accountability Act of 1996 (HIPAA). P.L. 104–191. In: 42 U.S.C. 1996.

Hripcsak G, Kuperman GJ, Friedman C. Extracting findings from narrative reports: software transferability and sources of physician disagreement. Methods Inf Med. 1998;37(1):1–7.

Hu M, Liu B. Mining and summarizing customer reviews. In: Proceedings of the tenth ACM SIGKDD international conference on knowledge discovery and data mining. Seattle: ACM; 2004. p. 168–77.

IBM – Watson. (n.d.) [cited 2012 April 5]. Available from: http://www-03.ibm.com/innovation/us/watson/index.html

Institute of Medicine (IOM). Initial national priorities for comparative effectiveness research [Consensus Report]. 2009 [cited 2012 Mar 19]. Available from: http://www.iom.edu/Reports/2009/ComparativeEffectivenessResearchPriorities.aspx

Institute of Medicine (IOM). The learning healthcare system in 2010 and beyond: understanding, engaging, and communicating the possibilities. [Workshop]. 2010 Apr [cited 2012 June 1]; Available from: http://www.iom.edu/Activities/Quality/VSRT/2010-APR-01.aspx

Jha AK. The promise of electronic records: around the corner or down the road? JAMA. 2011;306(8):880–1.

JULIE Lab. Jena University Language & Information Engineering Lab. [cited 2012 Mar 19]. Available from: http://www.julielab.de/

Jurafsky D, Martin JH. Speech and language processing: an introduction to natural language processing, computational linguistics, and speech recognition, Prentice Hall series in artificial intelligence. Upper Saddle River: Prentice Hall; 2000. xxvi, 934 p.

Kho AN, et al. Electronic medical records for genetic research: results of the eMERGE consortium. Sci Transl Med. 2011;3(79):79re1.

Kimia AA, et al. Utility of lumbar puncture for first simple febrile seizure among children 6 to 18 months of age. Pediatrics. 2009;123(1):6–12.

Kimia A, et al. Yield of lumbar puncture among children who present with their first complex febrile seizure. Pediatrics. 2010;126(1):62–9.

Kohane IS. Using electronic health records to drive discovery in disease genomics. Nat Rev Genet. 2011;12(6):417–28.

Kullo IJ, et al. Leveraging informatics for genetic studies: use of the electronic medical record to enable a genome-wide association study of peripheral arterial disease. J Am Med Inform Assoc. 2010;17(5):568–74.

Lexical Systems Group. Specialist NLP Tools. [cited 2012 June 1]. Available from: http://lexsrv3.nlm.nih.gov/Specialist/Home/index.html

Liao KP, et al. Electronic medical records for discovery research in rheumatoid arthritis. Arthritis Care Res (Hoboken). 2010;62(8):1120–7.

Lin C, et al. Feature engineering and selection for rheumatoid arthritis disease activity classification using electronic medical records. In: Proceedings of the 29th international ICML conference Workshop on Machine Learning for Clinical Data; 2012; Edinburgh, Scotland.

Lindberg DA, Humphreys BL, McCray AT. The unified medical language system. Methods Inf Med. 1993;32(4):281–91.

Liu B. Sentiment analysis and opinion mining. In: Paper presented at the twenty-fifth conference on artificial intelligence (AAAI-11 Tutorial); 2011; San Franciso. p. 1–99.

Lucene. Apache Lucene Core. [cited 2012 Mar 13]. Available from: http://lucene.apache.org/core/

Mack R, et al. Text analytics for life science using the unstructured information management architecture. IBM Syst J. 2004;43(3):490–515.

Manning CD, Schütze H. Foundations of statistical natural language processing. 2nd printing, with corrections. ed. Cambridge, MA: MIT Press; 2000. xxxvii, 680 p.

McCarty C, et al. The eMERGE Network: a consortium of biorepositories linked to electronic medical records data for conducting genomic studies. BMC Med Genomics. 2011;4(1):13.

Meystre S, Haug PJ. Evaluation of medical problem extraction from electronic clinical documents using MetaMap transfer (MMTx). In: Proceedings of MIE2005 – the XIXth international congress of the European Federation for Medical Informatics. Amsterdam: Ios Press; 2005. p. 823–8.

Meystre SM, et al. Extracting information from textual documents in the electronic health record: a review of recent research. Yearb Med Inform. 2008;2008:128–44.
Meystre SM, et al. Automatic de-identification of textual documents in the electronic health record: a review of recent research. BMC Med Res Methodol. 2010;10:70.
MIST: The MITRE identification scrubber toolkit. [cited 2012 June 4]. Available from: http://mist-deid.sourceforge.net/
Murphy SN, et al. Serving the enterprise and beyond with informatics for integrating biology and the bedside (i2b2). J Am Med Inform Assoc. 2010;17(2):124–30.
National Centre for Text Mining (NaCTeM). [cited 2012 Mar 19]. Available from: http://www.nactem.ac.uk/index.php
Neamatullah I, et al. Automated de-identification of free-text medical records. BMC Med Inform Decis Mak. 2008;8:32.
Nielsen RD, et al. An architecture for complex clinical question answering. In: Proceedings of the 1st ACM international health informatics symposium. Arlington: ACM; 2010. p. 395–9.
Online Colleges.net. Standford introducing five free online classes by Anna Schumann. 2012 [cited 2012 June 1]. Available from: http://www.onlinecolleges.net/2012/03/07/stanford-introducing-five-free-online-classes/
OpenNLP Tools 1.5.0 API: sentence boundary detector. [cited 2012 June 4]. Available from: http://opennlp.sourceforge.net/api/index.html
Palmer M, Gildea D, Kingsbury P. The proposition Bank: an annotated corpus of semantic roles. Comput Linguist. 2005;31(1):71–106.
Pestian JP, et al. Using natural language processing to classify suicide notes. AMIA Annu Symp Proc. 2008;2008:1091.
Pestian JP, et al. Sentiment analysis of suicide notes: a shared task. Biomed Inform Insights. 2012;5 Suppl 1:3–16.
Riviello Jr JJ, et al. Practice parameter: diagnostic assessment of the child with status epilepticus (an evidence-based review): report of the Quality Standards Subcommittee of the American Academy of Neurology and the Practice Committee of the Child Neurology Society. Neurology. 2006;67(9):1542–50.
Ruch P, et al. Medical document anonymization with a semantic lexicon. Proc AMIA Symp. 2000;2000:729–33.
Savova GK, et al. Mayo clinical Text Analysis and Knowledge Extraction System (cTAKES): architecture, component evaluation and applications. J Am Med Inform Assoc. 2010a;17(5): 507–13.
Savova GK, et al. Discovering peripheral arterial disease cases from radiology notes using natural language processing. AMIA Annu Symp Proc. 2010b;2010:722–6.
Savova GK, et al. Automated discovery of drug treatment patterns for endocrine therapy of breast cancer within an electronic medical record. J Am Med Inform Assoc. 2012;19(e1)83–9.
Sebastiani F. Machine learning in automated text categorization. ACM Comput Surv (CSUR). 2002;34(1):1–47.
Singh RK, et al. Prospective study of new-onset seizures presenting as status epilepticus in childhood. Neurology. 2010;74(8):636–42.
Sohn S, et al. Classification of medication status change in clinical narratives. AMIA Annu Symp Proc. 2010;2010:762–6.
Solti I. Increasing clinical trial enrollment: a semi-automated patient centered approach. NIH Project No: 5R00LM010227-04. [cited 2012 June 1]. Available from: http://projectreporter.nih.gov/project_info_description.cfm?aid=8215715&icde=12435657
Solti I, et al. Automated classification of radiology reports for acute Lung injury: comparison of keyword and machine learning based natural language processing approaches. Proceedings (IEEE Int Conf Bioinformatics Biomed). 2009;2009:314–19.
Stein SC, Hurst RW, Sonnad SS. Meta-analysis of cranial CT scans in children. A mathematical model to predict radiation-induced tumors. Pediatr Neurosurg. 2008;44(6):448–57.
Szarvas G, Farkas R, Busa-Fekete R. State-of-the-art anonymization of medical records using an iterative machine learning framework. J Am Med Inform Assoc. 2007;14(5):574–80.

Taira RK, Bui AA, Kangarloo H. Identification of patient name references within medical documents using semantic selectional restrictions. Proc AMIA Symp. 2002;2002:757–61.

U.S. Department of Health and Human Services (HHS). Secretary Sebelius announces final rules to support 'Meaningful Use' of electronic health records [News Release]. 2010 [cited 2012 Mar 19]. Available from: http://www.hhs.gov/news/press/2010pres/07/20100713a.html

U-Compare. [cited 2012 Mar 19]. Available from: http://u-compare.org/index.en.html

UIMA (Unstructured Information Management Applications). Apache UIMA. [cited 2012 June 4]. Available from: http://uima.apache.org/

Uzuner O, Luo Y, Szolovits P. Evaluating the state-of-the-art in automatic de-identification. J Am Med Inform Assoc. 2007;14(5):550–63.

Uzuner O, et al. A de-identifier for medical discharge summaries. Artif Intell Med. 2008;42(1): 13–35.

Warren Z, et al. Therapies for children with autism spectrum disorders. comparative effectiveness review, AHRQ, Number 26. 2011 [cited 2012 June 1]. Available from: http://www.effectivehealthcare.ahrq.gov/ehc/products/106/656/CER26_Autism_Report_04-14-2011.pdf

Weber GM, et al. The Shared Health Research Information Network (SHRINE): a prototype federated query tool for clinical data repositories. J Am Med Inform Assoc. 2009;16(5):624–30.

Weiming W, et al. Automatic clinical question answering based on UMLS relations. In: Proceedings of the third international conference on semantics, knowledge and grid. Washington, DC: IEEE Computer Society; 2007. p. 495–8.

Wellner B. Sequence models and ranking methods for discourse parsing. Waltham: Brandeis University; 2009.

Wilke RA, et al. The emerging role of electronic medical records in pharmacogenomics. Clin Pharmacol Ther. 2011;89(3):379–86.

Wolf MS, Bennett CL. Local perspective of the impact of the HIPAA privacy rule on research. Cancer. 2006;106(2):474–9.

Yu H, Cao YG. Automatically extracting information needs from Ad Hoc clinical questions. AMIA Annu Symp Proc. 2008;2008:96–100.

Zeng QT, et al. Extracting principal diagnosis, co-morbidity and smoking status for asthma research: evaluation of a natural language processing system. BMC Med Inform Decis Mak. 2006;6:30.

Chapter 11
Informatics and Decision Support

Judith W. Dexheimer, Laurie H. Johnson, Imre Solti, Dominik Aronsky, and John P. Pestian

Abstract Clinical decision support systems encompass both computerized and paper-based tools, which aid clinicians in decision-making. Decision support systems are frequently integrated with electronic medical records to guide treatment decisions and to aid in decision-making at the point of care. The decision support can be delivered in a variety of ways from medication suggestions and warnings to guideline recommendations. They can be built using simplistic rule-based techniques or more complex artificial intelligence methodologies. Improvement and evaluation of clinical decision support systems can be accomplished using translational informatics. This chapter uses examples in the management of pediatric asthma and pediatric epilepsy to discuss how informatics is being applied to improve clinical decision support systems.

J.W. Dexheimer, Ph.D. (✉)
Department of Pediatrics, University of Cincinnati College of Medicine, Cincinnati, OH, USA

Division of Emergency Medicine and Biomedical Informatics, Cincinnati Children's Hospital Medical Center, MLC 2008, 3333 Burnet Avenue, Cincinnati, OH 45229-3039, USA
e-mail: Judith.Dexheimer@cchmc.org

L.H. Johnson, M.D.
Department of Pediatrics, University of Cincinnati College of Medicine, Cincinnati, OH, USA

Division of Emergency Medicine, Cincinnati Children's Hospital Medical Center, MLC 2008, 3333 Burnet Avenue, Cincinnati, OH 45229-3039, USA

I. Solti, M.D., Ph.D., M.A. • J.P. Pestian, Ph.D.
Department of Pediatrics, University of Cincinnati College of Medicine, Cincinnati, OH, USA

Division of Biomedical Informatics, Cincinnati Children's Hospital Medical Center, MLC 2008, 3333 Burnet Avenue, Cincinnati, OH 45229-3039, USA

D. Aronsky, M.D., Ph.D., FACMI
Departments of Biomedical Informatics and Emergency Medicine, Vanderbilt University School of Medicine, Nashville, TN, USA

J.J. Hutton (ed.), *Pediatric Biomedical Informatics: Computer Applications in Pediatric Research*, Translational Bioinformatics 2, DOI 10.1007/978-94-007-5149-1_11,

11.1 Decision Support

Computerized decision support systems integrate with electronic health records (EHRs) to support treatment decisions at the point of care. They are frequently built upon evidence-based clinical guidelines that represent expert consensus about the best ways to manage patients under specific circumstances (Bakken et al. 2004). Decision support systems have demonstrated positive effects on patient outcomes (Dexter et al. 2004). For this reason, providers, payors, federal agencies, healthcare institutions, and patient organizations support the development and use of decision support, whose data substrate are clinical guidelines. However, there are many barriers that limit the implementation and integration of these into clinical practice.

Clinical decision support systems provide reminders, current knowledge, and alerts about many elements of patient care. Warnings about potential drug-drug interactions, dosing suggestions, alerting about an abnormal laboratory value, recommending a preventive care measure or other care suggestions are common. At the simplest level, decision support systems, e.g., medication warnings and alerts, provide reminders that encourage "good" care of an individual patient and help increase the standard of care provided by the organization as a whole. These decision support elements are virtually invisible to the end-user because they are integrated with the software design of an electronic health record (EHR).

Decision support provides users with passive or active alerts. Passive alerts are reminders that appear as part of general care such as a drug-dosing suggestion or displaying allergy information. Active alerts provide a prompt or pop-up to the user that must be acted upon (modal dialogue), such as when a new medication interacts with existing medications with a high probability of an adverse event. Active alerts are frequently time-dependent and strive to fit into the clinical workflows to display the right information at the right time to the right person. A potential issue with these alerts is alert-fatigue (Ash et al. 2007). Care must be taken within the design and implementation of alerts to include only those essential to providing good patient care and avoiding adverse events.

Clinical decision support systems are based on EHR data and generally require a knowledge base to represent the domain information. In machine learning this knowledge base is called a corpora. An inference mechanism, such as a rule engine, is also needed. This engine combines the patient's clinical characteristics with those of the knowledge base. An example is avoidance of adverse events by prospectively searching for potential drug-drug interactions when a new medication is ordered in the EHR. The decision support system combines a patient's existing medication list with the new medication and executes rules that evaluates whether an interaction representation exists for the new medication and any of the other medications.

Decision support can be built and integrated with the medical record, using rule-based or more complex techniques such as applying artificial intelligence methods (Friedman and Frank 1983). Artificial intelligence techniques such as Artificial Neural Networks, Support Vector Machines, Bayesian networks, or Natural Language Processing can be applied in domains where more complex and multi-dimensional problems exist, such as classifying patients or suggesting likely diagnoses or treatment decisions. Many approaches to decision support are sound theoretically, but are challenging to integrate in real-time patient care settings. Aspects such as workflow integration, alert-fatigue, specificity of alerts, and human-computer interaction all influence clinicians' acceptance of clinical decision support (Aronsky et al. 2001).

Artificial intelligence as a decision support methodology has been used in the emergency department (ED) to predict, for example, patient admission (Leegon and Aronsky 2006) and conditions such as gastrointestinal bleeding (Chu et al. 2008). Artificial neural networks are probabilistic networks that function as "black boxes" that can act as predictive models in clinical decision support systems; their "reasoning" can be difficult to interpret, but they can be leveraged as a tool for decision making (Green et al. 2009). Support vector machines are non-probabilistic and can be used to classify data into discrete categories; they can also be used for decision making (Dexheimer et al. 2007). Bayesian networks are probabilistic or belief networks that can be used to predict events, Sect. 11.2.1 provides an example of implementation of a Bayesian network. Natural language processing (NLP) is the process of extracting meaningful information from text and is discussed more fully in Chaps. 9 and 10. Research has shown that using more training data helps improve the accuracy of decision support methodologies (Leegon and Aronsky 2006).

Knowledge bases (KBs) support decision support queries. The knowledge base typically contains up-to-date information needed to identify cases of disease, to recommend ancillary studies to support the diagnoses, to recommend therapy, and to train and evaluate the machine learning algorithms. The rules in a rule-based system function can function as the knowledge base (KB). These rules need to be re-evaluated and updated as new information becomes available. For this reason, rule-based systems are costly to maintain (Giuse et al. 1995). Automated knowledge base updates are more efficient, but some level of oversight is needed. Clinical knowledge management is an integral part of developing an integrated informatics system and several tools can help (Sittig et al. 2010; Leegon and Aronsky 2006).

Improvement and evaluation of clinical decision support systems can be accomplished using translational informatics. We will show how informatics is being applied to improve clinical decision support systems for asthma management in pediatric emergency medicine and selection of optimal treatment for epilepsy. NLP will be necessary to interpret text-based data such as clinical notes and histories as discussed more fully in Chaps. 9 and 10. In the following sections we provide examples of clinical decision support based on NLP methods.

11.2 Pediatric Emergency Medicine

Pediatrics is one of the oldest medical specialties (American Board of Medical Specialties (ABMS)) and the subspecialty of pediatric emergency medicine has been in existence for at least two decades. Dedicated fellowship training programs began in 1980 (Babl et al. 2005) and pediatric emergency medicine was approved as a subspecialty by the American Board of Medical Specialties in the early 1990s (Strange et al. 2009).

Between 1996 and 2006, the number of pediatric ED visits increased by an average of 3% per year (Pitts et al. 2008). During this same time period, the overall population-based ED utilization rate increased by 18% (Pitts et al. 2008). Pediatric visits to emergency departments comprise approximately 25% of all emergency visits (Luo et al. 2003) and range in acuity from life-threatening medical illnesses and injuries to emergent/urgent medical issues and injuries to minor acute or non-acute problems/medical complaints. Patients seeking care specifically for acute injuries comprise approximately one-third of pediatric visits to an emergency department (Owens et al. 2008). Cases deemed to be "non-emergent" comprise up to 30% of pediatric emergency visits, and can be defined as "those in which immediate medical care would not be required within 12 hours" based on the patient's age, presenting complaints, symptoms, vital signs, and past medical history (Owens et al. 2008). Patient caregivers' perspectives on reasons for non-emergent ED visits include lack of access to physicians during non-standard working hours and lack of health insurance, while primary care physician and ED personnel perspectives for those visits was lack of knowledge of what is a "true emergency" (Ben-Isaac et al. 2010).

In 2010, an estimated 2.8% of children's hospitals had a comprehensive EHR and 17.9% had some form of basic EHR system (Nakamura et al. 2010). Of these, 34.3% had Computerized Provider Order Entry (CPOE) and 26.7% had clinical guideline support (Nakamura et al. 2010). Approximately 46% of emergency departments have EHRs (Geisler et al. 2010). These numbers will quickly outdate because EHRs are rapidly replacing paper-based processes for charting, ordering, patient tracking, and many other healthcare documentation and decision-making functions. Based on EHR data, computer-based decision support can be provided to caregivers to support clinical decision making and standardize patient care. Computerized decision support systems are integrated with EHRs to guide treatment decisions and to aid the decision-making process at the point of care. Decision support can be delivered in a variety of ways, such as suggesting a specific medication, warning about a potential adverse drug event, providing recommendations from a computerized guideline, alerting about an abnormal laboratory value, recommending a preventive care measure or other care suggestions.

11.2.1 Decision Support: Asthma Management in the Emergency Department

Many techniques have been used to identify patients eligible for clinical decision support systems. These techniques are generally developed in the "in silico lab"

(Sanders and Aronsky 2006a, b; Sanders et al. 2007) and then pilot tested in a real time clinical environment before implementation throughout a healthcare delivery system (Dexheimer et al. 2011). Ideally, measurement of performance is an integral part of piloting. A good example of this approach is the design, implementation and testing of a technique to identify asthma patients in a pediatric ED.

Asthma is the most common chronic childhood disease, affecting nine million children (12.5%) under 18 years of age in the United States (Centers for Disease Control and Prevention (CDC); Centers for Disease Control and Prevention (CDC) 2005), and is a significant burden on the healthcare system. Approximately four million children experience an asthma exacerbation annually; they account for an estimated 14 million missed school days (Centers for Disease Control and Prevention (CDC) 2005) and more than 1.8 million emergency department visits (Allergy & Asthma Advocate 2004). Uncontrolled asthma requires the patient to seek immediate care, which frequently leads to hospitalizations that account for more than 60% of asthma-related costs (Lozano et al. 1999; Eder et al. 2006).

In the emergency department early identification, accurate assessment of the severity of airway obstruction, and response to therapy are fundamental to this urgent care. The severity of the patient's asthma should be rated every one to two hours, using either an asthma scoring metric (Qureshi et al. 1998), peak flow measurement, or oxygen saturation. After each assessment, care decisions should be adjusted to the new severity level leading to a disposition decision within four to six hours. Providing standardized care is challenging in the fast paced emergency department of an academic-medical center. Several asthma guidelines exist to help support clinicians in providing appropriate treatment including the guideline from the National Heart Lung and Blood Institute (NHLBI) (1997). These guidelines are general rules concerning medications and treatments that clinicians should follow for optimal care.

Adherence to asthma guidelines improves patient care (Grimshaw et al. 2002; Scribano et al. 2001). Despite the wide dissemination of guidelines and scoring systems, practice variation continues to affect treatment decisions. The NHLBI guidelines direct treatment through peak flow readings, forced expiratory volume, and oxygen saturation. They include an asthma flow diagram that outlines general principles, but customization is required to account for variations in individual emergency department practices, including medications and scoring. The NHLBI guidelines emphasize early recognition and treatment stratified by severity (NHLBI 1997). The temporal evolution of the patient's condition and the complex environment of the emergency department make written guidelines difficult to follow. The challenges associated with care of asthmatic patients in an emergency room, then, provide an ideal opportunity for development and implementation of a well-designed computerized decision support system.

11.2.2 Decision Support: Workflow-Embedded Computerized Detection and Reminder System

The goal of this project was to design and implement a workflow-embedded computerized detection and reminder system for asthma care in the pediatric emergency department. The system design and prompts were influenced by the team-oriented workflow and current asthma care in the pediatric emergency department and the local adaptation of the national guidelines. The design of an integrated system included: developing a conceptual framework, establishing guideline eligibility using a Bayesian Network (BN) system (Sanders and Aronsky 2006a; Sanders et al. 2007), and providing an electronic protocol for management of the patient.

The ideal guideline management system should take advantage of the existing information system infrastructure of the ED to allow for a fully computerized solution with minimal or no prompting for additional information. By supporting the emergency department staff, the integrated approach helps to facilitate guideline adherence with limited additional work. This provides the basis for a computerized and fully integrated approach. Relevant information should be available at the right time, for the right individuals, and at the right place. Requirements for additional data entry should be minimal. To help ensure it is user-centric, providers should be able to opt out at any time within the system. To reach the design objectives, a multidisciplinary team was gathered to develop the system, including emergency department nurses, physicians, respiratory therapists, and leadership members from the emergency department and the hospital.

In the pediatric ED, approximately 10% of the patients present with asthma exacerbations (Sanders and Aronsky 2006a). An existing paper-based guideline, including a validated asthma severity metric (Qureshi et al. 1998) was available for several years prior and included reassessment and treatment suggestions; however, chart review suggested the guideline was used in only 7–10% of the asthma cases (Sanders and Aronsky 2006a).

The information system infrastructure in the pediatric ED at Vanderbilt University Medical Center, where the system was developed and implemented, includes five information systems that were used to develop and integrate the computerized guideline system. The five information systems include the EHR (Miller et al. 2005), the computerized triage application (Levin et al. 2006), computerized physician order entry (CPOE) (Miller et al. 2005), the computerized whiteboard application (Aronsky et al. 2008), and the computerized respiratory therapy documentation system. The longitudinal EHR includes patient information acquired from 1994 to the present (Giuse 2003). It represents the primary repository for all patient visit information, including the problem list, clinic notes, procedure notes, scanned documents, exam reports, and caregiver team communications.

The computerized triage system (Levin et al. 2006) captures triage data in a mostly coded format. It captures the triage information including past medical history, vital signs, acuity, any additional information and the nurse assigns the patient a coded chief complaint. Immediately following triage, the BN system calculates an asthma probability. If the patient has findings compatible with asthma, the nurse is prompted to perform an initial asthma score. An automatic page is sent to respiratory therapy containing the asthma score to alert them that a potential asthma patient has arrived.

The CPOE system is used throughout the hospital (Miller et al. 2005) and more than 90% of all medication orders are entered by physicians. The computerized asthma management system is triggered automatically by the BN detection system. The CPOE system provides a scoring page for physicians containing the patient's most recent asthma score, previous orders, and local guideline recommendations for order or discharge decisions. The physician is required to score the patient to move ahead into order-entry or can opt out of the asthma prompts. Orders suggested by the emergency department management are pre-selected for mild, moderate or severely scored patients.

The respiratory therapy documentation system is integrated with the EHR and the respiratory therapist documents the patient's asthma score (Qureshi et al. 1998) using a semi-structured format and notes about the patient's breathing or auscultations.

The computerized whiteboard provides an "at-a-glance" overview of the emergency department's clinical and operational statistics (Aronsky et al. 2008). The whiteboard displays emergency department patients and relevant information about orders and labs. It can also be used to access the patient's EHR or CPOE system. Because the whiteboard plays an integral role in communication in the emergency department, it is used to alert respiratory therapists and physicians of the need for reassessment of a patient.

The asthma detection algorithm is a BN (Fig. 11.1) (Sanders and Aronsky 2006a; Dexheimer et al. 2009) integrated within the ED information systems and uses data found in the EHR and electronic triage application. The algorithm makes an asthma exacerbation prediction in real-time using the information available in triage and past medical history. Upon the completion of triage, a scoring page is displayed requiring the nurse to perform the initial asthma scoring.

The asthma management system was built with workflow design and consideration for the pediatric ED, where it was to be deployed. While this limits the generalizability of the implementation, it helps to ensure the best-fit application for the environment. Patients identified by the BN and subsequently scored during triage have prompts for the computerized management system for their care throughout their stay in the emergency department. The management system uses the electronic whiteboard to display the current asthma scores, a trend graph of the scores throughout their treatment, and passive reminders for ongoing scoring and evaluation by the respiratory therapist and the physician. By graphing the trend of scores throughout the patient's stay, the clinicians are provided a concise view of what has occurred during the treatment process.

The CPOE system prompts are designed to streamline guideline adherence through order sets. Pre-selected orders allow for one-click ordering of

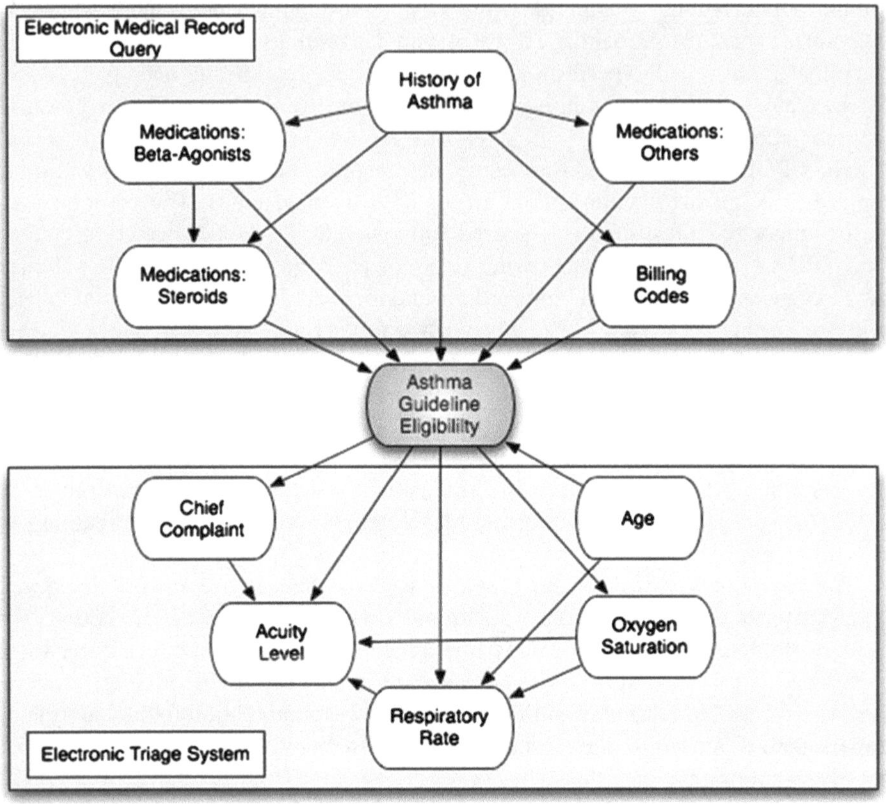

Fig. 11.1 Bayesian Network used for determining asthma guideline eligibility (Taken with the author's permission from Sanders and Aronsky 2006a; Dexheimer et al. 2009)

guideline-compliant asthma medications and treatments, but are easily modified to fit different clinical cases. Additional orders and modifications can always be performed after accepting or rejecting the order set. Reminders that disposition decisions need to be made also help the clinicians move patients through the ED. Disposition rules provide general guidelines for admitting or discharging a patient, based on previous experience with care.

The asthma management system has been well received and remains in use throughout the pediatric ED. Although no change has been shown in the length of stay after implementation (Dexheimer et al. 2011), the system integrates patient and clinician data to help the care team communicate more effectively. Everyone can easily follow the patient's visit based on the graph of scores displayed. The computerized asthma management system represents a workflow oriented, sustainable approach to improving patient care in a challenging environment.

11.3 Epilepsy

Epilepsy is defined as having two or more uncontrolled seizures. More than three million people in North America suffer from this disease. The incidence of 50/100,000 per year is highest for children below 5 years of age and among the elderly. Epilepsy is a common neurologic condition with significant personal, societal, medical, and economic load. Despite this, considerable gaps exist in the quality of care delivered to patients with epilepsy. Measuring quality is the first step toward eliminating these gaps and improving overall care.

11.3.1 Improvements in Quality of Clinical Notes

In 2010, the American Academy of Neurology (AAN) proposed eight measures of quality of care, based on 160 relevant recommendations culled from publications (Fountain et al. 2011). These measures include: reporting seizure type and frequency, etiology of the epilepsy, finding on EEG and neuro-imaging, side effects of antiepileptic drugs, safety issues, referral for neurosurgical evaluation of patients with medically refractory epilepsy, and issues for women of childbearing potential.

A team of investigators at Cincinnati Children's Hospital Medical Center and the University of Cincinnati (the Epilepsy Team including T. Glauser, K. Holland, P. Matykiewicz, D. Morita, S. Sandridge, and J. Pestian—an author of this chapter) has built a linguistic corpus around these measures, as required for development of algorithms using semi-automated natural language processing to analyze free text from the electronic health records of patients. The first step was to annotate these data. This requires manually assigning epilepsy related medical terms to progress notes. The following terms were annotated: "epileptic" or "non-epileptic events", "seizure frequency", "epilepsy classification", "1989 epilepsy etiology", "2010 epilepsy etiology", "epilepsy syndromes", "current treatment", "current treatment daily dose", "reason for change in treatment", "planned treatment", "planned treatment daily dose". These terms expand upon the AAN recommendations and include those of the International League Against Epilepsy and the National Institutes of Health.

Web-based software was developed for tracking progress of annotation and resolving issues that occur. Three individuals annotated the medical records from 51 epileptic patients. The records included 260 progress notes and 41,444 words. Approximately 140 inconsistencies between annotators were resolved by majority vote (see Chap. 9 for a discussion of annotation and creation of gold standards). Inter-annotator agreement was 0.86, as measured by Krippendorff's alpha coefficient. This means that annotators agreed 86% of the time. Information about epileptic and non-epileptic events that would be required, if the notes were of high quality, was present in 91% of progress notes, i.e. missing in 9%. The Epilepsy

Team, using NLP, uncovered deficiencies in the neurology department's documentation of patients' present history. Because of this, the neurology department modified requirements for data entry into the electronic medical record by adding form fields to be completed during each patient visit.

11.3.2 Decision Support: Identification of Medically Intractable Epilepsy

Nearly 70% of seizures in new onset epilepsy patients can be controlled with proper pharmacological treatment. The remaining patients are considered to have medically intractable epilepsy and should be considered for neurosurgical evaluation. The International League Against Epilepsy (ILAE) recommends that surgery be considered as soon as intractability is diagnosed (ILAE). Deciding who is a surgical candidate requires answers to several questions: Is the epilepsy truly intractable? Has the child received the correct medications in optimal dosage? What are the parents' wishes? What environmental factors have played a role? What is the surgical risk for a particular child? Can advanced technology identify surgical candidates? The question then becomes whether an automated process can be developed to examine a patient's complex history and provide an alert to the health care team when the question of surgery should be raised.

Testing this hypothesis requires selecting the correct features from the patients' electronic medical records, then examining these features to assemble an automated intractability classification system. To develop and test such an NLP-based algorithm, the Epilepsy Team selected an appropriate group of medical records for annotation and initial analyses. Epileptologists assigned patients and their medical records to appropriate groups to create a "gold standards" corpus. This corpus was used to train a supervised machine learning algorithm (see Sect. 9.3). Ultimately, the Epilepsy Team developed machine-learning algorithms that could classify surgical patients with at least 90% accuracy. The machine-learning algorithm has been moved from development to testing in an actual clinical care setting. Ultimately the algorithms will be integrated into the clinical workflow where it will generate the appropriate clinical alerts.

11.3.3 Multi-institutional Pediatric Epilepsy Decision Support

The Multi-Institutional Pediatric Epilepsy Decision Support (MiPeds) project aims to create a learning health care system focused on the care of children with epilepsy. The goal is to develop decision support tools that predict appropriate therapy for individual patients and to benchmark outcomes among centers. The participation of multiple institutions is necessary to accrue sufficient numbers of patients to permit analyses with high levels of statistical power. Experts in epilepsy (domain experts)

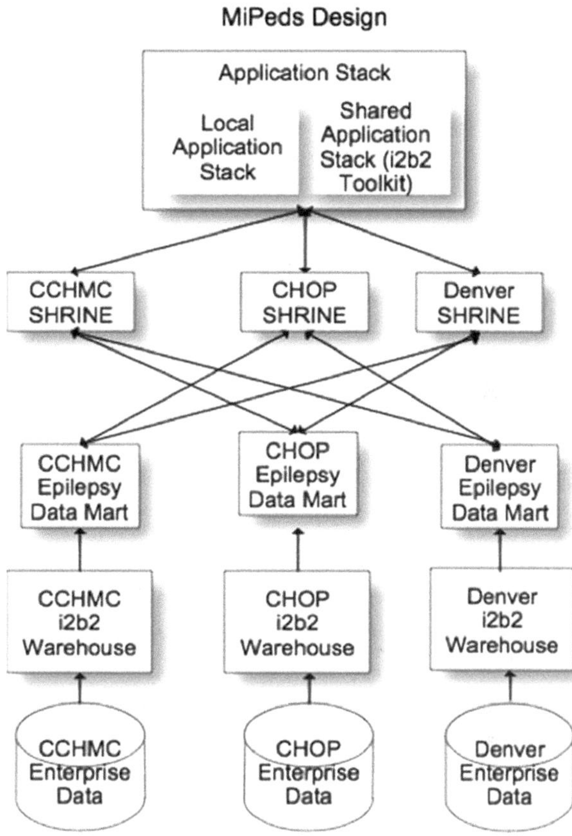

Fig. 11.2 MiPeds system design. Participating institutions are: Cincinnati Children's Hospital Medical Center (*CCHMC*), Children's Hospital of Philadelphia (*CHOP*) and the Children's Hospital of Colorado (Denver)

will develop consensus measures for defining the quality of clinical care provided to patients with epilepsy and integrate these into clinical practices of participating institutions. Homogeneous cohorts of patients will be identified to develop and test algorithms that capture data from electronic medical records using machine learning and natural language processing and feed the data into clinical decision support algorithms. This will test predictive models versus empiric therapy based on a physician's best judgment. The plan is to capture data in data repositories at each participating institution and link the repositories in a federated data-sharing network. In a federated network, each individual site retains administrative control of its data, but sites trust one another for certain types of data sharing. They expose the information needed for agreed upon analyses. The three participating sites, Cincinnati Children's Hospital Medical Center (CCHMC), Children's Hospital of Philadelphia (CHOP) and the Children's Hospital of Colorado (Denver), have a common goal: to increase quality by sharing data. To do this, the MiPeds project will employ the i2b2 and SHRINE data warehousing technology as illustrated in Fig. 11.2 (For detailed information, see Chap. 8, Sect. 8.3, Data Sharing within Distributed Research Networks and Natter et al. 2012). A SHRINE network is

Table 11.1 Indicators of outcomes of treatment of patients with epilepsy

Indicator	Delphi score	Operational definition	Classification
Quality of life score after any medical treatment	5	Quality of life score = PedsQL total score Medical treatment = anticonvulsant or ketagenic diet	Site specific
Side effects after any medical treatment	5	Side effects = PESQ total score after taking anticonvulsant Medical treatment = anticonvulsant or ketagenic diet	Treatment outcome

Operational definitions of the indicators specify how they are to be measured to avoid ambiguities in data collection and analysis among members of the networkAbbreviations: *PedsQL* Pediatric Quality of Life Inventory, *PESQ* Perceptual Evaluation of Speech Quality

composed of a single Query Aggregator and a set of SHRINE adaptors, one for each i2b2 data repository in the network. Authorized researchers can query the network for the total number of patients at participating institutions who meet a given set of criteria—a combination of demographics, diagnoses, medications and laboratory tests. Because counts are aggregate, patient privacy is protected.

Which features of the data should be collected is an important question. One often-used approach is to the Delphi method, a systematic approach that pools the collective expertise into a centralized repository (Linstone and Turoff 1975). Using this approach with a group of experts in care of patients with epilepsy, a series of measurement indicators of outcomes were created with operational definitions. Table 11.1 shows examples of these. SHRINE supports cross-institutional queries and permits analysis of measurement indicators in individual institutions as well as for the group as a whole. Displaying the results of these queries should promote knowledge. For example, Fig. 11.3 shows the top five side effects of the drug amantadine and the top five procedures that cause cerebral spinal fluid leaks.

11.3.4 Decision Support: Expert Systems

Clinicians want to improve the quality of the decisions they make, while caring for individual patients. Clinical decision support systems are widely used in the clinical setting to organize and present information to optimize the caregiver's clinical decisions. At the core of these clinical decision support systems is an expert system. Expert systems are designed to emulate the behavior of human experts in care of patients with specific disorders. These software-encoded agents provide both benefits and drawbacks. Some benefits include ability to incorporate best practices

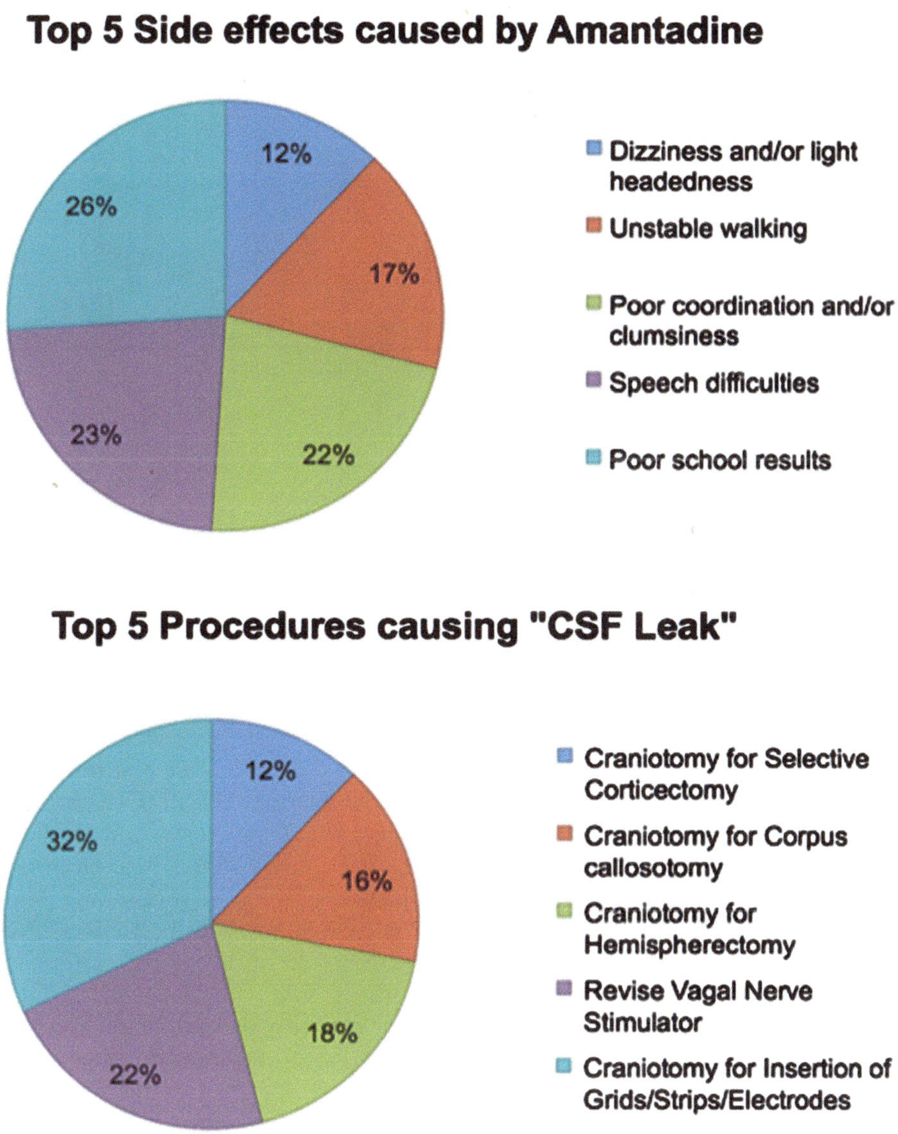

Fig. 11.3 Measurements of outcomes of treatment of patients with epilepsy. The *top panel* shows the top five side effects of the drug amantadine and the *bottom panel* the top five procedures that cause cerebral spinal fluid leaks

derived from analysis of huge amounts of literature, reproducibility, efficiency, and consistency. A substantial drawback of individual expert systems is their lack of generalizability beyond a specific domain like oncology or neurology, and their inability to link to other domains.

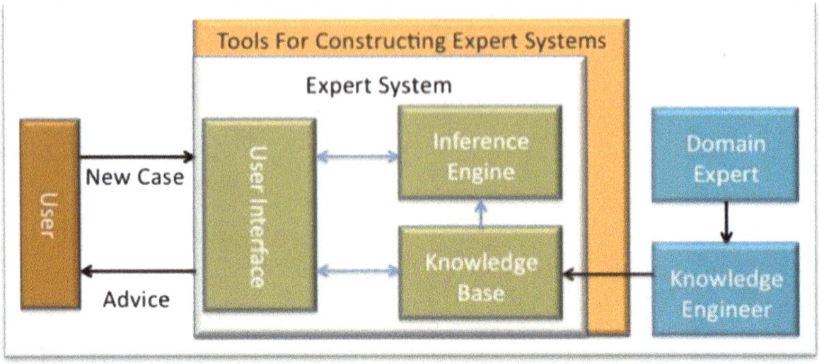

Fig. 11.4 Components of the CHRISTINE system

Expert clinical systems can improve patient safety, quality of care, and efficiency (Coiera 2003). At the authors' institution, a system termed "CHRISTINE" provides a framework for developing expert modules for decision support in care of patients with epilepsy, but the framework is generalizable. CHRISTINE integrates patient specific factors (seizure type, demographics, history, genetics, co-medications, co-morbidities, environment) with drug specific factors and then queries a knowledge base of expert consensus of best therapy for different classes of epileptic patients (Pestian et al. 2009).

Figure 11.4 shows the individual components of CHRISTINE and how they collaborate to provide clinical decision support. At its core, CHRISTINE provides an environment for developing and implementing modular expert systems. It is built using open source software, which ensures it can be distributed without licensing fees. The development language is a combination of Perl, JavaScript and Ajax and the data reside in a MySql database. Each module within the CHRISTINE framework is an expert system and can be extended to other clinical conditions, such as risk of suicide. Thus far at Cincinnati Children's Hospital, modular expert systems have been developed for: epilepsy, epilepsy annotation, suicide note annotation, identification of emotions in suicide notes, and suicide assessment. While all of these expert systems share CHRISTINE's common framework, each one contains its own domain-specific knowledge base. This approach allows fully supervised, semi-supervised, and automated analysis. For example, analyzing the free-text in medical records of epileptic patients along with the drugs selected provides a complete picture of drug utilization in patients with different types of seizures, as show in Fig. 11.5. To develop these capabilities in CHRISTINE is a complex task that requires processing of both free-text and discrete data. To do this requires an appropriately annotated corpus, tools for natural language processing, tools for generation of reports, and user training.

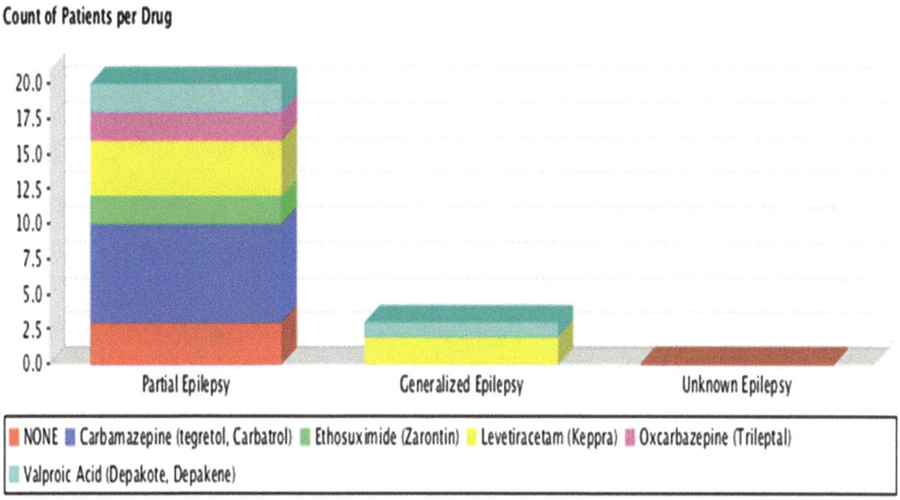

Fig. 11.5 Drugs utilized in treatment of epilepsy

References

Akinbami LJ, Moorman JE, Liu X. Asthma prevalence, health care use, and mortality: United States, 2005–2009. National health statistics reports; no 32. Hyattsville: National Center for Health Statistics; 2011.

American Board of Medical Specialties (ABMS). Approval of new member boards. [cited 2012 Apr 2]. Available from: http://www.abms.org/About_ABMS/ABMS_History/Extended_History/Approving_New_Boards.aspx

Aronsky D, Chan KJ, Haug PJ. Evaluation of a computerized diagnostic decision support system for patients with pneumonia: study design considerations. J Am Med Inform Assoc. 2001;8(5):473–85.

Aronsky D, et al. Supporting patient care in the emergency department with a computerized whiteboard system. J Am Med Inform Assoc. 2008;15(2):184–94.

Ash JS, et al. Some unintended consequences of clinical decision support systems. AMIA Annu Symp Proc. 2007;2007:26–30.

Babl FE, et al. Advanced training in pediatric emergency medicine in the United States, Canada, United Kingdom, and Australia: an international comparison and resources guide. Ann Emerg Med. 2005;45(3):269–75.

Bakken S, Cimino JJ, Hripcsak G. Promoting patient safety and enabling evidence-based practice through informatics. Med Care. 2004;42(2 Suppl):II49–56.

Ben-Isaac E, et al. National profile of nonemergent pediatric emergency department visits. Pediatrics. 2010;125(3):454–9.

Centers for Disease Control and Prevention (CDC). QuickStats: percentage of children aged <18 years who have ever had asthma diagnosed, by age group—United States, 2003. MMWR. 2005;54(16):412. [cited 2011 Mar 9]. Available from: http://www.cdc.gov/mmwr/preview/mmwrhtml/mm5416a5.htm

Centers for Disease Control and Prevention (CDC). FastStats: asthma. [cited 2009 Mar 9]. Available from: http://www.cdc.gov/nchs/fastats/asthma.htm/

Chu A, et al. A decision support system to facilitate management of patients with acute gastrointestinal bleeding. Artif Intell Med. 2008;42(3):247–59.
Coiera E. Interaction design theory. Int J Med Inform. 2003;69(2–3):205–22.
Dexheimer JW, et al. Comparing decision support methodologies for identifying asthma exacerbations. Stud Health Technol Inform. 2007;129(Pt 2):880–4.
Dexheimer JW, et al. Development of an asthma management system in a pediatric emergency department. AMIA Annu Symp Proc. 2009;2009:142–6.
Dexheimer JW, et al. A randomized controlled trial of a computerized asthma management system in a pediatric emergency department. AMIA Annu Symp Proc. 2011;2011:1740.
Dexter PR, et al. Inpatient computer-based standing orders vs physician reminders to increase influenza and pneumococcal vaccination rates: a randomized trial. JAMA. 2004;292(19): 2366–71.
Eder W, Ege MJ, von Mutius E. The asthma epidemic. N Engl J Med. 2006;355(21):2226–35.
Fountain NB, et al. Quality improvement in neurology: AAN epilepsy quality measures: report of the quality measurement and reporting subcommittee of the American Academy of Neurology. Neurology. 2011;76(1):94–9.
Friedman RH, Frank AD. Use of conditional rule structure to automate clinical decision support: a comparison of artificial intelligence and deterministic programming techniques. Comput Biomed Res. 1983;16(4):378–94.
Geisler BP, Schuur JD, Pallin DJ. Estimates of electronic medical records in U.S. emergency departments. PLoS One. 2010;5(2):e9274.
Giuse DA. Supporting communication in an integrated patient record system. AMIA Annu Symp Proc. 2003;2003:1065.
Giuse DA, Giuse NB, Miller RA. Evaluation of long-term maintenance of a large medical knowledge base. J Am Med Inform Assoc. 1995;2(5):297–306.
Green M, et al. Exploring new possibilities for case-based explanation of artificial neural network ensembles. Neural Netw. 2009;22(1):75–81.
Grimshaw JM, et al. Changing physicians' behavior: what works and thoughts on getting more things to work. J Contin Educ Health Prof. 2002;22(4):237–43.
International League Against Epilepsy (ILAE). Surgery in epilepsy: guidelines for the management of epilepsy in India. [cited 2012 June 25]. Available from: http://www.ilae.org/visitors/Documents/15-surgeryinepilepsy.pdf
Leegon J, Aronsky D. Impact of different training strategies on the accuracy of a Bayesian network for predicting hospital admission. AMIA Annu Symp Proc. 2006;2006:474–8.
Levin S, et al. The effects of computerized triage on nurse work behavior. AMIA Annu Symp Proc. 2006;2006:1005.
Linstone HA, Turoff M. The Delphi method: techniques and applications. Reading: Addison-Wesley Pub. Co., Advanced Book Program; 1975.
Lozano P, et al. The economic burden of asthma in US children: estimates from the National Medical Expenditure Survey. J Allergy Clin Immunol. 1999;104(5):957–63.
Luo X, et al. Children's health insurance status and emergency department utilization in the United States. Pediatrics. 2003;112(2):314.
Miller RA, et al. The anatomy of decision support during inpatient care provider order entry (CPOE): empirical observations from a decade of CPOE experience at Vanderbilt. J Biomed Inform. 2005;38(6):469–85.
Nakamura MM, et al. Electronic health record adoption by children's hospitals in the United States. Arch Pediatr Adolesc Med. 2010;164(12):1145–51.
National Heart Lung and Blood Institute (NHLBI). Expert Panel Report 2: guidelines for the diagnosis and management of asthma. 1997. Available from: http://www.nhlbi.nih.gov/guidelines/archives/epr-2/index.htm
Natter MD, Quan J, Ortiz DM, et al. An i2b2-based, generalizable, open-source, self-scaling chronic disease registry. J Am Med Inform Assoc. 2012; doi:10.1136/amiajnl-2012-001042

Owens PL, et al. Annual report on health care for children and youth in the United States: focus on injury-related emergency department utilization and expenditures. Ambul Pediatr. 2008;8(4):219–240.e17.

Pestian J, et al. Personalizing drug selection using advanced clinical decision support. Biomed Inform Insights. 2009;2009:23–9.

Pitts SR, et al. National Hospital Ambulatory Medical Care Survey: 2006 emergency department summary. Natl Health Stat Rep. 2008;7:1–38.

Qureshi F, et al. Effect of nebulized ipratropium on the hospitalization rates of children with asthma. N Engl J Med. 1998;339(15):1030–5.

Sanders DL, Aronsky D. Detecting asthma exacerbations in a pediatric emergency department using a Bayesian network. AMIA Annu Symp Proc. 2006a;2006:684–8.

Sanders DL, Aronsky D. Prospective evaluation of a Bayesian Network for detecting asthma exacerbations in a Pediatric Emergency Department. AMIA Annu Symp Proc. 2006b;2006:1085.

Sanders DL, Gregg W, Aronsky D. Identifying asthma exacerbations in a pediatric emergency department: a feasibility study. Int J Med Inform. 2007;76(7):557–64.

Scribano PV, et al. Provider adherence to a clinical practice guideline for acute asthma in a pediatric emergency department. Acad Emerg Med. 2001;8(12):1147–52.

Sittig DF, et al. The state of the art in clinical knowledge management: an inventory of tools and techniques. Int J Med Inform. 2010;79(1):44–57.

Strange GR. Pediatric emergency medicine a comprehensive study guide, McGraw-Hill's access emergency medicine. 3rd ed. New York: McGraw-Hill; 2009.

Chapter 12
Support of Perinatal and Neonatal Research

Eric S. Hall

Abstract The emergence of the field of biomedical informatics has enabled the development of computerization within the Neonatal Intensive Care Unit (NICU); however, the unique population encountered in this setting requires special consideration beyond minor adaptations from an adult Intensive Care Unit (ICU) or even a Pediatric Intensive Care Unit (PICU). Deficiencies in data capture of the traditional Electronic Health Record must be addressed to support research of NICU populations. Additional precision is required for the measurement of data elements such as age and weight where day-to-day changes can be clinically relevant. Obtaining gestational age, a key measure in neonatal research, has its own unique array of challenges. Although gestational age has a precise definition, numerous opportunities for error are introduced in documenting the variable, as measures are necessarily based upon estimates. Another challenge to neonatal data collection is the integration of information from parental medical records. The prenatal record and maternal medical history, though pertinent to the neonate, are contained in the mother's medical chart. Access to prenatal data may be limited by barriers between institutions where care was provided, the transition between types of care providers (obstetrics to neonatology), concerns for the mother's privacy, and the absence or unreliability of traditional identifiers used in record linking such as name and social security number. The development of neonatal terminologies allows for standardization in the description and communication of concepts specific to neonatology, and thereby facilitates research networks that study rare neonatal conditions across national and international landscapes.

E.S. Hall, Ph.D. (✉)
Department of Pediatrics, University of Cincinnati College of Medicine, Cincinnati, OH, USA

Perinatal Institute & Division of Biomedical Informatics, Cincinnati Children's Hospital Medical Center, 3333 Burnet Ave, ML 7009, Cincinnati, OH 45229-3039, USA
e-mail: eric.hall@cchmc.org

J.J. Hutton (ed.), *Pediatric Biomedical Informatics: Computer Applications in Pediatric Research*, Translational Bioinformatics 2, DOI 10.1007/978-94-007-5149-1_12,

12.1 The Sub-specialty of Neonatology

Although neonatology is a relatively new sub-specialty of pediatrics, advances in the assessment and care of newborn infants along with technological innovations have been responsible for drastic reductions in infant mortality over the past two centuries. In the early 1860s the availability of vital statistics data indicating an infant mortality rate hovering around 45% of all live births led the British medical community to search for the causes of their high infant mortality (Baines 1862). Since that time, vital records have played a key role in measuring the state of worldwide neonatal health. Near the turn of the nineteenth century, French obstetricians with a special interest in fetal and perinatal development made one of the first major breakthroughs in advancing neonatal health by introducing the first incubators. Now termed isolettes, the climate-controlled bassinets enable regulation of humidity and temperature, while limiting exposure to sources of infection. In spite of reducing infant deaths, incubators were slow to be adopted in the United States due to a lack of convincing data to support their effectiveness (Baker 1991). However, isolettes have become a mainstay of the modern NICU. In the 1950s Dr. Virginia Apgar introduced the Apgar score, which has become a simple, repeatable worldwide standard method to assess newborn infants (Finster and Wood 2005). By the 1970s the increased requirements for monitoring of ventilated infants led to the formation of newborn intensive care units (NICUs) as dedicated units within hospitals. In 1975, the American Board of Pediatrics established certification for neonatology, which now requires the completion of a fellowship program following training as a pediatrician (Pearson et al. 2001). The development of pulmonary surfactant replacement therapy in the 1980s improved the survival prospects for extremely preterm infants, who previously had not been considered viable. By the 1990s, as a result of widespread adoption of surfactant therapy, infants born as early as 24 weeks of gestation had more than a 50% chance of survival, though neonatologists suspected they had reached the limits of viability (Allen et al. 1993). Table 12.1 demonstrates 5-year averages of worldwide infant mortality from 1950 to 2010 (United Nations 2010). Corresponding with advances in neonatal care, all world regions have a experienced steady decline in infant mortality over the past several decades; though none are as dramatic as the 90% reduction which has occurred in the more developed regions of the world.

Since the 1990s, an increased focus has been placed on predicting and achieving long-term survival and healthy development of extremely premature infants (Saigal and Doyle 2008). And while neonates are technically just young children, there are a number of important factors that differentiate the practice of neonatology from general pediatrics ranging from the typical care setting and workflow to the types of clinical problems that are managed.

Along with the adult intensive care unit (ICU) or pediatric ICU (PICU), there is great potential for the NICU to benefit from computerization and decision support. While clinicians in all ICU settings could benefit from computer aided monitoring, data assimilation, and tools which aid and promote patient safety, fundamental differences in the NICU require special consideration.

Table 12.1 Global 5 year average infant mortality rates (infant deaths per 1,000 live births) since 1950

World region	1950–1955	1955–1960	1960–1965	1965–1970	1970–1975	1975–1980	1980–1985	1985–1990	1990–1995	1995–2000	2000–2005	2005–2010
World	133	122	114	94	86	80	71	64	60	56	51	46
More developed regions	60	43	33	26	22	18	15	13	11	9	7	6
Less developed regions	151	140	130	105	96	89	79	71	66	62	56	50
Least developed countries	192	176	164	152	150	136	125	116	109	100	89	80
Less developed regions, excluding least developed countries	145	134	125	98	86	80	70	62	56	52	46	41
Less developed regions, excluding China	165	149	135	123	113	101	90	81	74	69	61	55

12.2 NICU Care and the Neonatal Electronic Health Record

NICUs may be located either within a tertiary center in proximity to a Labor and Delivery unit where infants are born or within a quaternary facility specializing in pediatric care. Although neonates may be admitted to a NICU because of an emerging condition such as hyperbilirubinemia or sepsis after having gone home, it is more common for a neonate to be transferred to a NICU from a lower acuity unit such as a normal newborn or special care nursery. When complications are expected, it is desirable for laboring mothers to deliver at a center housing a NICU in order to avoid the transfer of a fragile or unstable newborn and to limit physical separation of mother and baby.

Many disorders treated within the NICU are unique to the care setting, including preterm birth and congenital anomalies along with their associated complications. Thus, the NICU setting provides an unparalleled opportunity for advancing research efforts related to the better understanding and treatment of these conditions. Specialized monitoring and sensitive instruments are also required for managing this high acuity population.

An infant born prior to the 37th week of gestation is considered preterm, though the earlier an infant is delivered, the greater the likelihood the infant will suffer complications. Immature lungs may lead to an array of respiratory complications often requiring mechanical ventilation or other respiratory support. Premature infants are also more susceptible to infection, including pneumonia and sepsis, as well as tissue death of the bowels known as Necrotizing Enterocolitis (NEC). Immature digestive systems among preterm infants necessitate nutritional supplementation through intravenous feeding called Parenteral Nutrition (PN). Immature blood vessels may be compromised by changes in circulation leading to bleeding in the brain or intraventricular hemorrhage. Other common conditions encountered in the NICU among preterm infants include inability to regulate body temperature, failure of closure of the neonate's ductus arteriosus, and retinopathy of prematurity, which can lead to blindness. Hyperbilirubinemia or jaundice resulting from bilirubin is also commonly managed in the NICU though it affects both term and preterm infants. Also, no respecter of age is neonatal abstinence syndrome which results when a neonate withdrawals from drugs or medications to which he was prenatally exposed. Finally, congenital anomalies requiring immediate surgical management account for another sizeable fraction of the NICU census. Although neonatal care continues to progress, there is still a dearth of understanding related to the causes of preterm birth generally, as well as in the best ways to manage or prevent complications of preterm birth.

It is insufficient to merely install an Electronic Health Record (EHR) system designed for an adult or even general pediatric population into the NICU setting. In particular, the NICU EHR must support a higher level of granularity and precision of measurements than is traditionally available for more mature populations. Whereas in the adult ICU, an approximate patient age in years may suffice, neonatologists concern themselves with infant age in days and weeks of life. Similarly, an EHR capturing only the admission weight or periodically updated patient weight may lack the precision necessary to accommodate rapid infant growth. Particularly when medication dosages and nutritional orders are based upon the infant's weight, it is imperative that the EHR support and accurately represent the neonate's ever fluctuating weight. An opportunity for improving the usefulness of the EHR within the NICU lies in the better integration of newborn screening examinations and reporting. The screens, including both dried blood spot and hearing exams, are intended to identify disorders that are difficult to detect clinically.

12.3 Perinatal Data Sources

A broad range of perinatal data is of interest to researchers and includes everything from patient-specific measures to representations of worldwide health and everything in between. Individual patient data may be captured by a number of sources, including care providers, billers, insurance providers, and quality improvement teams. These data include diagnoses, procedures, laboratory results, demographics and other characteristics, outcomes, results of biosample and DNA analyses, medication orders, and treatments. Although the categories of interest are not unique to the neonatal population, the most commonly desired elements for capture are representative of the specific patient population. For example, infant gestational age and birth weight are particularly relevant to the study of neonates. Also, the sub-populations of newborns that attract the greatest research focus are defined by diagnoses and treatments not prevalent in other pediatric care domains.

Aggregated patient data is also of interest to researchers (see Fig. 12.1). Individual NICU investigators may be interested in studying patient costs, outcomes, or disease prevalence within their unit. Likewise, aggregations of data at the regional, national, or global level may provide important insight into assessing regional health quality and costs, identifying hotspots of disease activity or risk factors, or in guiding policy and lawmakers. Various levels of data aggregation provide the opportunity to combine health data with other regional datasets, including vital statistics, census, environmental, or other registries that may support the investigation of population health hypotheses. In particular, community level

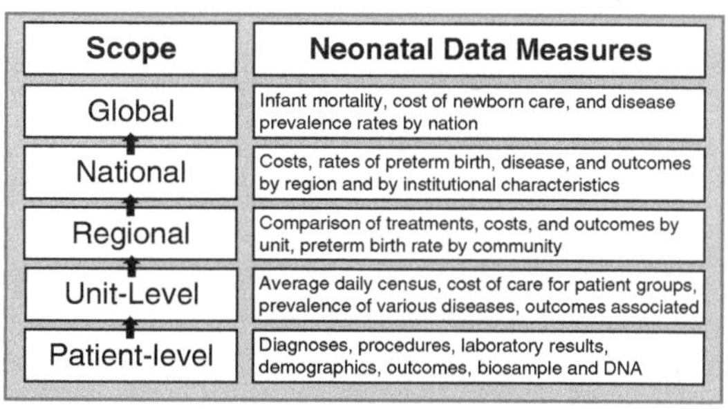

Fig. 12.1 Examples of neonatal data at various levels of aggregation, from patient-specific to global in scope

perinatal studies may be strengthened by associating hospital acquired data with community datasets collected during follow-up care or by home visitation programs.

Quality of data and the purpose for which data are collected are important considerations when selecting which elements to include in analyses. For example, ICD-9-CM (International Classification of Diseases, Ninth Revision, Clinical Modification) codes are generated from discharge summaries and other clinical notes by non-clinical personnel primarily for billing and other administrative purposes, not to support clinical care. Despite numerous analyses demonstrating the deficiencies in completeness, accuracy, granularity, precision, and content of ICD-9-CM codes, they continue to be commonly used in research because they are discrete and easily acquired (Iezzoni et al. 1992; Hsia et al. 1988).

More specific to perinatal research are the limitations of birth certificate or vital statistics data. In the United States, birth certificates are the only population based, national data source of measures such as gestational age (Wier et al. 2007). Therefore, because the birth certificate data set captures more than 99% of births annually, it remains an important resource in studying the health of maternal and newborn populations in spite of data quality concerns which arise from unreliable source data and variation in data collection (Schoendorf and Branum 2006). The birth certificate data set contains large numbers of incomplete records particularly for records associated with high-risk women and vulnerable infants (Gould et al. 2002). While some birth certificate fields such as maternal risk factors and pregnancy complications are often under reported, other measures such as pregnancy history, outcomes, and demographics have been found to be reliable data sources (Vinikoor et al. 2010).

12.4 Measurement of Gestational Age

One of the measurements most commonly used for categorization of neonatal cohorts is gestational age with the approximate normal gestational length lasting 40 weeks (see Fig. 12.2). Gestational age is determined by the elapse of time following the onset of the mother's Last Menstrual Period (LMP) prior to becoming pregnant. Although precisely defined, obtaining accurate measures of gestational age remains elusive. Some studies use infant birth weight as a proxy for gestational age with ranges of birth weights assigned to various cohorts. Birth weights are nearly universally obtained and much less prone to error than gestational ages within a hospital setting; however, while birth weight is also an important data point to capture, there are limitations to using birth weight to represent fetal development. Infants who are either smaller or larger than appropriate for their gestational age may be at increased risk for poor outcomes; however, in the absence of gestational age information the birth weight loses some predictive power. Furthermore, gestational age is a measure not merely of size, but of the level of infant maturity and physiological development. Thus, gestational age becomes a useful guide in the appropriate prescription of medications and administration of therapies. Preterm

Fig. 12.2 Stages of pregnancy in weeks of gestation

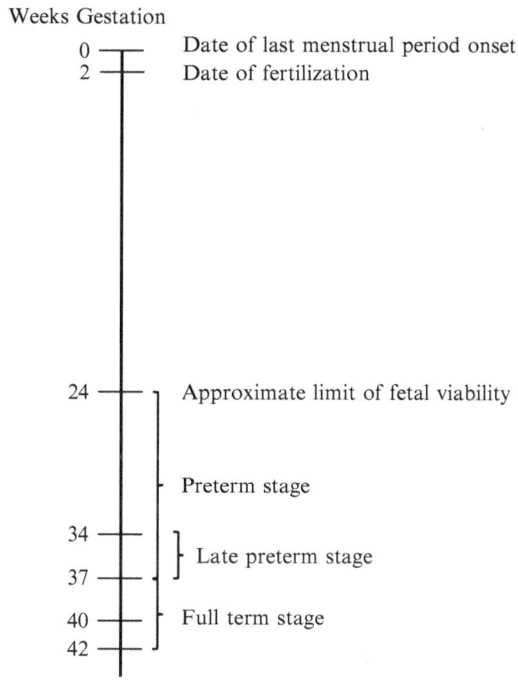

birth status is also based on gestational age rather than birth weight and is useful in anticipating potential neonatal complications. Even babies classified as late preterm infants (34–36 weeks gestation) are at increased risk for neonatal morbidity and mortality and incur greater cost than their full term counterparts (McIntire and Leveno 2008; Kramer et al. 2000; Wang et al. 2004). Thus, gestational age is also a critical risk adjuster in calculating infant mortality and other neonatal outcomes.

Several methods are available to estimate gestational age, yet the most commonly used is dating based upon the mothers LMP. Although defined by LMP onset, menstrual-based gestational age dating is error prone due to irregularities in women's cycle length and errors in accurate recall of related dates (Ananth 2007). Obstetricians are particularly interested in obtaining accurate gestational age estimates to provide care appropriate to the level of fetal development. Obstetric estimates of gestational age are often based upon early ultrasound examinations. Within the first trimester, fetal growth is remarkably consistent, allowing for accurate estimation of gestational age based on ultrasound appearance. As pregnancy progresses beyond the first trimester, growth is less consistent, making gestational age assignment based upon infant appearance on ultrasound less reliable. When maternally reported LMP is dramatically different from ultrasound-based estimates of gestational age, the obstetrician must determine which estimate most likely represents the "true" gestational age of the infant. The obstetric estimate is typically recorded in the mother's medical record.

The Ballard or Dubowitz examinations can be performed on a newborn during the first day of life to estimate gestational age based on appearance, skin texture, motor function and reflexes. Although physical examinations have been validated for the entire newborn population including extremely preterm infants, the examinations are not routinely performed on full term infants (Ballard et al. 1991). Additionally, variation of 1–2 weeks of true gestational age is common, therefore physical examinations are more often used to confirm than to reassign obstetric estimates of gestational age. Estimates derived from physical examinations will typically be recorded in the infant's medical record as they are obtained during postnatal care. Variations in timing and method may result in several discrepant estimates of a single infant's gestational age stored in different charts or in different hospital units. To further complicate matters, United States vital statistics data also contain another representation of gestational age generated using an algorithm that mediates differences between LMP and obstetrician-based estimates using birth weight measures (Martin et al. 2002). Thus investigators planning to incorporate gestational age measures in study analysis should be aware of variations in representations of the data point and should aim for consistency in which representation is utilized.

12.5 Infant Mortality and Preterm Birth

Reduction of infant mortality remains a primary neonatal research objective. Although rates of infant mortality in the United States declined steadily during the 1900s from approximately 100 deaths per 1,000 live births to 6.89 per 1,000 live births, the rate remained steady between 2000 and 2005 much to the chagrin of policy makers and the befuddlement of researchers (Macdorman and Mathews 2008). Furthermore, infant mortality in the United States remains considerably higher than rates in other developed nations. Differences in reporting criteria alone do not explain the United States' relatively low international ranking. Furthermore, considerable disparity exists between Infant Mortality Rates (IMRs) among various racial and ethnic groups with non-Hispanic black women experiencing over double the IMR of the national average. In 2007, the five leading primary causes of infant death were (from greatest to least) congenital malformations, prematurity, sudden infant death syndrome, effects of maternal complications of pregnancy, and accidents (Mathews and MacDorman 2011). While preterm birth may not have been the chief contributor for all cases, 36.0% of infant deaths listed preterm-related causes. Additionally, the leveling off of the United States IMR can be largely explained by the combined effects of an increase in the percentage of preterm infants of any gestational age born during the last decade and the lack of decline in IMR among early preterm infants. Thus preterm birth, one of the most commonly managed conditions in NICU's remains a prime research focus.

Although the quality of global data has much room for improvement, worldwide, an estimated 13 million infants are born preterm each year (9.6% of all births)

(Simmons et al. 2010). Preterm birth ranks first among causes of neonatal death globally, reported as a direct cause in 27% of neonatal deaths. Despite its devastating impact on infant mortality, the causes of preterm birth remain poorly understood. In fact, preterm birth itself is not a singular disease, but is better understood as the common outcome resulting from the convergence of multifactorial etiologic pathways related to biology, genetics, behavior, and socioeconomics (Behrman et al. 2007; DeFranco et al. 2007, 2008; Kogan 1995) Although a number of studies have demonstrated a disproportional occurrence of preterm birth among racial minorities and the socioeconomically disadvantaged, both globally and within the United States, it is a problem that transcends race and socioeconomic status (Goldenberg et al. 2008). Clearly there is a need for additional high quality data ranging from the patient specific, atomic-level to the population level in order to better understand and combat preterm birth.

Along with infant mortality, long-term developmental outcomes are important measures of NICU care. Survival prospects for extremely low birth weight infants (<1,000 g) have improved; however, very preterm infants are at an increased risk for neurologic impairment at school age, as well as for cerebral palsy, blindness, and hearing impairment compared to term infants (Bhutta et al. 2002; Marlow et al. 2005; Ward and Beachy 2003). Long-term outcomes for infants with congenital abnormalities or surgical interventions are also important considerations. There is great opportunity to study the long-term impact of changes in neonatal care on infant and child development as well as to improve the mechanisms by which parents and clinicians are informed of long-term prognoses for preterm infants receiving care. Linking follow-up data to neonatal and obstetrical information would be a powerful tool to in understanding the implications of preterm birth and congenital anomalies and the treatment of these conditions. Currently, long-term follow-up is performed on a limited scale, with a narrow disease or range of birth weight focus. A large-scale data collection effort over a more diverse set of conditions and treatment centers would aid in developing a better understanding of natural neonatal disease histories.

12.6 Neonatal Informatics

Many of the breakthrough advances in neonatology occurred prior to the integration of computerization and medicine. Although a specialized subfield of neonatal or perinatal informatics has been slow to emerge, the hope by many is that the adoption of Information Technology (IT) systems, particularly the EHR and Radiology information systems will "save babies" (Miller and Tucker 2011). There remains great potential for neonatologists and other perinatal care providers to build upon core concepts of Biomedical Informatics to improve perinatal care (Drummond 2009). The American Medical Informatics Association defines Biomedical Informatics as, "the interdisciplinary field that studies and pursues the effective uses of biomedical data, information, and knowledge for scientific inquiry, problem solving and decision making, motivated by efforts to improve human health"

(Shortliffe 2010). Thus, neonatal or perinatal informatics would likewise concern itself with those same efforts related to the domain of infant health spanning from the molecular to the population level.

12.6.1 Computerization of Neonatal Care

The last few years have produced numerous neonatal endeavors building on core informatics concepts. Tools have been integrated into the NICU EHR to improve safety and communication including computerized physician order entry and tools supporting physicians at hand-off (Palma et al. 2011a, b). Other efforts to improve the EHR have focused on human factors, seeking to improve the presentation of EHR data to better aid in the physician decision-making process (Brown et al. 2011). Using metadata produced by the EHR, others have sought to model and improve patient care teams (Gray et al. 2011). Clinical models have been developed to predict mortality among preterm infants (Medlock et al. 2011). At the population health level, global infant trends and projections in mortality rates have been a focus of study (Oestergaard et al. 2011). While tools are being integrated into hospital-based EHRs, other efforts have focused on web-based dissemination of decision support to standardize care processes and to improve the continuity of newborn care (Longhurst et al. 2009; Thornton et al. 2007). Signal analysis has been applied to studying the effects of ground and air transport on infants by interpreting heart rate variability (Karlsson et al. 2012). Additionally, the analyses of heart rate characteristics and EEG have aided in the prediction of neurodevelopmental outcomes among some groups of high acuity newborns (Addison et al. 2009; Spitzmiller et al. 2007). Along with improving care quality and outcomes, it is hoped that technological advances will support cost savings.

With origins in classical epidemiology, spatial analysis is another opportunity for studying and improving perinatal health at a population level (Rushton and Lolonis 1996). Traditionally, spatial analysis techniques have been utilized to identify geographical disease outbreaks. Population-based interventions may be designed around specific disease profiles in areas exhibiting excess disease prevalence (Caley 2004; Richards et al. 1999). Geographical analysis has also been used to study patterns related to pregnancy outcomes including preterm birth (English et al. 2003). Other prime applications for spatial analyses lie in the investigation of associations between environmental or socio-demographic factors and the occurrence of birth defects as well as other perinatal outcomes.

12.6.1.1 A Computerized NICU Tool Implementation

A research project at the Cincinnati Children's Hospital Medical Center (CCHMC) NICU aids in illustrating the implementation process of an automated decision support tool within the inpatient setting. An initial literature review offered scant evidence to support diagnostic or treatment standards for neonates with Tongue-Based Airway Obstruction (TBAO) (Bookman et al. 2012). Motivated to decrease health care costs and improve patient and family satisfaction for the affected population, a principal investigator recognized an opportunity to champion an IT solution in order to aid in timely identification of infants with TBAO. The primary delay in diagnosis resulted from delayed referral of potentially affected neonates to an Ear Nose and Throat (ENT) physician for evaluation. It was recognized that clues for the necessary ENT referral were often embedded in admission notes, which were included in the CCHMC EHR system. Although computerized, the note components were not coded and free-text content did not support a simple querying mechanism in order to identify patients with a suspected TBAO diagnosis. The completed notes, however, were easily obtainable by querying the EHR database for all admission notes associated with NICU patients. Additionally, diagnosis codes indicating TBAO along with dates of admission, referral to ENT specialist, and diagnosis resided in the EHR as coded elements and were easily obtained.

A team of clinical specialists was assembled to generate a starter set of terms for construction of a Natural Language Processing (NLP) paradigm that would trigger suspicion of TBAO (requiring an ENT referral for evaluation) in a neonate. The preliminary NLP tool retrospectively parsed thousands of transcribed admission notes generated during infant NICU admission for patients whose ultimate TBAO status had already been determined. The set of clinically identified terms were ranked according to their positive predictive value in identifying neonates who were diagnosed with TBAO. The NLP algorithm was subsequently refined until an optimal set of search terms had been identified at which point plans were made to automate and integrate the algorithm within the CCHMC research data warehouse for testing and validation. Upon final validation, the pilot NLP tool will be integrated into the clinical workflow resulting in the following processes: (1) A neonate is admitted to the NICU, (2) within 12 h of admission, an admission note will be generated and signed by the attending physician within the EHR, (3) the scheduled automated alerting tool will be executed and will scan each NICU admission note, (4) a report of patients recommended by the tool for specialist referral will be generated, (5) each referred patient will be evaluated for TBAO and when present, the TBAO diagnosis will be documented. The cycle time between admission and diagnosis for the TBAO population using the new process will be compared to the cycle time pre-implementation.

12.6.1.2 A Perinatal Spatial Analysis

An example of an analysis conducted in Hamilton County, Ohio demonstrates how spatial analysis techniques can be applied to the perinatal realm of the public health domain. Initiatives to reduce preterm birth are often inefficient with broad, rather than geographically focused programmatic implementations of specific interventions. Hamilton County, Ohio, reports an approximate preterm birth rate of 14%, however the distribution and risk factors for preterm birth are not

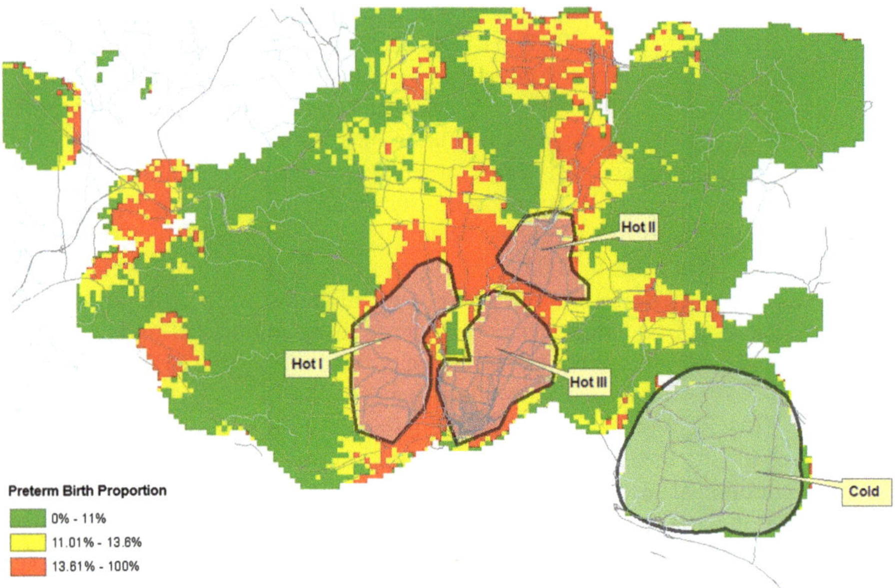

Fig. 12.3 Spatial analysis of preterm birth in Hamilton County, Ohio indicating three hotspots of activity with variation in preterm birth risk profiles

uniform across the metropolitan area. A spatial analysis was conducted to identify regions with the highest rates of preterm birth as well as regions of low preterm birth. Additionally, risk factors in each of the identified regions were compared in order to aid in the implementation of efficient preterm birth intervention strategies (South et al. 2012). Vital statistics data including preterm birth status, demographics, and maternal clinical characteristics were collected by birth hospitals and reported to the Ohio Department of Health. For births that occurred in the State of Ohio, each mother's reported residence address was geocoded into a latitude-longitude coordinate pair. These data were obtained by investigators and linked with publicly available socioeconomic data obtained from the decennial census. Socioeconomic measures that represented the census block of residence as determined by the corresponding geocoded location were linked to each birth record.

Figure 12.3 represents three regions of high preterm birth and one region of low preterm birth within Hamilton County. Subsequent analysis of each of the four regions identified preterm birth risk factors including advanced maternal age, diabetes, less than high school education, chronic hypertension, pregnancy induced hypertension, previous preterm birth, short inter-pregnancy interval, low income, low maternal weight, smoking, and African American race. Attributable risk analysis highlighted variation in the number of cases of preterm birth resulting from the identified risk factors. For example, although the risk associated with the occurrence of a previous preterm birth contributed to the greatest number of preterm birth cases in the county as a whole, two of the high-risk regions were most impacted by maternal smoking, while the third was most impacted by maternal underweight status. The preterm births occurring in the region of low risk were mostly attributed to advanced maternal age.

Particularly in geographic regions with patterns of uneven preterm birth distribution, the spatial analysis approach to preterm birth analysis supports the targeting of interventions within highly affected neighborhoods and the prioritization of risk factors to target. In general, the application of a spatial analysis approach potentially enables tailored interventions resulting in the development of more efficient and effective public health strategies.

12.6.2 Integration of Perinatal Data

A primary challenge to perinatal research is integrating relevant data sources. Although pertinent data exist at the time of birth, the newborn infant is essentially issued a blank medical record with a brand new medical record number (MRN). Maternal and fetal care provided up until birth, as well pregnancy complications, are recorded in the mother's prenatal record. Information related to the mother's birth history and general medical history may be summarized in the prenatal record, but may also be detailed in separate maternal charts. Likewise, the paternal medical history is stored in the father's various medical records. A number of barriers prevent the seamless aggregation of data into the newborn's medical record including the transition between care institutions, transition between provider types, concerns for privacy, the absence of an established medical home, and technical limitations (see Chap. 1, Pediatric Electronic Health Records and Research and Chap. 5, Protecting Privacy in the Child's Electronic Health Record).

While information pertaining to the newborn delivery is captured at the birth hospital, prenatal data is typically collected in a clinic setting. Although prenatal data may be obtained to guide delivery care, it is often in a non-computerizable (i.e. faxed, photocopied, or scanned) or summarized state, which does not facilitate integration with the hospital EHR. Transfers of a newborn to a higher acuity care center may further complicate data integration, particularly when the birth facility and transfer facility belong to different care networks or do not share an EHR system resulting in incongruent infant medical record numbers. Even the use of the same base EHR system by two facilities does not guarantee seamless exchange of patient data, even from a technological perspective. Additionally, large quantities of data may be missing and unobtainable in medical records belonging to infants of mothers who received scant or no prenatal care.

During pregnancy, care is predominantly delivered by an obstetrician or midwife. Although the health of both fetus and mother are of concern of the care provider, data capture is associated with the mother's MRN. Following delivery, newborn care delivered by neonatologists and pediatricians is recorded in the infant's own medical record represented by an MRN assigned to the infant. The transition of care providers accompanying the transition from fetal to infant stage results in a separation of data, although some modern EHRs are built to facilitate the manual migration of relevant data from the maternal to infant chart during the course of care.

As described in Chap. 5 on protecting privacy, concerns for maintaining data confidentiality must be satisfied when integrating newborn data with parental health information, including data pertaining to the newborn's prenatal care. While confidentiality remains a consideration during the course of administering clinical care, considerations for privacy are particularly stringent during the course of conducting research.

A final challenge to integration of perinatal data is the absence of a newborn's established medical home, which complicates the acquisition of follow-up data related to ultimate outcomes. Although the birth hospital or NICU care providers may inquire as to the intended primary care provider for the soon to be discharged infant, that patient provider relationship has often not been established. Furthermore, data integration once again is challenged by transitions in care providers and institutions.

Even if all other barriers are mitigated, linking newborn records remains a particular challenge, as many of the identifiers traditionally used by record matching algorithms are absent or unreliable in the newborn medical record. The newborn does not have a unique social security number or any other unique identifier that can be used in linking, nor in many cases does the infant have a name. Often the name listed on the newborn's medical record is represented by the same last name as the infant's mother and "baby boy" or "baby girl", if a first name has not yet been chosen. The infant name may be subsequently modified by providing a first name or by adjusting the last name, which further complicates linking. Probabilistic and deterministic matching of newborn records as well as newborn-to-mother records are still possible using characteristics such as gender, date of birth/date of delivery, birth/delivery hospital, insurance type, and address elements such as street number and zip code. Nevertheless, in cases of multiple gestation births, detailed characteristics such as infant birth weight may be necessary to break ties between potentially matched records.

12.6.3 Perinatal Population Systems

An example of a successful implementation for supporting population health research is the Pregnancy to Early Life Longitudinal (PELL) data system. The public-private collaborative partnership between the Maternal and Child Health Department at the Boston University School of Public Health, the Massachusetts Department of Public Health, and the Centers for Disease Control and Prevention has enabled researchers in Massachusetts to investigate a broad range of topics related to pregnancy, premature birth, and infant mortality (Barfield et al. 2008). The PELL system is a longitudinally linked data set of mothers and their children that allows researchers to track mothers and children over time. The core of the data system is a deterministically and probabilistically linked public health data set consisting of vital statistics, health services utilization data, and program participation data (Shapiro-Mendoza et al. 2006). The data core can subsequently be linked to additional resources such as an early intervention services database to evaluate eligibility and referrals of children with developmental delay to the early intervention program (Clements et al. 2008). In addition to monitoring preterm birth rates and associated neonatal morbidity, the system has been used to conduct a variety of analyses including the evaluation of early intervention programs, to evaluate maternal outcomes following elective Cesarean delivery, and for surveillance of autism. PELL also offers a conceptual model for other states hoping to conduct a similar range of analyses, including the opportunity to perform analysis related to specific mothers over multiple subsequent pregnancies.

In Wisconsin, a different approach to supporting perinatal research has taken the form of PeriData.Net. The web-based database was developed in partnership between the Wisconsin Association for Perinatal Care and the University of Wisconsin Milwaukee/Center for Urban Population Health. Over 200 hospitals utilize the system to house hospital-supplied data corresponding to the state birth certificate fields. In addition to enabling electronic submission of birth certificate data, the system allows institutions to benchmark patient outcomes against other participants and to supplement the standard set of data fields with custom fields intended to support specific initiatives (Costakos 2006). Additionally, the system provides for standardized data capture facilitating the potential coordination of perinatal focused efforts between participating partners.

Pediatrix Medical Group, a national provider of maternal-fetal, newborn, and pediatric subspecialty physician services, has established another model with a shared clinical information system. Their large data system representing over 100 NICUs has enabled reports on national NICU medication use, morbidity and mortality associated with intrauterine growth restriction, and has supported quality improvement in areas such as reducing catheter related bloodstream infections (Clark et al. 2006; Garite et al. 2004).

Another national scale database including data from over 4,000 hospitals is the Kids Inpatient Database (KID). KID was established by the Healthcare Cost and Utilization Project (HCUP) sponsored by the Agency for Healthcare Research and Quality (AHRQ) to include a broad range of data collected during children's inpatient hospital encounters. KID includes between two and three million hospital discharges for children 20 years old and younger, a significant proportion of whom were newborns. The data including diagnoses, procedures, patient demographics, charges, length of stay, and hospital characteristics have been used to for analysis of national health care utilization, charges, quality, and outcomes (HCUP Kids' Inpatient Database (KID)).

12.7 Neonatal Research Networks

Many of the most clinically significant and costly conditions to treat in the neonatal population are relatively rare. For example, approximately 2.0% of babies in the United States are very preterm, born at less than 32 weeks gestation. Necrotizing enterocolitis (NEC) affects about 10% of the infants in the very preterm category. A single clinical site may not manage enough cases of NEC to power a statistically significant analysis of care interventions. For this purpose, multi-center research networks have been established to conduct clinical trials and observational studies in neonatal medicine.

The Vermont Oxford Network (VON) founded in 1986 and National Institute of Child Health and Human Development (NICHD) Neonatal Research Network founded in 1988 are two of the longest running quality improvement and research networks. VON maintains a database of information regarding high-risk newborn care and outcomes contributed by its more than 900 participating NICUs worldwide. The NICHD network also maintains a registry referred to as the generic database (GDB) which collects more detailed data than the VON database. In addition, the NICHD network coordinates research protocols for the 17 participating centers around drug therapies and therapies to manage neonatal conditions including sepsis, intraventricular brain hemorrhage, and chronic lung disease. Both networks focus primarily on very preterm infants with birth weights less than 1,000–1,500 g, as well as select other infants with specific conditions of high interest to researchers. These network databases enable the monitoring of disease trends and changes in neonatal survival and outcomes, while providing data for hypothesis generation and sample size calculation.

12.8 Neonatal Terminologies

Although the NICHD and VON network collect similar measures from similar populations, the data dictionaries used by each database are not compatible. For example, both networks request information about retinopathy of prematurity (ROP). Vermont Oxford asks if a retinal exam was done and if so, requests the worst ROP stage documented on a scale of 0–5. Subsequently, yes or no responses are requested to inquiries of whether or not the ROP was treated with an anti-WEGF drug or surgery. The GDB also asks if ROP was diagnosed; however, rather than asking the ROP stage, the GDB requests to know if ROP reached stage 3 or worse. In addition to requesting information about treatment with anti-WEGF drugs, information about the specific intervention therapies of retinal ablation, scleral buckle, or vitrectomy is collected. For another example, both networks request data about necrotizing enterocolitis (NEC). The Vermont Oxford network asks for a yes or no response to the question of whether necrotizing enterocolitis occurred and if NEC was present, if surgery was performed. The GDB data requests NEC status as a single question including the options of "Absent/Suspect," "Proved, no surgery," or "Proved, surgery." In cases where definitions of the two networks are approximate, even in the absence of exact one-to-one mappings between data elements, it may be possible to translate data from one format to the other; however, incongruent granularity may result in information loss.

Although clinical terminologies such as SNOMED CT were designed to enable data standardization and exchange, existing terminologies do not adequately represent the pediatric domain, particularly concepts unique to neonatology. Insufficient established standard terminology has lead individual research networks to establish their own data definitions to achieve their research objectives. As a consequence the ability to pool or share data between research networks is severely limited. Furthermore, leveraging complimentary datasets such as the outcomes focused Children's Hospitals Neonatal Database (CHND), which captures therapies and outcomes of infants referred to tertiary centers, is also frustrated.

Development of the Neonatal Research Networks Terminology (NRNT) has begun in order to enable the exchange of perinatal data among research networks (Padula 2012). Modeled after the structure of SNOMED CT and developed by subject matter experts, NRNT contains mappings to clinical findings, observations, and procedures used in current neonatal research networks. Serving as a hub, NRNT will facilitate meta-analyses as well as studies spanning multiple research networks. By providing the structure of a central hierarchy of neonatal concepts NRNT will allow the translation of data from one network to another, as well as the aggregation of data for higher powered studies.

References

Addison K, et al. Heart rate characteristics and neurodevelopmental outcome in very low birth weight infants. J Perinatol. 2009;29(11):750–6.

Allen MC, Donohue PK, Dusman AE. The limit of viability–neonatal outcome of infants born at 22 to 25 weeks' gestation. N Engl J Med. 1993;329(22):1597–601.

Ananth CV. Menstrual versus clinical estimate of gestational age dating in the United States: temporal trends and variability in indices of perinatal outcomes. Paediatr Perinat Epidemiol. 2007;21 Suppl 2:22–30.

Baines MA. Excessive infant-mortality; how can it be stayed? A paper contributed to the National Social Science Association, London meeting. To which is added a short paper, reprinted from the Lancet [1861] on Infant-alimentation; or, Artificial feeding, as a substitute for breast-milk, considered in its physical and social aspects. London: Churchill; 1862. 20 p.

Baker JP. The incubator controversy: pediatricians and the origins of premature infant technology in the United States, 1890 to 1910. Pediatrics. 1991;87(5):654–62.

Ballard JL, et al. New Ballard Score, expanded to include extremely premature infants. J Pediatr. 1991;119(3):417–23.

Barfield WD, et al. Using linked data to assess patterns of early intervention (EI) referral among very low birth weight infants. Matern Child Health J. 2008;12(1):24–33.

Behrman RE, Butler AS, Institute of Medicine (U.S.). Committee on Understanding Premature Birth and Assuring Healthy Outcomes. Preterm birth: causes, consequences, and prevention. Washington, DC: National Academies Press; 2007. xvi, 772 p.

Bhutta AT, et al. Cognitive and behavioral outcomes of school-aged children who were born preterm: a meta-analysis. JAMA. 2002;288(6):728–37.
Bookman LB, et al. Neonates with tongue-based airway obstruction: a systematic review. Otolaryngol Head Neck Surg. 2012;146(1):8–18.
Brown P, et al. Variations in faculty assessment of NICU flowsheet data: implications for electronic data display. Int J Med Inform. 2011;80(7):529–32.
Caley LM. Using geographic information systems to design population-based interventions. Public Health Nurs. 2004;21(6):547–54.
Clark RH, et al. Reported medication use in the neonatal intensive care unit: data from a large national data set. Pediatrics. 2006;117(6):1979–87.
Clements KM, et al. Maternal socio-economic and race/ethnic characteristics associated with early intervention participation. Matern Child Health J. 2008;12(6):708–17.
Costakos DT. Of lobsters, electronic medical records, and neonatal total parenteral nutrition. Pediatrics. 2006;117(2):e328–32.
DeFranco E, Teramo K, Muglia L. Genetic influences on preterm birth. Semin Reprod Med. 2007;25(1):40–51.
DeFranco EA, et al. Area-level poverty and preterm birth risk: a population-based multilevel analysis. BMC Public Health. 2008;8:316.
Drummond WH. Neonatal informatics–dream of a paperless NICU. Part one: the emergence of neonatal informatics. NeoReviews. 2009;10:e480–7.
English PB, et al. Changes in the spatial pattern of low birth weight in a southern California county: the role of individual and neighborhood level factors. Soc Sci Med. 2003;56(10):2073–88.
Finster M, Wood M. The Apgar score has survived the test of time. Anesthesiology. 2005;102(4):855–7.
Garite TJ, Clark R, Thorp JA. Intrauterine growth restriction increases morbidity and mortality among premature neonates. Am J Obstet Gynecol. 2004;191(2):481–7.
Goldenberg RL, et al. Epidemiology and causes of preterm birth. Lancet. 2008;371(9606):75–84.
Gould JB, et al. Incomplete birth certificates: a risk marker for infant mortality. Am J Public Health. 2002;92(1):79–81.
Gray JE, et al. Using digital crumbs from an electronic health record to identify, study and improve health care teams. AMIA Annu Symp Proc. 2011;2011:491–500.
HCUP Kids' Inpatient Database (KID). Healthcare Cost and Utilization Project (HCUP) 2006 and 2009. Rockville: Agency for Healthcare Research and Quality. www.hcup-us.ahrq.gov/kidoverview.jsp
Hsia DC, et al. Accuracy of diagnostic coding for Medicare patients under the prospective-payment system. N Engl J Med. 1988;318(6):352–5.
Iezzoni LI, et al. Comorbidities, complications, and coding bias. Does the number of diagnosis codes matter in predicting in-hospital mortality? JAMA. 1992;267(16):2197–203.
Karlsson BM, et al. Sound and vibration: effects on infants' heart rate and heart rate variability during neonatal transport. Acta Paediatr. 2012;101(2):148–54.
Kogan MD. Social causes of low birth weight. J R Soc Med. 1995;88(11):611–15.
Kramer MS, et al. The contribution of mild and moderate preterm birth to infant mortality. Fetal and Infant Health Study Group of the Canadian Perinatal Surveillance System. JAMA. 2000;284(7):843–9.
Longhurst C, Turner S, Burgos AE. Development of a Web-based decision support tool to increase use of neonatal hyperbilirubinemia guidelines. Jt Comm J Qual Patient Saf. 2009;35(5):256–62.
Macdorman MF, Mathews TJ. Recent trends in infant mortality in the United States. NCHS Data Brief. 2008;9:1–8.
Marlow N, et al. Neurologic and developmental disability at six years of age after extremely preterm birth. N Engl J Med. 2005;352(1):9–19.
Martin JA, et al. Births: final data for 2001. Natl Vital Stat Rep. 2002;51(2):1–102.
Mathews TJ, MacDorman MF. Infant mortality statistics from the 2007 period linked birth/infant death data set. Natl Vital Stat Rep. 2011;59(6):1–30.

McIntire DD, Leveno KJ. Neonatal mortality and morbidity rates in late preterm births compared with births at term. Obstet Gynecol. 2008;111(1):35–41.
Medlock S, et al. Prediction of mortality in very premature infants: a systematic review of prediction models. PLoS One. 2011;6(9):e23441.
Miller AR, Tucker CE. Can health care information technology save babies? J Pol Econ. 2011;119(2):289–324.
Oestergaard MZ, et al. Neonatal mortality levels for 193 countries in 2009 with trends since 1990: a systematic analysis of progress, projections, and priorities. PLoS Med. 2011;8(8):e1001080.
Padula M. Neonatal research network terminology harmonization: a formative research initiative of the National Children's Study. National Children's Study Metadata Repository Workshop; 2012.
Palma JP, et al. Neonatal informatics: computerized physician order entry. NeoReviews. 2011a;12:393–6.
Palma JP, Sharek PJ, Longhurst CA. Impact of electronic medical record integration of a handoff tool on sign-out in a newborn intensive care unit. J Perinatol. 2011b;31(5):311–17.
Pearson HA, et al. Committee report: American Pediatrics: milestones at the millennium. Pediatrics. 2001;107(6):1482–91.
Richards TB, et al. Geographic information systems and public health: mapping the future. Public Health Rep. 1999;114(4):359–73.
Rushton G, Lolonis P. Exploratory spatial analysis of birth defect rates in an urban population. Stat Med. 1996;15(7–9):717–26.
Saigal S, Doyle LW. An overview of mortality and sequelae of preterm birth from infancy to adulthood. Lancet. 2008;371(9608):261–9.
Schoendorf KC, Branum AM. The use of United States vital statistics in perinatal and obstetric research. Am J Obstet Gynecol. 2006;194(4):911–15.
Shapiro-Mendoza CK, et al. Risk factors for neonatal morbidity and mortality among "healthy," late preterm newborns. Semin Perinatol. 2006;30(2):54–60.
Shortliffe EH. Biomedical informatics in the education of physicians. JAMA. 2010;304(11): 1227–8.
Simmons LE, et al. Preventing preterm birth and neonatal mortality: exploring the epidemiology, causes, and interventions. Semin Perinatol. 2010;34(6):408–15.
South AP, et al. Spatial analysis of preterm birth demonstrates opportunities for targeted intervention. Matern Child Health J. 2012;16(2):470–8.
Spitzmiller RE, et al. Amplitude-integrated EEG is useful in predicting neurodevelopmental outcome in full-term infants with hypoxic-ischemic encephalopathy: a meta-analysis. J Child Neurol. 2007;22(9):1069–78.
Thornton SN, et al. Neonatal bilirubin management as an implementation example of interdisciplinary continuum of care tools. AMIA Annu Symp Proc. 2007;2007:726–30.
United Nations, Department of Economic and Social Affairs, Population Division. World population prospects: the 2010 revision, CD-ROM edition; 2011.
Vinikoor LC, et al. Reliability of variables on the North Carolina birth certificate: a comparison with directly queried values from a cohort study. Paediatr Perinat Epidemiol. 2010;24(1): 102–12.
Wang ML, et al. Clinical outcomes of near-term infants. Pediatrics. 2004;114(2):372–6.
Ward RM, Beachy JC. Neonatal complications following preterm birth. BJOG. 2003;110 Suppl 20:8–16.
Wier ML, Pearl M, Kharrazi M. Gestational age estimation on United States livebirth certificates: a historical overview. Paediatr Perinat Epidemiol. 2007;21 Suppl 2:4–12.

Chapter 13
Informatics Support of Outcome Measures

Keith Marsolo

Abstract Outcome measures enable the comparison of a patient's health status before and after an intervention. Such interventions can include surgical procedures, drug treatments, vaccines or changes in nursing care. Outcome measures are used heavily in support of quality improvement initiatives and comparative effectiveness research. Quality improvement involves the setting of goals and then comparing the performance of that organization (or clinic or provider) with those goals, identifying gaps in expectations. The knowledge of existing gaps is used to improve the processes of care. With comparative effectiveness research, the outcomes of several interventions are compared across large populations. Examples include the study of efficacy of two or more drugs and the comparison of frequency of adverse reactions to a drug in patients with different genotypes for drug metabolizing enzymes. To allow the widespread use of outcome measures in the age of the electronic health record, it is necessary to automate the process of computing them. This task is more challenging than it may seem. To accurately compute an outcome measure, it is necessary to create an operational definition that states precisely the algorithm used to compute the value of the measure. This involves the clear definition of specified data elements, data sources, and logic on how to handle unexpected conditions. It also involves close consultation with clinicians in order to ensure that the measure is actually calculating something meaningful. Only then will they be useful for quality improvement and comparative effectiveness research.

K. Marsolo, Ph.D. (✉)
Department of Pediatrics, University of Cincinnati College of Medicine, Cincinnati, OH, USA
e-mail: keith.marsolo@cchmc.org

J.J. Hutton (ed.), *Pediatric Biomedical Informatics: Computer Applications in Pediatric Research*, Translational Bioinformatics 2, DOI 10.1007/978-94-007-5149-1_13,

13.1 Introduction

Chronic illness in children can negatively affect their current health and development, as well as their future health as adults (Forrest and Riley 2004). Since these conditions occur uncommonly, most clinical research questions must be addressed by networks of health care providers in order to accrue enough patients to conduct a study. This is particularly true of outcomes and comparative effectiveness research where clinicians identify critical healthcare knowledge gaps, design the studies, and utilize the findings to improve care of their patients. It is advantageous, from a cost and resource perspective, if the data to support this research can be collected in the patients' electronic health record during the process of clinical care.

Although currently optimized for clinician productivity over data accuracy, electronic health record (EHR) interfaces have many of the real-time data validation capabilities that are found in dedicated clinical trials electronic data entry systems—e.g., range checks and mandatory entry requirements. In addition to capturing more clinical data in coded format directly at the point of care, natural language processing can extract information from dictated reports and notes, such as discharge and procedure notes, pathology and radiology reports, and even death certificates (Chapman and Cohen 2009) (see Chap. 10, Natural Language Processing – Applications in Pediatric Research). This leads to the vision of capturing high-quality, research-grade data during standard clinical care (data-in-once) that can be shared over a secure, HIPAA-compliant collaborative research network for a wide range of secondary uses, including comparativeness effectiveness research and continuous clinical quality improvement. Unfortunately, evidence exists that current EHR documentation practices may not result in complete, consistent, and unbiased data collection (Weiner and Embi 2009; Aronsky and Haug 2000; Brennan and Stead 2000; Hogan and Wagner 1997; Stein et al. 2000).

To combat this, there are six attributes necessary for a data system to enable learning from every patient in a study: (1) a structured process to define and develop measures; (2) a variety of ways to move data in and out of the data system, ensuring adaptability to variations in work processes; (3) standards for content and messaging (Chap. 2, Interfaces and Data Transmission Standards); (4) standardized and user-configurable reports; (5) assurance of data quality, including data validity checks and skip checks, auditing features to track data entry, change and deletion events, measures to assess the completeness of forms and of population data, and procedures for error correction; and, (6) security and confidentiality, including features to guard data integrity, and policies and procedures to control and audit access to protected health information. Chapter 8 (EHR-linked Registries for Studies of Populations) discusses design and implementation of registries that have many of these attributes and are supported by networks of health care providers. Particular attention is paid to methods of data entry, data quality control, and data sharing among members of a network. This chapter focuses on uses of registries to support the definition and calculation of outcome measures, which underlie comparative effectiveness research and quality improvement (QI) initiatives.

13.2 Measure Development and Data Collection

Outcome measures support quality improvement initiatives and comparative effectiveness research. Outcome measures allow the comparison of a patient's health status before and after an intervention, which can range from the implantation of a medical device, to the administration of a new drug or to a change in a patient's after-visit instructions. QI requires the setting of targets, or goals. In a QI initiative, the performance of the organization, clinic or provider (the unit performing the work) is compared against the specified target. Gaps in performance are identified and the knowledge of those gaps is used to improve the processes of care. Comparative effectiveness research, on the other hand, is used to compare the outcomes of more than one intervention in large patient populations. Examples include comparisons of efficacies of two or more drugs and comparisons of frequency of adverse reactions to a drug in patients with different genotypes for drug metabolizing enzymes.

In order to properly and reliably calculate an outcome measure, it is necessary to have an operational definition that states precisely the algorithm needed to compute it. Before an operational definition can be created, the relevant patient population, data elements, data sources, etc., all must be identified. As a result, the process for development of measures, where the EHR is the primary data source, tends to have the following flow:

1. Identify the condition to be improved or intervention to be tested. Define the outcomes of interest as well as the population to be studied.
2. Identify the clinical workflow that will generate the data needed to compute the outcome measure. Within the EHR, identify the corresponding electronic workflow that will be used to capture the relevant data.
3. If the EHR screens needed to collect the data do not exist, it is necessary to create them.
4. Develop the technical documentation that defines the algorithm needed to compute the outcome measure. Include the required data elements, data source(s), location of the data elements in those data sources, etc. Also include logic on how to handle unexpected conditions. Code and validate the algorithm.
5. Put the measure into production, publishing and/or visualizing the results.

From an informatics perspective, steps 2 and 5, which go hand in hand, are often the most challenging. The EHR typically provides a number of different ways to document the same information. In order to generate accurate and reliably results, it is necessary to either (a) develop an algorithm that covers all of these different alternatives, which can be time consuming and expensive to maintain, or (b) work with clinicians to standardize on a single workflow for entering the required data. While it requires changes in the behavior of clinicians, standardizing the workflow is usually a more tractable approach to the problem.

The above process is based on an approach promoted by James and others (James 2003), that is employed to first develop outcome measures and then to

Table 13.1 Example measures related to model care that are reported monthly by the ImproveCareNow network

Domain	Measures
Model care	% of patients with a documented visit within the last 6 months
	% of visits where, when treatment with thiopurine is started, TPMT has been measured
	% of patients treated with infliximab, where the dose of infliximab prescribed is at least the dose recommended by the infliximab protocol
	% of visits where, when disease activity is moderately or severely active and the patient is being treated with a thiopurine, the 6-TGN level has been measured within the previous 90 days
	% of patients treated with infliximab (at least 5 mg/kg)
	% of visits where, when disease activity is moderately or severely active and the patient is being treated with infliximab, the infliximab trough level has been measured within the previous 180 days from infusion date
	% of patients treated with methotrexate where the dose is at least 10 mg/m^2 per week

implement a system to capture the discrete data elements needed to calculate these outcomes. This approach starts with desired end-result and works backward to front-line data collection and data flow. The highly structured process results in a parsimonious set of data, containing only those data elements that lead to actionable reports that comprehensively cover the entire process of care. Measures to monitor the process and outcomes of care are developed based on evidence-based recommendations. From those recommendations about care delivery, a list of outcome measures is defined. Table 13.1 shows a set of example measures that are reported monthly to participants in the ImproveCareNow network. As discussed in Chap. 8, this network of pediatric gastroenterologist aims to improve the care of children with inflammatory bowel disease and is comparing the effectiveness of various treatments.

Robust operational definitions are crucial in order to successfully automate the calculation of an outcome measure. While it is relatively easy for a human to reason through an ambiguous definition to develop a conclusion, it is difficult, if not impossible for a computer to do so. As a result, operational definitions should be as specific as possible. At a minimum, they should include the following:

- Patient population and subset of interest for the measure (e.g., population – all patients with Inflammatory Bowel Disease; subset of interest – patients with a visit in the past 30 days)
- Method for determining the subset of interest (e.g., how to determine whether a patient had a visit in the past 30 days)
- Data source and data elements needed to compute the measure
- Formula for computing the measure (numerator and denominator)
- Explanation on how to deal with missing or extraneous data elements

Table 13.2 Typical elements of an operational definition

1. Measure
2. Description and rationale
3. Population definition (inclusions/exclusions)
4. Data source(s)
5. Data collection plan
6. Numerator
7. Denominator
8. Formula to compute the measure
9. Analysis plan and frequency of reporting
10. Reporting venues
11. Limitations
12. Experts/Resources
13. Revision history

Table 13.2 provides another listing of the elements that are typically included in an operational definition. This process is relevant to development of computable outcome measures to support both comparative effectiveness research and quality improvement.

13.2.1 Designing Data Capture and Monitoring Data Entry Compliance

Tools for the collection and import of clinical data are discussed in Chap. 8. The focus of this section is on the capture of outcomes data in the EHR. Simultaneous advances in EHRs' functionality, graphical user interface designs, and mobile/wireless hardware have expanded the sophistication of tools that support direct entry of detailed clinical data into the EHR as discrete coded data elements (Los et al. 2005; Venema et al. 2007). As was demonstrated in Chap. 8, EHR screens can be designed to do this.

Even with screens that allow data to be captured discretely, it is still necessary to monitor data entry compliance to ensure that the forms are actually being completed and the necessary data are being captured. Otherwise, it is difficult, if not impossible to compute the outcome measures. Many tools for electronic data capture that are used for clinical research have mandatory fields that must be completed before a form can be submitted. With EHRs, however, mandatory fields are not typically used. If a patient left suddenly in the middle of a visit before all of the necessary data was collected, having a mandatory field could prevent the encounter from being closed, which could potentially affect clinical operations. To improve data entry compliance, it is important to provide timely feedback to users. The data are extracted from the EHR and metrics are computed to determine information such as the percentage of all visits with a completed record, where "completeness" is dictated by the operational definition. By evaluating the data entry compliance, analysts can work with the clinic to determine whether the lack of data entry is due to factors like inefficient clinical flow or an unclear operational definition.

13.2.2 Clinical Care Process Alignment

It is extremely important to incorporate registry data collection into routine clinical workflow and to use an approach that links outcome measures to data elements that are collected during clinical care (James 2003). It is advisable for sites within a network to adopt a structured clinical encounter form similar to the form shown in Fig. 13.1. Data elements and the design of the form to collect them are specific for each disease or project. The elements in Fig. 13.1 are collected by the ImproveCareNow network discussed in Chap. 8.

Simply adopting structured clinical forms is not sufficient to achieve highly reliable data collection. Achieving this goal requires changes in the behavior of physicians and other health care professionals. In this case, a desired process outcome would be identified – appropriate documentation collected at >98% of all visits, for instance. Drivers of this outcome would be identified, such as the creation of a standard process for collecting the data and making clinicians aware of its existence. Interventions to support these drivers would include education and training, a visual cue to complete the form during the clinic visit, and a process to identify incomplete forms. This approach provides a conceptual framework that relates the desired outcome, which is that appropriate documentation is collected at >98% of visits, to changes designed to support reliable use of the encounter forms. Annotated run charts can be created to show the relationship of changes in workflow to process performance, illustrating that the workflow changes successfully achieved the desired results.

13.2.3 Data Quality

Data quality is a function of completeness and accuracy. Data completeness can be measured as the proportion of patients included or as a proportion of required data elements that are entered. It is important that data are present for all variables, visits, and patients. It is also important that the data items are valid entries. Systematic error may occur, for example, if a data element does not have an operational definition that is commonly agreed upon, accepted, or used. Random error may occur through inaccurate data entry. Observations with larger error terms can be detected as outliers during data analysis. Software modules can be developed to clean and validate study data that have been uploaded to a registry or data warehouse.

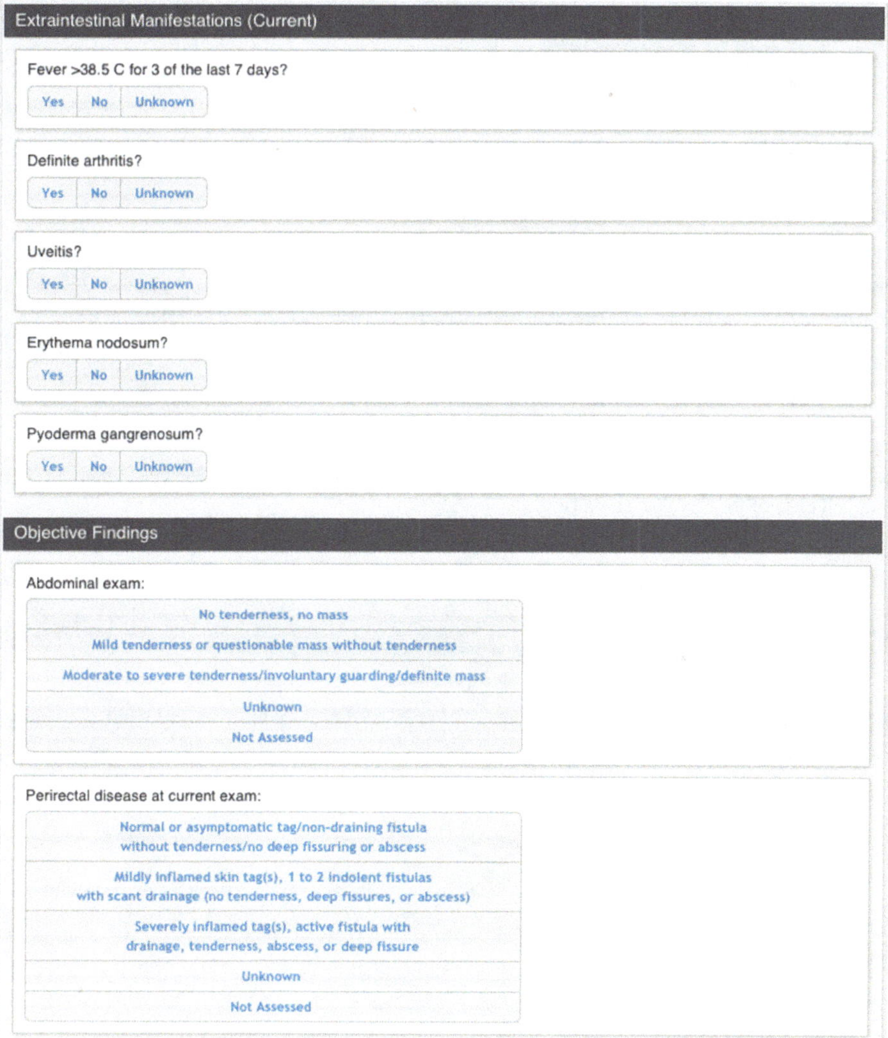

Extraintestinal Manifestations (Current)

Fever >38.5 C for 3 of the last 7 days?
Yes | No | Unknown

Definite arthritis?
Yes | No | Unknown

Uveitis?
Yes | No | Unknown

Erythema nodosum?
Yes | No | Unknown

Pyoderma gangrenosum?
Yes | No | Unknown

Objective Findings

Abdominal exam:
No tenderness, no mass
Mild tenderness or questionable mass without tenderness
Moderate to severe tenderness/involuntary guarding/definite mass
Unknown
Not Assessed

Perirectal disease at current exam:
Normal or asymptomatic tag/non-draining fistula without tenderness/no deep fissuring or abscess
Mildly inflamed skin tag(s), 1 to 2 indolent fistulas with scant drainage (no tenderness, deep fissures, or abscess)
Severely inflamed tag(s), active fistula with drainage, tenderness, abscess, or deep fissure
Unknown
Not Assessed

Fig. 13.1 A section of a clinic form focused on patients with Inflammatory Bowel Disease. The form is intended to be incorporated into the EHR in order to collect the data elements necessary to calculate outcome measures for patients with IBD

13.2.4 Reporting and Visualization

Feedback to the end-user on an outcome measure can consist of exception reports and summary data in tabular format, as well as visualization tools such as run charts and bar charts. This is often dependent on the needs of the user. The results for a particular measure or outcome are displayed on a chart, optionally with

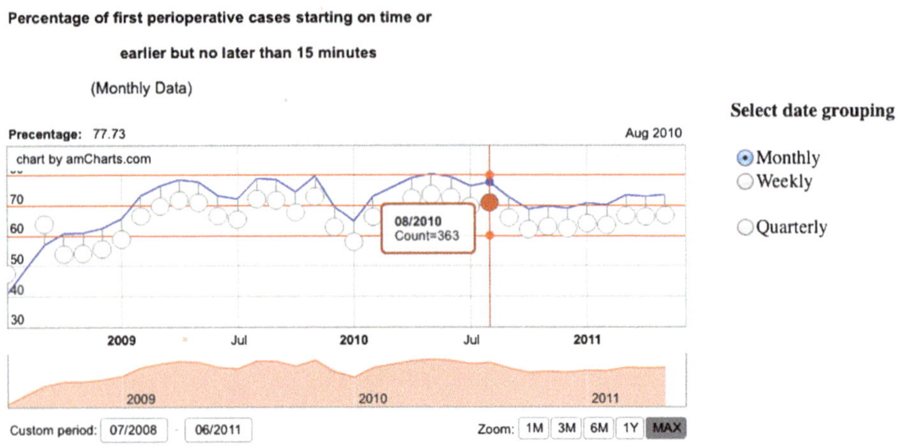

Fig. 13.2 A run chart of an outcome measure. This measure is used to track whether the first surgical case of the day started on time

Fig. 13.3 A run chart that shows an outcome measure stratified by an attribute, in this case Room Type. These attributes will be specific to the outcome being measured

population-level statistics as well as comparators for the individual site. This enables each measure to be tracked over time, with comparisons of single or multiple measures against the overall group or a subset of sites. Figure 13.2 shows an example of a run chart, which, depending on the system, may allow users to drill down into a point to see patient-level data or allow them to group the values based on different time periods (monthly, weekly, quarterly, etc.). Figure 13.3 shows an example of a run chart with values split by specific attributes, in this case, Room Type.

For multi-center or network-based projects, charts may also visualize the results for the user's center against the network as a whole. Multi-center projects can

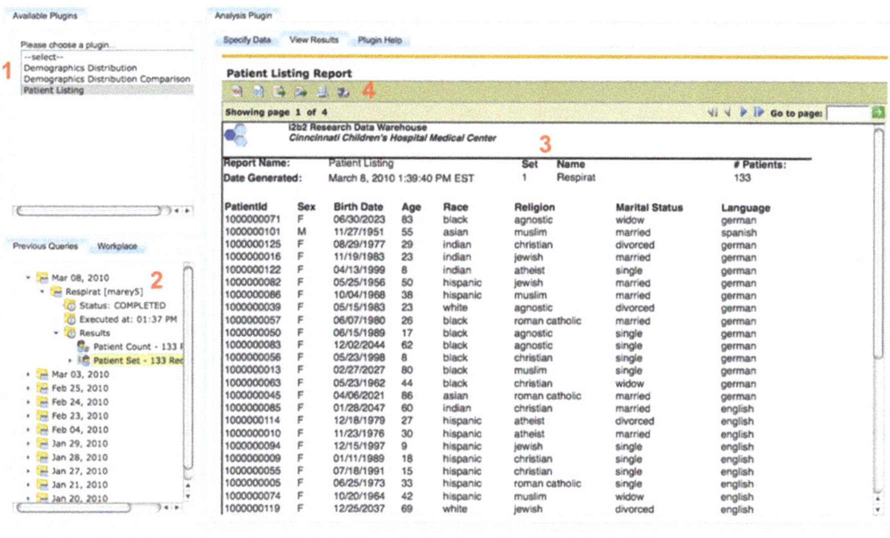

Fig. 13.4 Screenshot of the Analysis/Reporting section of the CCHMC i2b2 workbench. A user chooses one a report (*1*) and applies it to a patient cohort (*2*). The results appear in the reporting window (*3*) and the user can export the data or the report (*4*)

be supported by data sharing within a distributed research network. Examples are described in Chap. 8, where local EHR-linked i2b2 research patient data warehouses are federated for data sharing using SHRINE. If the study is SHRINE-enabled, the aggregate EHR-linked data will be queried from the other SHRINE sites and can be aggregated to produce reports. The governance of the project will determine whether users at one center can see the patient-level results of another center. The EHR-linked, i2b2-based registry includes an integrated reporting module based on the open-source BIRT toolkit (Business Intelligence and Reporting Tools) as well as the ability to display stand-alone visualizations created in languages like Javascript or HTML5. An example report using the open-source BIRT module (http://www.eclipse.org/birt/phoenix/) is shown in Fig. 13.4. The system supports both pre-defined and ad hoc reports that are generally defined by data analysts. Reports can be created for the purposes of population and patient management, detailing adherence, risk stratification and response to therapy. Population management is used to stratify patient populations into risk categories and deliver recommendations on what interventions to provide those patients between visits. An example population management report is shown in Fig. 13.5. For a given measure, clinicians can view the breakdown of their patients, either by age, diagnosis, risk, etc., depending on what makes the most sense (top image). They can then drill down into the measure to view the patient-level data for the patients in the particular breakdown (bottom). Patient management or pre-visit planning reports are typically used to plan a patient's upcoming visit and make sure they receive the recommended care while they are in the clinic.

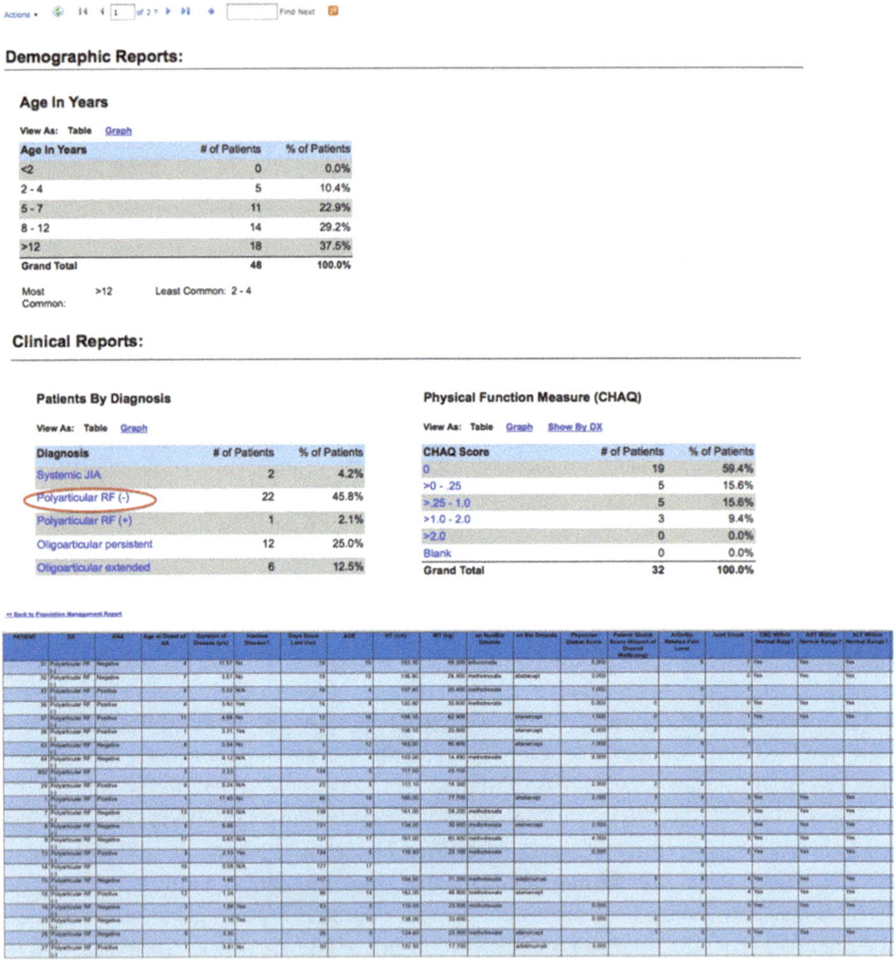

Fig. 13.5 An example population management report. The *top image* shows a set of measures, along with a breakdown by diagnosis. The *bottom image* shows a patient-level breakdown of patients matching specific criteria. In the case of this particular example, it is all patients with a diagnosis of Polyarticular RF(−)

References

Aronsky D, Haug P. Assessing the quality of clinical data in a computer-based record for calculating the pneumonia severity index. J Am Med Inform Assoc. 2000;7(1):55–65.

Brennan P, Stead W. Assessing data quality: from concordance, through correctness and completeness, to valid manipulatable representations. J Am Med Inform Assoc. 2000;7(1):106–7.

Chapman WW, Cohen KB. Current issues in biomedical text mining and natural language processing. J Biomed Inform. 2009;42(5):757–9.

Forrest CB, Riley AW. Childhood origins of adult health: a basis for life-course health policy. Health Aff (Millwood). 2004;23(5):155–64.
Hogan W, Wagner M. Accuracy of data in computer-based patient records. J Am Med Inform Assoc. 1997;4(5):342–55.
James B. Information system concepts for quality measurement. Med Care. 2003;41(1 Suppl): 171–9.
Los R, van Ginneken A, van der Lei J. Open SDE: a strategy for expressive and flexible structured data entry. Int J Med Inform. 2005;74(6):481–90.
Stein H, et al. Exploring the degree of concordance of coded and textual data in answering clinical queries from a clinical data repository. J Am Med Inform Assoc. 2000;7(1):42–54.
Venema A, et al. Is openSDE an alternative for dedicated medical research databases? An example in coronary surgery. BMC Med Inform Decis Mak. 2007;7:31.
Weiner MG, Embi PJ. Toward reuse of clinical data for research and quality improvement: the end of the beginning? Ann Intern Med. 2009;151(5):359–60.

Chapter 14
Patient Safety – Automated Detection and Reporting

Eric S. Kirkendall

Abstract The release of the 1999 Institute of Medicine report *To Err is Human* brought focus on iatrogenic harm in healthcare and spotlighted technology's potential role in mitigating adverse events. However, the use of health information technology to report on safety events and patient harm is a relatively nascent collaboration. To date, many healthcare organizations still rely on traditional methods of harm detection such as voluntary reporting or manual chart review. Automated reporting of safety events through the use of electronic "triggers" confers many advantages over those methodologies. Accurate, timely and complete reporting of adverse events allows institutions to analyze the quality of care they provide and identify any weaknesses in the care delivery system that may contribute to patient harm. Pediatric populations are particularly vulnerable to certain types of harm such as adverse drug events and have higher event rates than adults. The judicious use of electronic reporting systems in children can be used to detect past safety events as well as predict which patients are at risk for future events. This chapter will begin with a brief overview of pediatric safety events and reporting mechanisms including automated systems integrated with the electronic health record. The advantages, disadvantages, and challenges of electronic reporting will then be discussed, including research opportunities in informatics. Challenges with automated event detection specifically related to the pediatric population will be addressed, including examples from the author's institution. In closing, we will discuss other related topics and future areas of research in this rapidly evolving field.

E.S. Kirkendall, M.D., MBI, FAAP (✉)
Department of Pediatrics, University of Cincinnati College of Medicine, Cincinnati, OH, USA

Divisions of Hospital Medicine and Biomedical Informatics, Cincinnati Children's Hospital Medical Center, 3333 Burnet Avenue, ML 3024, Cincinnati, OH 45229-3039, USA
e-mail: eric.kirkendall@cchmc.org

J.J. Hutton (ed.), *Pediatric Biomedical Informatics: Computer Applications in Pediatric Research*, Translational Bioinformatics 2, DOI 10.1007/978-94-007-5149-1_14,

14.1 Patient Safety Surveillance and Harm Detection; Why It Is Important

The Institute of Medicine's oft-cited report *To Err is Human* presented the case for under-recognized high levels of harm. In that report, the authors state that between 44,000 and 98,000 preventable deaths and over 1,000,000 injuries occur per year in hospitals in the United States (Kohn et al. 2000). While some researchers have questioned the statistics of the IOM report, other studies have highlighted high rates of harm in different patient populations, demonstrating that iatrogenic harm is a common phenomenon (McDonald et al. 2000; Leape 2000). A 2008 report from the Office of the Inspector General found that after surveying the records of almost one million Medicare patients, 27% of them had had an adverse drug event (Levinson 2008). Other high profile, high sample number studies have shown similar high rates of adverse drug events, between 25 and 33% of all hospital admissions (Landrigan et al. 2010; Classen et al. 2011).

In a follow-up report to *To Err is Human*, the IOM report *Crossing the Quality Chasm* cited health information technology as one potential mechanism for mitigating harm (Institute of Medicine (U.S.). Committee on Quality of Health Care in America 2001). The IOM's 2011 report, *Health IT and Patient Safety: Building Safer Systems for Better Care,* summarized literature concerning health IT's effects on patient safety. The committee noted that while health IT has the capability to promote patient safety, large knowledge gaps exist and that realizing widespread gains will take participation from both the public and private sectors (Institute of Medicine (U.S.). Committee on Patient Safety and Health Information Technology 2011). At this point in time, much has been published about healthcare IT systems' ability to decrease medication-related prescribing errors, but a paucity of data exists showing convincing evidence that it improves outcome, rather than process measures (Kim et al. 2006; Ammenwerth et al. 2008). For instance, many studies demonstrate a reduction in prescribing errors, but none have convincingly shown a translation to better clinical outcomes such as fewer adverse drug events (ADEs) (McKibbon et al. 2012).

The potential use of computers to detect and prevent iatrogenic safety events related to healthcare has long been recognized. Significant advancements, however, have only been realized over the past several decades as technology has become more powerful, cheaper, easier to use, and more ubiquitous. Recent U.S. Government incentives to increase adoption of health IT such as the Health Information Technology for Economic and Clinical Health (HITECH) portion of the American Recovery and Reinvestment Act will expedite adoption by many organizations and increase the amount of electronically-stored healthcare data. These data repositories will become new targets of safety event detection because rare events will become easier to detect through data mining, natural language processing, and other reporting techniques. Interest in detecting and preventing these events will also increase as the U.S. healthcare delivery system will be incentivized, and in some cases, required to participate in national quality and safety initiatives such

Table 14.1 Factors that increase the complexity of prescribing medications to a pediatric population (McPhillips et al. 2005; Kaushal et al. 2004; Koren et al. 1986; Perlstein et al. 1979; Zandieh et al. 2008; Rieder et al. 1988; Schirm et al. 2003)

Factor	Notes
Weight-based dosing	*Children are usually dosed in mg drug/kilogram weight as opposed to "fixed" absolute dosing in adults*
Varying drug metabolism & physiology	*As children develop, their physiologic properties change, which affects the pharmacokinetics and pharmacodynamics of drug metabolism*
Increase off-label use	*50–75% of medications are labeled as having insufficient info for pediatric use* (Schirm et al. 2003)
Accurate/changing weight in growing children	*Weight often changes rapidly in children, which affects weight-based dosing*
Conversion of pounds to kilos	*Most parents relate their children's weight in the English system of weights (lbs. and ounces). Many prescribers and pharmacists must then convert that weight to the metric system (kilograms and grams). This conversion can create calculation errors.*
Many formulations, preparations, concentrations	*Medications often come in many different formulations with differing strengths. Oral forms often have several concentrations to consider.*
Total daily dose divided into multiple doses	*Pediatric dosing recommendations are often stated in total daily dosing, divided by a frequency of administration. This introduces chances of calculation errors.*
Tenfold errors can occur easily	*Pharmacists are less likely to recognize due to being used to adult doses* (Koren et al. 1986; Perlstein et al. 1979; Rieder et al. 1988)
Providers must know pediatric and adult dosing	*Older and/or larger children may need to be prescribed an adult-like absolute dose if the calculated weight-based dose exceeds a maximum amount.*

as meaningful use, accountable care organization standards, and other regulatory standards from oversight groups such as the Joint Commission (JCAHO). Other pressures to improve safety initiatives and be transparent with outcomes will come from comparative surveys such as those lead by U.S. News and World Report and the Leapfrog Initiative.

Pediatric patients are particularly vulnerable to iatrogenic harm. Several studies have shown higher rates of medical errors and ADEs in children than have been detected in adult populations. The reasons for this are numerous and are largely based on factors relating to physiologic growth and development. Some of the specific factors cited as increasing prescribing complexity are shown in Table 14.1. Due to the propensity for errors in this particular population, safety detection and error prevention are especially paramount.

In order to set the stage for detection, it is very important that one first understands what types of safety events and errors are common and most detrimental in pediatrics. The following section will serve to outline the broad types of events that pediatric safety researchers have identified. In many cases, there is an opportunity for research in informatics to improve detection and prevention.

14.2 A Brief Overview of Safety Events

The World Health Organization and the Institute for Healthcare Improvement both define patient harm as:

> *Unintended physical injury resulting from or contributed to by medical care that requires additional monitoring, treatment, or hospitalization, or that results in death.* (Griffin and Resar 2009)

Safety events that lead to patient harm are often grouped into acts of commission (an act of healthcare led to the event occurring) or acts of omission (care was not performed appropriately that could have avoided the event) and are referred to as adverse events (AEs). The categories of adverse events that occur in pediatric care are similar to those in adult patients, although the frequency of occurrence often varies substantially from their adult counterparts. Table 14.2 displays several categories of adverse events, but is not by any means an all-inclusive list, nor are the categories mutually exclusive.

Medication-related safety events, commonly known as adverse drug events (ADEs), are a subset of all AEs and are one of the most common types of errors. Due to their high frequency, they warrant further discussion. ADEs are often assigned to a phase of the clinical medication-patient process. There are five phases of the medication process: (Aspden and Institute of Medicine (U.S.). Committee on Identifying and Preventing Medication Errors 2007).

- Ordering/prescribing
- Transcribing and verifying
- Dispensing and delivering
- Administering
- Monitoring and reporting

Most of the errors in pediatrics are related to the first phase, when care providers are prescribing medications. Drug-dosing makes up the largest proportion of prescribing errors, due in large part because doses in children are based on their weight (such as 10 mg/kg of drug X) as opposed to absolute dosing used for most doses of adult medications (100 mg of drug X). As such, most of the literature on clinical decision support (CDS) in EHRs revolves around dosing support. One systematic review of the literature on CDS with medication dosing support showed that 95.7% of studies pertained to either the prescribing or the monitoring phase (McKibbon et al. 2012). See Chap. 1, Pediatric Electronic Health Records and Research, for further discussion of issues related to CDS of drug dosing within EHRs.

Many of the laboratory-based safety events fall into the monitoring-based phase of the medication process. Many drug levels can be followed and, when exceeded, may demonstrate unsafe monitoring frequency or dosing. Laboratory values can also be followed that may indicate many other underlying mechanisms of error. One such example is abnormal coagulation results such as an elevated d-dimer, which may indicate a deep venous thromboembolism. The other categories of safety events will not be covered in detail here, but examples for each are shown in Table 14.2.

Table 14.2 Safety event (adverse event) categories and examples

Safety event category	Example	Potential etiology of the harm/event
Medication-related (ADEs)	Naloxone use for oversedation	Patient was oversedated with narcotics; naloxone is a reversal agent
Laboratory result-based	Gentamicin levels out of range	Inappropriate dosing or monitoring of gentamicin use
Peri-operative events	Unplanned and/or return to surgery	Failure to perform procedure correctly necessitates return trip to the operating room and subsequent procedures and risks
Admit/Discharge/Transfer events	Unplanned readmissions	Previous hospital discharge occurred too early, patient returned to hospital
Clinical deterioration	Transfer to higher level of care (such as to an intensive care unit)	Failure to monitor patient closely or recognize deterioration earlier
Hospital-acquired/ nosocomial infections	Ventilator-associated pneumonia	Failure to perform known best practices that reduce infections in patients on ventilators
Care-related	Pressure ulcer development in chronic patients	Inappropriate positioning of the patient during long stays or inspection for skin problems

14.3 Many Different Types of Safety Event Reporting

The mainstay of safety reporting for most healthcare organizations is voluntary reporting, which relies on providers or staff to provide details of potential safety events to a central repository. Voluntary reporting has many pitfalls and has been shown to be woefully underutilized, often times only catching 5–15% of all events (Ferranti et al. 2008; Naessens et al. 2009; Olsen et al. 2007). The reasons for the low yield are numerous: failure to recognize events as errors, additional workflow requiring time during already heavy workloads, fear of punitive actions by organizations, and fear of legal reprisal are some major detractors.

Most healthcare institutions have instituted a local voluntary reporting system. These systems are usually protected from discovery by external entities in order to foster reporting to improve local processes and prevent future errors. Other voluntary reporting systems are more widespread, to either local or national levels. Several states have mandated reporting of medical errors. An example of a government-sponsored nationwide effort is the Federal Drug Administration's Adverse Event Reporting System, or AERS (Food and Drug Administration 2012). Despite the robust nature of these various systems and mandated reporting in some cases, it is still widely known that these systems only catch a small minority of safety events that occur in US hospitals and care facilities.

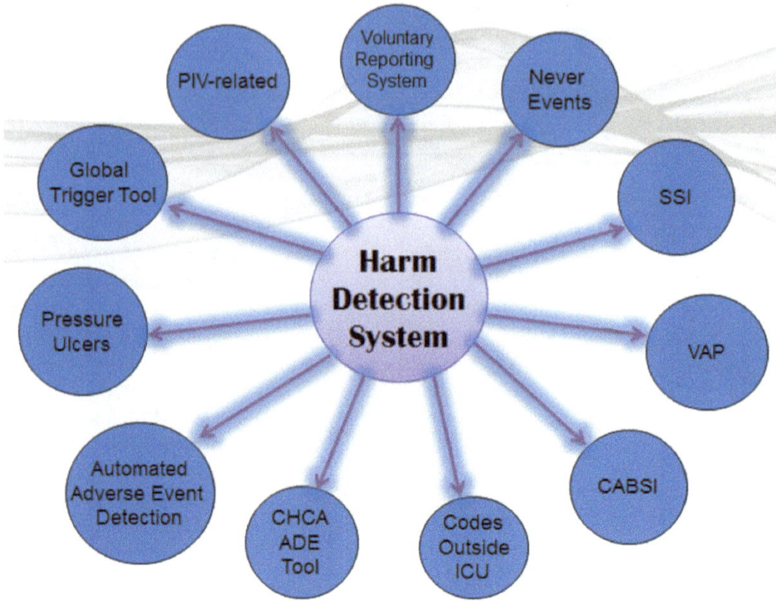

Fig. 14.1 Example of an Organization-wide Harm Detection System. Each spoke and node represents a different type of mechanism for detecting harm. The different types of detection systems are designed to find and report different types of safety events. Abbreviations *SSI* Surgical Site Infections, *VAP* Ventilator-Associated Pneumonia, *CABSI* Catheter-Associated BloodStream Infections, *ICU* Intensive Care Unit, *CHCA* Child Health Corporation of America, *ADE* Adverse Drug Event, *PIV* Peripheral IntraVenous (catheter)

Due to the shortcomings of voluntary reporting systems, various other safety reporting systems have been used to capture safety events. Multiple studies have shown that reporting via different mechanisms has a high yield of detection of different types of events with very little overlap or redundancies (Ferranti et al. 2008; Naessens et al. 2009; Olsen et al. 2007; Weissman et al. 2008). Many institutions rely on several, if not many, different kinds of reporting. An example construct of reporting systems is shown in Fig. 14.1, Example of an Organization-wide Harm Detection System.

Administrative and billing claims data are often the next system implemented by reporting groups. These data have the advantage of being the by-product of modern care delivery and as such, are readily available and typically do not rely on additional work by staff as voluntary reports do. The major downside to using billing and administrative data is inherent in the properties of the data collected; by nature it is not meant to be used for safety reporting and as such, is often not granular enough or, at best, can only serve as a proxy for an event. Two potential sources for mining of this data include the use of e-codes and the Pediatric Health Information System (PHIS) database. E-codes are a subset of widely used International Classification of Diseases (ICD) billing codes that almost all American healthcare organizations use to submit reimbursement claims to payors. The ICD system is maintained and updated by the World Health Organization. The e-code subset of the ICD system is the grouping of codes that represent external causes of injuries such as trauma, and more pertinent to safety reporting, adverse effects of drugs and other medicinal or biologic substances (codes e930-949 in the ICD-9-CM system, Y40-59 in ICD-10). The version that most healthcare institutions utilize, ICD-9-CM, is under federal mandate to be replaced by ICD-10. The change to ICD-10 will increase the granularity and detail of the coding system, expanding the 17,000 codes in ICD-9-CM to about 68,000 in ICD-10 and providing an opportunity for the reporting of more detail surrounding adverse events.

The PHIS database is an administrative database managed by the Child Healthcare Corporation of America (recently renamed the Children's Hospital Association, Shawnee Mission, KA) (Kittle et al. 2002). Forty-three of the largest pediatric hospitals in the US contribute administrative data to this database and each member can, under a use agreement, extract limited data sets for research. To date, dozens of studies have been published that have used PHIS aggregated data. In May, 2011, the author retrieved 32 articles in Medline with a simple search for the database. One collaborative project currently in progress and titled PHIS^{+} aims to add more clinical data to PHIS in order to overcome the difficulties inherent with using administrative data for clinical research (Narus et al. 2011). The PHIS^{+} project will add laboratory, radiology and microbiology study results from six pilot institutions to augment the current database and will conduct four comparative effectiveness research studies to demonstrate viability. As was the case with the expansion of the ICD coding system, the changes to PHIS will increase the potential for detection of safety events. PHIS is just one example of many available administrative databases that can be used for research. The Healthcare Cost and Utilization Project (H-CUP) family of databases are also available online. These include the Kids' Inpatient Database (KID), the Nationwide Inpatient Sample (NIS), and the Nationwide Emergency Department Sample (NEDS). More information is available at http://www.hcup-us.ahrq.gov/databases.jsp.

One widely used strategy for estimation of safety event rates in institutions is to conduct manual chart reviews. Manual chart reviews usually involve taking a small sample of patient charts and applying a systematic review process to those charts to gather adverse event data and rates. The sample rates are then extrapolated to patient census data from the reviewing institution to estimate rates of harm across the entire population. Most techniques for manual chart review involve identifying "triggers" in the clinical care documents of a patient. Clinical Triggers are defined by the Agency of Healthcare Research and Quality as data flags identifying patients who may be at risk of harm or clinical situations that have the potential for harm (Triggers and Targeted Injury Detection Systems (TIDS) 2012). Trigger tools are usually groups of like or unlike triggers, combined into an instrument designed to detect different types of events. Resar and Rozich first described modern trigger tool methodologies in 2003 in several descriptive papers (Resar et al. 2003; Rozich et al. 2003). Subsequently, their work led to the development of the best known manual chart review tool used for detecting safety events, the Institute of Healthcare

Table 14.3 Comparison of Adverse Event (*AE*) and Adverse Drug Event (*ADE*) rates in pediatric studies

Study	AEs per 100 patients	ADEs per 100 patients	AEs per 1,000 patient days	ADEs per 1,000 patient days
Matlow et al. (2005)	15.1			
Takata et al. (Phase 1) (2008)		9.8		13.4
Takata et al. (Phase 2) (2008)		11.1		15.7
Holdsworth et al. (2007)		6.0		
Kaushal et al. (2001b)		2.3		

Improvement's Global Trigger Tool (IHI GTT) (Griffin and Resar 2009). The GTT was developed by the IHI as an adult tool and has been used in many studies to document and confirm high levels of iatrogenic harm. The GTT has been used to demonstrate much higher levels of harm than other detection modalities including voluntary reporting and the use of billing data such as ICD-9 codes (Classen et al. 2011; Naessens et al. 2009). One study in 2011 found ten times the number of safety events via use of the GTT than was documented through voluntary reporting and the use of administrative data (Classen et al. 2011). The GTT has also been found to detect high rates of harm (2–3 times previously published rates) when applied to the pediatric population. Table 14.3 demonstrates previously published pediatric harm rates.

The GTT and other trigger tool methodologies have since been used or adapted for many different patient populations and settings including the inpatient hospitalized (Landrigan et al. 2010; Classen et al. 2011; Naessens et al. 2009, 2010; Franklin et al. 2010; Good et al. 2011; Matlow et al. 2005, 2011; Sharek et al. 2011; Takata et al. 2008), outpatient ambulatory care (Singh et al. 2009; de Wet and Bowie 2009; Kaafarani et al. 2010; Mull et al. 2011), intensive care units (Agarwal et al. 2010; Larsen et al. 2007; Resar et al. 2006; Stockwell and Kane-Gill 2010; Sharek et al. 2006), surgery (Kaafarani et al. 2010; Griffin and Classen 2008; Lander et al. 2010), and nursing homes/extended care facilities (Handler and Hanlon 2010). Several of these studies were focused on detecting harm to children (Matlow et al. 2005, 2011; Takata et al. 2008; Agarwal et al. 2010; Larsen et al. 2007; Stockwell and Kane-Gill 2010; Sharek et al. 2006; Lander et al. 2010). Notable pediatric trigger tools that have been developed include the CHCA Trigger Tools (Pediatric Adverse Drug Event, PICU, NICU), The Canadian Paediatric Trigger Tool, and the United Kingdoms' National Health Service Paediatric Trigger Tool.

The downside to manual chart review is the time involvement in determining whether or not a safety event occurred, as well as to gather additional information regarding the circumstances of the event (the characteristics of the AE). By leveraging technologic resources, however, this time commitment can be minimized.

The remainder of this chapter will discuss the automation of safety event reporting; what it is, what are the pros and cons of using it, how it fits in with other surveillance systems, what can be done with the data it produces, and how informatics might be applied to improve detection and reporting in the future.

14.4 Automated Reporting

Automated safety reporting generally refers to the identification of triggers within an electronic health record or other digital source of healthcare information. Electronic triggers or trigger tools are, in concept, basically digital equivalents of the manual tools discussed in the previous section. Different types of safety or healthcare events are variably amenable to automation, the details of which will be discussed in the other sections of this chapter. Electronic safety triggers confer several attractive advantages over manual trigger tool methodologies. Automated reports are typically timelier, more efficient, and are better able to survey entire patient populations than the methodologies used in their non-electronic trigger counterparts.

Whether automated safety reports are generated ad hoc/on-demand or scheduled periodically, trigger reports can be easily configured to allow near-real time information to be delivered to safety surveillance personnel. Real-time reporting is also possible, although much less common. In most cases, institutions are either trying to gather data about past events to help prevent future occurrences or are trying to inform improvement teams with metrics of interest. Real-time reporting, in some clinical cases, allows early identification of an event and can assist in mitigating or eliminating harm. This is accomplished by notifying appropriate healthcare personnel (such as a Patient Safety Officer) who then react to take corrective actions. In contrast, most manual trigger tool methodologies call for review of the chart to take place after a healthcare encounter has ended; such as an entire hospital stay, after the patient has been discharged and left the hospital. Manual review in these cases may take place days to years after a potential or real safety event has occurred. This delayed retrospective review may lead to suboptimal event characteristic data collection. Most improvement initiatives use staff interviewing techniques to gather more detail and insight than documentation affords, and delayed interviews weeks removed from an event (or longer) can introduce memory or recall issues and biases. Electronic reports can, of course, be configured to report events from distant times as well, but the real value of the reports (in regards to timing) is the ability to quickly identify and report potential events.

Automated reporting also allows organizations to survey entire populations of patients for safety events. In large populations, it is not practical to manually review each patient's medical record. Automated reports can overcome this challenge. The electronic storage of healthcare data allows computer queries to be constructed that can target specific areas of a database or data repository to expedite reporting. In much the same way operating system search functions can locate a file on a hard drive faster than a person could manually, automated reports efficiently survey large populations of patients with large amounts of electronic data stored about aspects of their healthcare.

Despite the many advantages to automated safety reporting, there are also some challenges and limitations to their effectiveness that can be addressed through informatics research. The primary barrier to effective report generation, technical and resource requirements aside, is the reliance on good data to use as the basis for a trigger. Discrete data (data with some structure, organization, or of finite number of values) are the best candidates for triggers. An example of a common trigger that is based on discrete data is the administration of an antidote to a patient. If drug X is given to a patient and the most common use of drug X is as a rescue medication against drug Y, then the documentation of administration of drug X would be a good proxy (or trigger) to the safety event. The electronic health record stores the administration of drugs in a structured format such that the system "knows" drug X from drug Y and drug Z. It is a much more difficult task to query unstructured data such as free narrative text, the basis for most clinical notes at this time. Advanced artificial intelligence techniques such as natural language processing (NLP) use advanced computational algorithms to try to parse meaning from unstructured data (NLP is discussed in Chaps. 9 and 10). NLP has the potential to extract what the author of the text was trying to communicate and imparts computational "meaning" to it. Revisiting our example above, in theory NLP could be used to scan clinical notes for evidence that drug X was given to a patient (perhaps written in a note that the patient took the medication at an outside hospital). There are many different ways of writing that drug X was given to a patient and success in this use case would require that the NLP program be able to correctly identify and classify the correct text as a trigger. The complexities, idiosyncratic nature, and non-standard conventions of the English language are challenges that researchers in NLP are working to overcome.

While NLP holds much promise, it is largely a research tool at this time and progress towards clinical applications has been characterized as slow by many prominent informatics leaders, including NLP experts (Chapman et al. 2011). Despite the perceived slow progression, NLP holds a lot of promise for applications in quality and safety operations and research. A more detailed discussion and overview of NLP is presented in Chap. 9, while specific applications in pediatrics are discussed in Chap. 10.

Another challenge to automating the safety event review process is that automation can only go so far. Many of the event characteristics that must be gathered for a quality review require human assessment or adjudication. Coupled with an increased number of potential events that are usually detected by automated reports, the resource requirements to support robust safety event detection systems can be considerable. In this respect, implementing technology to support automated detection systems is like implementation of any other healthcare information technology.

Most triggers that are implemented by healthcare institutions are used to gather information and statistics about the amount of harm that has occurred. This information is used to inform interventions and prevent future occurrences. In this way, the triggers can be characterized as reactive or retrospective event reporting triggers. In contrast, predictive or preventative triggers are proxies to potential or evolving harm. As an example, consider triggers that have been developed and put into use to detect acute kidney injury, or AKI. AKI has been shown to cause increased morbidity and mortality in children and is being increasingly recognized as a predecessor to developing lifelong kidney disease (Goldstein and Devarajan 2008, 2010, 2011; Koralkar et al. 2011; Askenazi et al. 2009). AKI triggers usually are based on changes in creatinine clearance (CrCl). Creatinine is a byproduct of muscle metabolism that is excreted by the kidneys. As kidney function declines, creatinine clearance decreases and creatinine levels in blood increase. Absolute values of CrCl below a certain level or relative increases of 50% or more in serum creatinine indicate that a kidney injury has occurred. Detecting either indicates a potential safety event that could be due to inappropriate use of medications harmful to the kidneys, improper fluid or electrolyte management, or failure to adjust doses of nephrotoxic drugs. Reacting to triggers may help mitigate further damage, but it cannot prevent the injury from occurring in the first place. Predictive AKI triggers, on the other hand, detect patients who may be susceptible to injury and have the potential to prevent iatrogenic harm. At the author's institution, we have created predictive triggers based on the number and duration of nephrotoxic medications a patient has or will be exposed to (submitted, but unpublished). The criteria for the triggers were based on previously published studies (Goldstein 2011; Moffett and Goldstein 2011; Zappitelli et al. 2011). Predictive triggers can then be used to identify patients at risk for injury and intervene to reduce the probability of harm occurring. Adverse event (AE) reporting triggers can then be used to detect the actual rates of harm occurring to judge whether interventions have been successful. The design and use of different types of triggers should be guided by the intent and end-goal of data collection; will the information be used to simply detect harm or be used to inform changes in the system to prevent injuries? For most institutions, the first step is to characterize the amounts and types of harm that are occurring inside their walls. Once harm has been identified and characterized, then targeted interventions can be designed and implemented to address the causative mechanisms of harm.

14.5 Challenges of Reporting

Creating automated adverse event reports is not without its own set of unique challenges. Researchers creating automated safety reports must consider how the reports will be adopted into or work with the other technologies at their institution. The time-consuming implementation of triggers by IT teams often contrasts with the immediacy of rapid-cycle improvement projects, requiring collaboration and patience on the part of informatics and safety experts. In addition, adverse event trigger criteria must be configured thoughtfully to meet the needs of the project. The need for real-time reporting is one of the forces of driving informatics research in the field.

We now take a deeper dive into these currently existing challenges and offer brief insight into how they may be managed or overcome in the future.

14.5.1 Underlying Architecture of Modern-Day EHRs

Most modern-day EHRs are based on transactional models of database operation, meaning that the underlying architecture of the database was designed to handle input and output of data in an operations-focused manner. As such, EHRs are optimally constructed to assist in the acute episodic care of patients, the "here and now" of delivering healthcare. This presents a problem when a researcher tries to answer questions such as "did this patient have a medication-related rash during his hospitalization" or "was an X-ray taken intra-operatively to look for a surgical instrument that was left in the patient." In both cases, the questions are very reasonable, but difficult to report if the EHR is not designed to capture that specific information or if providers do not reliably input that information into the system. EHRs and other care delivery HIT systems are generally designed with other priorities in mind such as administrative, billing, regulatory, and medico-legal documentation. While products of most large EHR vendors now have some functionality supporting quality improvement and patient safety activities, the onus largely remains on individual institutions to develop these tools. It can be quite difficult to develop robust, effective reports when current-day tools are rudimentary or not well supported. Proprietary limitations can also stymie the development of effective trigger tools. Despite historical limitations, many vendors have recently developed more safety and quality improvement (QI) tools because of the increased attention to safety and QI by the research community, increased federal reporting requirements (such as those mentioned in Sect. 14.1), and increased media coverage of adverse events in the last two decades. Advocacy on the part of customers and collaboration between customers and vendors can serve to guide this development of appropriate tools. In the meantime, many customers chose either to create reporting systems external to the proprietary software or to buy products from other vendors that have been designed to readily interface with EHRs and their data repositories. Examples of patient safety software include products such as Theradoc™,Visensia's Physiologic Monitoring System™, the Safety and Risk Management system from Quantros, and the healthcare application modules from The Patient Safety Company® just to name a few. A critical assessment of the characteristics of the candidate product and the needs and requirements of the organization must be carefully conducted before purchasing software.

14.5.2 Configuring Criteria for Triggering a Safety Alert

Setting the test properties of event triggers can also be difficult. In many ways event triggers are much like a diagnostic test. Diagnostic clinical tests have measurable properties such as rates for true positives (the proportion of cases where a trigger was reported and a true safety event occurred), false positives (a trigger fired, but did not represent a true safety event), false negatives, and true negatives. Adjusting trigger criteria will affect the report's output and the trigger's efficacy. Different institutions may identify a safety event via different criteria. For instance, a trigger could be designed several different ways to detect iatrogenic hypoglycemia (low blood sugar) events from inappropriate dosing of insulin (a sugar-lowering medication). The simplest method would be to report a low value of glucose. But how low would you set the threshold? A value of 60 mg/dL will be high enough to avoid reporting cases from other physiologic causes of hypoglycemia, such as vomiting and dehydration from gastroenteritis. The lower the threshold is set, the more likely it is that the hypoglycemia was medication-induced. However, if the threshold were lowered to a value like 40 mg/dL, true cases of insulin-induced hypoglycemia may be missed, if it did not result in a value under the cut-off. Triggers are more reliable if investigators use compound criteria, such as the combination of medication order and laboratory result, as query criteria. In the case of hypoglycemia, a trigger can fire if a blood glucose result is under 60 *and* the patient received insulin within the previous 24 h. Regardless, decisions such as these are common, must be considered, and final decisions on trigger criteria should be aligned with the safety detection philosophy of the organization.

14.5.3 Real-Time Reports

Another challenge with safety reporting is to generate the reports in a timely manner. As this has already been discussed earlier in the chapter, we will not expand on this aspect very much other than to say that implementing real-time alerting and reporting generally increases the complexity of technology and the supporting systems required to realize the advantages. Several major vendors of EHRs and HIT have developed or are currently developing more real-time functionality in their products, based on demand and the input of their customers.

14.5.4 Dynamics and Agility of Safety Research Methodologies

Developing technology in a timely manner to support quality improvement and patient safety initiatives can be quite difficult as many of the QI methodologies are based on small, but rapid tests of change. Much of the contemporary work in QI and patient safety is based on the Plan-Do-Study-Act cycle (the so-called Deming or Shewart Cycle) and methodologies that favor rapid iterations of plans that are not perfected (Ransom 2008; Carey and Lloyd 1995; Shaw and American Health Information Management Association 2010; Verzuh 2011). One of the tenants of good HIT project management is to avoid scope creep and changing requirements (Verzuh 2008; Wysocki 2011). Even the most agile IT developments must have sure footing and solid direction in order to be efficient with resources. With this in mind, conflicts can exist in projects that are not well defined or stable. Project managers who develop HIT to assist in patient safety initiative must be mindful of both the risks and benefits of rapid development in projects that do not yet have clearly defined methodologies. The development of automated safety event reports is not without many challenges. Careful assessment, planning, and implementation can overcome these difficulties and avoid pitfalls in automated reporting.

14.6 Pediatrics-Specific Considerations

Creating automated safety reports for the pediatric population differs from creating reports for the adult population in some respects. While the principles remain the same, there are several aspects to consider for this unique population. As the old pediatrics adage "children are not little adults" goes, so the saying also applies to automated reporting.

14.6.1 Differences in Disease Pathology

There are many different diseases and pathologies that affect adults and children differentially. Many of the most common illnesses in adults are either not present in pediatrics or are rare. In adults, many of the most common etiologies of disease develop over decades of life (acquired disease) and only rarely manifest in childhood. Triggers for diseases like thromboembolic strokes, heart attacks, and consequences of aging like traumatic falls and hip fractures are largely the domain of the adult and elderly. While there are some manifestations of these diseases in pediatrics, triggers not specific for children will lead to inefficiency and wasting of resources. For example, while a fall in an adult healthcare facility can be a sign of inappropriate monitoring on the part of staff, falls in infants and toddlers are much more likely to be developmental in nature and benign. Triggers designed to catch falls documented in the record will therefore be of little use in pediatrics and will lead to many false positive results.

14.6.2 Medication-Related Alerts in Pediatrics

Providing advanced clinical decision support (CDS) for medication prescribing in children necessitates a greater focus on drug-dosing than it does in adults. While drug-drug interactions are numerous in adults, pediatric prescribing tends to have more issues with drug dosing support. At the author's institution a 6 month log of medication alerts demonstrated that drug-dosing alerts were more numerous (773,721 alerts) than drug-drug interactions (669,884 alerts). As was shown in Table 14.1, there are many factors unique to pediatrics that complicates the prescribing of accurate doses to children. Other differences in frequency and distribution of types of alerts between adults and children are also to be expected; pediatric populations will generally have fewer drug-alcohol and pregnancy-related alerts.

Most large proprietary pediatric EHRs offer functionality that allows organizations to purchase drug-dosing rule databases to use in conjunction with the computerized provider order entry (CPOE) system. These database rules are typically based on FDA-approved dosing parameters. Since many medications are used in an "off-label" fashion in pediatrics, dosing parameters are often not available and not incorporated into the database products. This leads to an alert, "dose-checking is not available", and a gap in clinical decision support to the end user. If alerts are common enough in a CPOE system, alert fatigue can set in. Users will often then ignore further alerts, regardless of the content, and helpful clinical decision support may be bypassed. This may then lead to prescribing errors, and theoretically, adverse events that could have been prevented. A more thorough discussion of the sources of pediatric medication dosing errors is presented in an excellent paper written by McPhillips et al. (2005). There are many types of medication errors other than drug-dosing (Sharek et al. 2006; McPhillips et al. 2005; Folli et al. 1987; Kaushal et al. 2001a, b, 2004, 2007, 2010; Koren et al. 1986; Perlstein et al. 1979; Zandieh et al. 2008).

14.6.3 Medical Devices

As technology progresses and more and more medical devices are routinely used in healthcare, evidence is accumulating about safety concerns. While some devices are made specifically for children, many are adapted or borrowed from adult applications (Food and Drug Administration 2009; Institute of Medicine (U.S.). Committee on Postmarket Surveillance of Pediatric Medical Devices et al. 2006). The design of special devices for children must take into account the uniqueness of the population, for example, the smaller size of most children, the active nature of infants and children, and growth and development. Human factors engineering, the study of ergonomics, has long studied issues with technology and medical devices in the healthcare industry. Many researchers in this field have published articles and books describing high-consequence failures in medical devices (Carayon 2007; Wang et al. 2010). In a recent study using the PHIS database (see Sect. 14.3), researchers found that 2.5% of all medical visits (including inpatient and ambulatory encounters) had an adverse medical device event (ADME) associated with care.

Given the increasing ubiquity of medical devices in pediatric care, the U.S. Food and Drug Administration is championing several initiatives (including collaborations with the Institute of Medicine) to encourage the development of safe pediatric equipment and to increase the detection and reporting of safety events that occur when using medical devices. At this time, however, much remains unknown about harm rates and the overall safety of the devices already in use.

Medical device failures or safety issues can be insidious and very difficult to detect. Automated reporting can be used for timely detection of device failures or other forms of harm. An example of how automated reporting combined with quality improvement methodologies and timely surveillance can uncover a device failure is shown as a control chart. Figure 14.2 depicts a control chart, a quality control chart that displays an outcome of interest over time. Statistical techniques are applied to the chart to determine what limits represent normal variability of the outcome over time. In this example, the outcome (on the ordinate or y-axis) is a measure of Central Venous Catheter-Associated Laboratory-Confirmed Bloodstream Infections

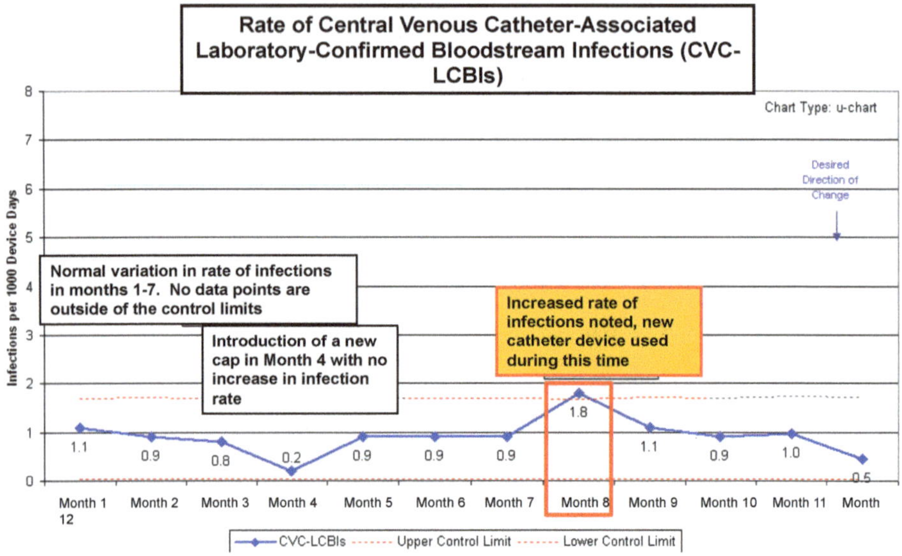

Fig. 14.2 Central Venous Catheter-Associated Laboratory-Confirmed Bloodstream Infections (CVC-LCBIs). Statistical Process Control chart for monitoring CVC-LCBI rates. The x-axis represents time, the y-axis is the number of infections normalized over 1,000 device days (the number of days patients had blood access catheters in place). The *dashed lines* represent the measurement limits one would expect for normal process variation. In Month 8 at the same time a new device was used, the rate of infections increased above the limit. The device was pulled from use and the rates of infection returned to within normal variation

(CVC-LCBIs). CVC-LCBIs are bacterial infections that occur while a vascular catheter was in place. These infections are harmful and are generally deemed preventable. The x-axis represents time. Rates of CVC-LCBIs are followed over time and normal variation between the statistically-defined limits is expected, if no changes to the system occur.

In Fig. 14.2, a decrease in the rate of infections has occurred since the monitoring began. The annotations on the control chart describe equipment and process changes that have occurred over time and the decrease in infections is very likely in response to these interventions. In Month 8 (red box) a higher than normal rate of infections was noticed via automated reporting. When plotted on the control chart, the rate exceeds the upper limit of what would be considered normal variation. When the catheter-patient care system was investigated, it was determined that the higher rates of infection coincided with a new cap for a type of commonly used catheter. Further investigation demonstrated suboptimal mechanics between the cap and the catheter. The cap was replaced with another device and subsequent rates of infection returned to baseline levels. In this case, automated reporting quickly discovered that something in the system had changed and new safety issues had arisen. Many of the care providers were unaware of the equipment change and would likely not have noticed a small, but significant change in the infection rate. In addition, the timeliness of the reporting allowed early discovery of an issue that could have perpetuated infinitely.

14.6.4 Pediatric Safety Networks and Collaborations

Up to this point, we have largely discussed automated safety reporting efforts on the local level. We will now turn the discussion towards the efforts of collaborations and networks at the regional and national level. There are many pediatric safety collaborations in existence, too many, in fact, to list here. The following descriptions will serve as examples of some of the more prominent pediatric safety reporting programs and initiatives in place in 2011.

There are several national automated safety reporting collaborations that focus on the pediatric population. The Automated Adverse Event Detection Collaborative (AAEDC) seeks to share learning and results from different organizations that are at varying development stages of their respective automated triggers programs (investigating, implementing, or analyzing/reporting results). Members of the collaborative use different EHRs at their institutions and vendors are often engaged on the monthly calls. Several vendors have developed automated reporting tools in response to this collaboration. The tools are in the form of software modules that can be downloaded, interfaced with existing model systems, and configured to meet the needs of the customer. Results and lessons learned from implementations are transparently shared among the participants. Publications describing the AAEDC's efforts will soon appear.

Another burgeoning national collaborative is the Children's Hospital Association-sponsored U.S. National Trigger Tool Working Group. The objective of this group is to define a trigger tool to screen for potential safety events that occur during healthcare delivery to children in the U.S. The working group is comprised of experts in trigger tools from the United States, Canada and the United Kingdom who have developed their own tools in their respective countries. The charge to the group is to come to a consensus about which triggers to include and the suggested query criteria for those triggers. Further work will delineate which triggers are amenable to automation and which triggers are likely to be more effective as manual triggers. The collaboration is working with both the Children's Hospital Association and federal entities such as the Agency for Healthcare Research and Quality to optimize both the development of the tools and the diffusion of innovation.

The Federal Drug Administration's Adverse Event Reporting System (FDA-AERS) is a database that accumulates post-market surveillance safety data on prior approved medications and biologic devices. Both healthcare providers and consumers can report adverse events. Reporting, for the most part, is not mandatory in the U.S. and is a limitation of the system. Despite this, the number of reports has increased steadily (over 750,000 reports in 2010) as reporting portals such as MedWatch have become available and have made reporting easier. The FDA uses the information gathered from AERS to inform them of potential new safety concerns involving drugs or medical devices. AERS data is in turn available to researchers through downloadable files, although relational database management skills are required to extract and use the data (Food and Drug Administration 2012). In addition, the FDA has held pediatric safety surveillance workshops that have brought together many patient safety and adverse event detection experts to further inform the division's efforts to increase detection of AEs (Critical Path Institute 2010).

The Leapfrog Group is a well-known not-for-profit group in the U.S. that provides assessment tools for pediatric medication ordering within Computerized Provider Order Entry (CPOE) systems (The Leapfrog Group: computerized physician order Entry). The annually-administered Leapfrog Survey contains several modules. The CPOE module is designed to assess the level of clinical decision support an organization's EHR/CPOE system offers when entering a battery of medication orders on pre-specified fake patients. Participating institutions receive instructions on patient demographics, which they enter into a test environment. A timed survey then starts during which test-takers enter a series of orders and report what kind of decision support, if any, their system offered (alerts, reminders, corollary orders/tests, etc.). The responses then generate a score, which is reported to both the institution and the Leapfrog Group. The Leapfrog Group uses the score combined with scores from the other modules to assess the quality and safety of the healthcare facility. While the Leapfrog survey does not have a component that directly relates to automated event reporting, its description is presented here because of its importance and relation to medication safety reporting.

The Leapfrog Group survey, the FDA-AERS, and the trigger tool cooperatives are all great examples of national efforts to disseminate knowledge about medication safety and/or automated reporting at the national level. Many smaller, regional efforts exist that seek to accomplish many of the same goals. The Ohio Children's Hospital Association's Solutions for Patient Safety (SPS) is a statewide initiative to spread knowledge and best practices across the State as well as to the national level (www.solutionsforpatientsafety.org). The eight major pediatric hospitals in the State of Ohio collaborate to share results and compare detection techniques and other methodologies. The hospitals openly disseminate safety rates, set statewide goals with aggregate measures, and work together to achieve these goals as a network. Accomplishments to date include a reduction of surgical site infections by 60%, as well as an overall adverse drug event rate reduction of 40%. To date the efforts of SPS have saved an estimated 7,700 children from iatrogenic harm and avoided $11.8 million in unnecessary healthcare costs in Ohio alone. Informatics, and in particular automated reporting, has greatly assisted in the success of initiatives such as SPS.

Collaborations such as the ones highlighted in this section are important because they not only perform groundbreaking work described, but because they also share the lessons learned and increase the efficiency and efficacy with which other organizations may improve the safety of the pediatric patients outside the walls of the pioneering institutions. Only through the transparency of such networks can the benefits of patient safety research be realized and disseminated quickly.

14.7 Future Directions

The use of automated safety triggers to detect harm in pediatric patients is still in its infancy. As more institutions adopt health information technology and EHRs and as adopted systems mature, more knowledge will be acquired through research and experience. The nascent field of automated patient safety reporting presents exciting opportunities for groundbreaking research. The maturation and expansion of regional and national collaborations will help to disseminate lessons learned. Vendor development of new software and applications, mediated through engagement with customers and automated trigger experts, will increase adoption of detection systems. The application of exciting artificial intelligence techniques such as natural language processing (Chaps. 9 and 10) will increase our ability to glean information from clinical notes, unlocking a wealth of additional safety data to examine and analyze. Continued development of predictive triggers will move detection upstream from adverse events to risk detection and prevention of harm. The same can be said for real-time detection systems, which, once developed and adopted, can alert response teams of events in progress or threatening provision of safe care.

It is exciting to be on the forefront of many great developments in automated adverse event detection, and it will be interesting to look retrospectively in 10, 20, or 50 years to see where we once stood and the gains made using informatics and automated reporting in pediatric safety.

Acknowledgements The author would like to thank the following individuals for their help and guidance in my career in Informatics and Patient Safety, as well as for their efforts in making Cincinnati Children's Hospital Medical Center (CCHMC) and the other pediatric institutions in the world safer:

S. Andrew Spooner, MD, Chief Medical Information Officer, CCHMC
Stephen Muething, MD, Vice President of Safety, CCHMC
Elizabeth Kloppenborg, MSN, BSN, Project Manager, CCHMC
Stuart Goldstein, MD, Director for the Chronic and Acute Care Nephrology Center, CCHMC
Patrick Brady, MD, Assistant Professor of Hospital Medicine, CCHMC
David Stockwell, MD, MBA, Medical Director of Patient Safety at Children's National Medical Center, Washington, DC

References

Agarwal S, et al. Prevalence of adverse events in pediatric intensive care units in the United States. Pediatr Crit Care Med. 2010;11(5):568–78.

Ammenwerth E, et al. The effect of electronic prescribing on medication errors and adverse drug events: a systematic review. J Am Med Inform Assoc. 2008;15(5):585–600.

Askenazi DJ, Ambalavanan N, Goldstein SL. Acute kidney injury in critically ill newborns: what do we know? What do we need to learn? Pediatr Nephrol. 2009;24(2):265–74.

Aspden P, Institute of Medicine (U.S.). Committee on Identifying and Preventing Medication Errors. Preventing medication errors, Quality chasm series. Washington, DC: National Academies Press; 2007. xvi, 463 p.

Carayon P. Handbook of human factors and ergonomics in health care and patient safety, Human factors and ergonomics. Mahwah: Lawrence Erlbaum Associates; 2007. xiv, 995 p.

Carey RG, Lloyd RC. Measuring quality improvement in healthcare: a guide to statistical process control applications. New York: Quality Resources; 1995. xix, 194 p.

Chapman WW, et al. Overcoming barriers to NLP for clinical text: the role of shared tasks and the need for additional creative solutions. J Am Med Inform Assoc. 2011;18(5):540–3.

Classen DC, et al. 'Global trigger tool' shows that adverse events in hospitals may be ten times greater than previously measured. Health Aff (Millwood). 2011;30(4):581–9.

Critical Path Institute. Pediatric safety surveillance workshop. 2010 [cited 2012 Mar 11]. Available from: http://www.c-path.org/Events/PediatricSafety.cfm

de Wet C, Bowie P. The preliminary development and testing of a global trigger tool to detect error and patient harm in primary-care records. Postgrad Med J. 2009;85(1002):176–80.

Ferranti J, et al. Reevaluating the safety profile of pediatrics: a comparison of computerized adverse drug event surveillance and voluntary reporting in the pediatric environment. Pediatrics. 2008;121(5):e1201–7.

Folli HL, et al. Medication error prevention by clinical pharmacists in two children's hospitals. Pediatrics. 1987;79(5):718–22.

Food and Drug Administration. Pediatric medical devices. 2009 [cited 2012 Mar 11]. Available from: http://www.fda.gov/MedicalDevices/ProductsandMedicalProcedures/ucm135104.htm

Food and Drug Administration. Adverse Event Reporting System (AERS). 2012 [cited 2012 Mar 11]. Available from: http://www.fda.gov/drugs/guidancecomplianceregulatoryinformation/surveillance/adversedrugeffects/default.htm

Franklin BD, et al. Testing a trigger tool as a method of detecting harm from medication errors in a UK hospital: a pilot study. Int J Pharm Pract. 2010;18(5):305–11.

Goldstein SL. Acute kidney injury in children: prevention, treatment and rehabilitation. Contrib Nephrol. 2011;174:163–72.

Goldstein SL, Devarajan P. Progression from acute kidney injury to chronic kidney disease: a pediatric perspective. Adv Chronic Kidney Dis. 2008;15(3):278–83.

Goldstein SL, Devarajan P. Pediatrics: acute kidney injury leads to pediatric patient mortality. Nat Rev Nephrol. 2010;6(7):393–4.

Goldstein SL, Devarajan P. Acute kidney injury in childhood: should we be worried about progression to CKD? Pediatr Nephrol. 2011;26(4):509–22.

Good VS, et al. Large-scale deployment of the Global Trigger Tool across a large hospital system: refinements for the characterisation of adverse events to support patient safety learning opportunities. Qual Saf Health Care. 2011;20(1):25–30.

Griffin FA, Classen DC. Detection of adverse events in surgical patients using the Trigger Tool approach. Qual Saf Health Care. 2008;17(4):253–8.

Griffin F, Resar R. IHI global trigger tool for measuring adverse events. 2nd ed. Cambridge, MA: Institute for Healthcare Improvement; 2009.

Handler SM, Hanlon JT. Detecting adverse drug events using a nursing home specific trigger tool. Ann Longterm Care. 2010;18(5):17–22.

Holdsworth M, et al. Impact of computerized prescriber order entry on the incidence of adverse drug events in pediatric inpatients. Pediatrics. 2007;120(5):1058–66.

Institute of Medicine (U.S.). Committee on Patient Safety and Health Information Technology. Health IT and patient safety: building safer systems for better care. Washington, DC: Institute of Medicine; 2011.

Institute of Medicine (U.S.). Committee on Postmarket Surveillance of Pediatric Medical Devices, Field MJ, Tilson HH. Safe medical devices for children. Washington, DC: National Academies Press; 2006. xxi, 457 p.

Institute of Medicine (U.S.). Committee on Quality of Health Care in America. Crossing the quality chasm: a new health system for the 21st century. Washington, DC: National Academy Press; 2001. xx, 337 p.

Kaafarani HM, et al. Development of trigger tools for surveillance of adverse events in ambulatory surgery. Qual Saf Health Care. 2010;19(5):425–9.

Kaushal R, Barker KN, Bates DW. How can information technology improve patient safety and reduce medication errors in children's health care? Arch Pediatr Adolesc Med. 2001a;155(9):1002–7.

Kaushal R, et al. Medication errors and adverse drug events in pediatric inpatients. JAMA. 2001b;285(16):2114–20.

Kaushal R, et al. Pediatric medication errors: what do we know? What gaps remain? Ambul Pediatr. 2004;4(1):73–81.

Kaushal R, et al. Adverse drug events in pediatric outpatients. Ambul Pediatr. 2007;7(5):383–9.

Kaushal R, et al. Medication errors in paediatric outpatients. Qual Saf Health Care. 2010;19(6):e30.

Kim GR, et al. Error reduction in pediatric chemotherapy: computerized order entry and failure modes and effects analysis. Arch Pediatr Adolesc Med. 2006;160(5):495–8.

Kittle K, et al. Using a pediatric database to drive quality improvement. Semin Pediatr Surg. 2002;11(1):60–3.

Kohn LT, Corrigan J, Donaldson MS. To err is human: building a safer health system. Washington, DC: National Academy Press; 2000. xxi, 287 p.

Koralkar R, et al. Acute kidney injury reduces survival in very low birth weight infants. Pediatr Res. 2011;69(4):354–8.

Koren G, Barzilay Z, Greenwald M. Tenfold errors in administration of drug doses: a neglected iatrogenic disease in pediatrics. Pediatrics. 1986;77(6):848–9.

Lander L, et al. A trigger tool fails to identify serious errors and adverse events in pediatric otolaryngology. Otolaryngol Head Neck Surg. 2010;143(4):480–6.

Landrigan CP, et al. Temporal trends in rates of patient harm resulting from medical care. N Engl J Med. 2010;363(22):2124–34.

Larsen GY, et al. Preventable harm occurring to critically ill children. Pediatr Crit Care Med. 2007;8(4):331–6.

Leape LL. Institute of Medicine medical error figures are not exaggerated. JAMA. 2000;284(1): 95–7.

Levinson DR. Adverse events in hospitals: overview of Key issues. Washington, DC: Department of Health and Human Services, Office of the Inspector General, U.S. Government; 2008.

Matlow A, et al. The development of the Canadian paediatric trigger tool for identifying potential adverse events. Healthc Q. 2005;8 Spec No:90–3.

Matlow AG, et al. Description of the development and validation of the Canadian Paediatric Trigger Tool. Qual Saf Health Care. 2011;20(5):416–23.

McDonald CJ, Weiner M, Hui SL. Deaths due to medical errors are exaggerated in Institute of Medicine report. JAMA. 2000;284(1):93–5.

McKibbon KA, et al. The effectiveness of integrated health information technologies across the phases of medication management: a systematic review of randomized controlled trials. J Am Med Inform Assoc. 2012;19(1):22–30.

McPhillips H, et al. Methodological challenges in describing medication dosing errors in children. In: Advances in patient safety: from research to implementation. Rockville: Agency for Healthcare Research and Quality; 2005.

Moffett BS, Goldstein SL. Acute kidney injury and increasing nephrotoxic-medication exposure in noncritically-ill children. Clin J Am Soc Nephrol. 2011;6(4):856–63.

Mull HJ, et al. Consensus building for development of outpatient adverse drug event triggers. J Patient Saf. 2011;7(2):66–71.

Naessens JM, et al. A comparison of hospital adverse events identified by three widely used detection methods. Int J Qual Health Care. 2009;21(4):301–7.

Naessens JM, et al. Measuring hospital adverse events: assessing inter-rater reliability and trigger performance of the Global Trigger Tool. Int J Qual Health Care. 2010;22(4):266–74.

Narus SP, et al. Federating clinical data from six pediatric hospitals: process and initial results from the PHIS + Consortium. AMIA Annu Symp Proc. 2011;2011:994–1003.
Olsen S, et al. Hospital staff should use more than one method to detect adverse events and potential adverse events: incident reporting, pharmacist surveillance and local real-time record review may all have a place. Qual Saf Health Care. 2007;16(1):40–4.
Perlstein PH, et al. Errors in drug computations during newborn intensive care. Am J Dis Child. 1979;133(4):376–9.
Ransom ER. The healthcare quality book: vision, strategy, and tools. 2nd ed. Chicago/Washington, DC: Health Administration Press/AUPHA Press; 2008. xvi, 495 p.
Resar RK, Rozich JD, Classen D. Methodology and rationale for the measurement of harm with trigger tools. Qual Saf Health Care. 2003;12 Suppl 2:ii39–45.
Resar RK, et al. A trigger tool to identify adverse events in the intensive care unit. Jt Comm J Qual Patient Saf. 2006;32(10):585–90.
Rieder MJ, et al. Tenfold errors in drug dosage. CMAJ. 1988;139(1):12–3.
Rozich JD, Haraden CR, Resar RK. Adverse drug event trigger tool: a practical methodology for measuring medication related harm. Qual Saf Health Care. 2003;12(3):194–200.
Schirm E, Tobi H, den Berg LTdJ-v. Risk factors for unlicensed and off-label drug use in children outside the hospital. Pediatrics. 2003;111(2):291–5.
Sharek PJ, et al. Adverse events in the neonatal intensive care unit: development, testing, and findings of an NICU-focused trigger tool to identify harm in North American NICUs. Pediatrics. 2006;118(4):1332–40.
Sharek PJ, et al. Performance characteristics of a methodology to quantify adverse events over time in hospitalized patients. Health Serv Res. 2011;46(2):654–78.
Shaw P, American Health Information Management Association. Quality and performance improvement in healthcare: a tool for programmed learning. 4th ed. Chicago: American Health Information Management Association; 2010. xxxiii, 490 p.
Singh R, et al. Experience with a trigger tool for identifying adverse drug events among older adults in ambulatory primary care. Qual Saf Health Care. 2009;18(3):199–204.
Stockwell DC, Kane-Gill SL. Developing a patient safety surveillance system to identify adverse events in the intensive care unit. Crit Care Med. 2010;38:S117–25.
Takata GS, et al. Development, testing, and findings of a pediatric-focused trigger tool to identify medication-related harm in US children's hospitals. Pediatrics. 2008;121(4):e927–35.
The Leapfrog Group: computerized physician order entry. [cited 2012 Apr 24]. Available from: http://www.leapfroggroup.org/for_hospitals/leapfrog_safety_practices/cpoe
Triggers and Targeted Injury Detection Systems (TIDS). 2012 [cited 2012 Apr 24]. Available from: http://www.ahrq.gov/qual/triggers/
Verzuh E. The fast forward MBA in project management, The portable MBA. 3rd ed. Hoboken: Wiley; 2008. xvii, 462 p.
Verzuh E. The fast forward MBA in project management. 4th ed. Hoboken: Wiley; 2011. xxiv, 519 p.
Wang C, et al. Emergency department visits for medical device-associated adverse events among children. Pediatrics. 2010;126(2):247–59.
Weissman JS, et al. Comparing patient-reported hospital adverse events with medical record review: do patients know something that hospitals do not? Ann Intern Med. 2008;149(2):100–8.
Wysocki RK. Effective project management: traditional, agile, extreme, sixth edition. 1st ed. Indianapolis: Wiley Pubishing, Inc; 2011.
Zandieh SO, et al. Risk factors in preventable adverse drug events in pediatric outpatients. J Pediatr. 2008;152(2):225–31.
Zappitelli M, et al. Acute kidney injury in non-critically ill children treated with aminoglycoside antibiotics in a tertiary healthcare centre: a retrospective cohort study. Nephrol Dial Transplant. 2011;26(1):144–50.

Chapter 15
Network Analysis in Translational Research

Minlu Zhang, Jingyuan Deng, Lirong Tan, Ye Chen, and Long Jason Lu

Abstract Networks, where nodes/vertices denote entities and links/edges denote associations, provide a unified representation for a variety of complex systems, from social relationships to molecular interactions. In general, network analysis has proved useful in applications such as prediction of protein function, guiding the design of wet-lab experiments, and discovery of biomarkers of disease. Driven by the availability of large-scale data sets and appropriate informatics' tools, the research community has begun to apply network analysis of these data to define underlying causes of pediatric diseases. This will almost certainly lead to more effective strategies for prevention and treatment. In this chapter, we will introduce classic and the state-of-the-art network analysis methodologies, approaches and their applications. We then provide three examples of recent research, where network analysis is being applied in pediatrics. These include identification of targets

M. Zhang, Ph.D. • J. Deng • L. Tan
School of Computing Sciences and Informatics, University of Cincinnati College of Engineering & Applied Science, Cincinnati, OH, USA

Division of Biomedical Informatics, Cincinnati Children's Hospital Medical Center, 3333 Burnet Avenue, MLC 7024, Cincinnati, OH 45229-3039, USA

Y. Chen
Division of Biomedical Informatics, Cincinnati Children's Hospital Medical Center, 3333 Burnet Avenue, MLC 7024, Cincinnati, OH 45229-3039, USA

School of Electronics and Computing Systems, University of Cincinnati College of Engineering & Applied Science, Cincinnati, OH, USA

L.J. Lu, Ph.D. (✉)
Division of Biomedical Informatics, Cincinnati Children's Hospital Medical Center, 3333 Burnet Avenue, MLC 7024, Cincinnati, OH 45229-3039, USA

Department of Pediatrics, University of Cincinnati College of Medicine, Cincinnati, OH, USA
e-mail: long.lu@cchmc.org

J.J. Hutton (ed.), *Pediatric Biomedical Informatics: Computer Applications in Pediatric Research*, Translational Bioinformatics 2, DOI 10.1007/978-94-007-5149-1_15,

for development of new drugs to treat *Pseudomonas* infections, protein interactions in high-density lipoprotein particles that underlie development of cardiovascular disease, and analysis of networks in brain that may underlie conditions such as Attention Deficit Hyperactivity Disorder.

15.1 Introduction

Networks or graphs provide a unified representation for a variety of complex systems, from social relationships, e.g., co-authorship of different authors, to associations between molecular entities. Molecular networks, where each node denotes a molecule like a protein or gene and each edge denotes an association, form the foundation of contemporary systems biology. To date, network models are applied to depict a variety of biological systems. For example, protein-protein interaction (PPI) networks or maps, where each node is a protein and each edge is an interaction, are adopted to describe physical interactions or attachments between protein pairs in sets of proteins such as those found within a cell organelle, or particle (Rual et al. 2005). Gene co-expression networks, where each node is a gene and each edge indicates gene expression similarity, are applied to summarize similarities in expression patterns between gene pairs (Stuart et al. 2003). Transcriptional regulatory networks, where each node is either a transcription factor protein or a target gene, depict transcriptional regulatory relationships between transcription factors and target genes (Guelzim et al. 2002). Other common biological networks include metabolic networks of metabolites and their chemical reactions and signal transduction pathways of multiple interacting genes and proteins (Zhang et al. 2010). Analyzing large-scale networks may provide a systems-level view of these biological systems and thus advance understanding of the underlying biological processes. In recent years, applications of network analysis have proved useful in guiding the design of experiments, uncovering novel and effective prognostic biomarkers of disease, and facilitating drug discovery. Some basic features of networks are illustrated in Fig. 15.1.

In this chapter, we will first introduce tools and processes for state-of-the-art biological network analyses and provide a few examples of their application to basic research into causes of pediatric diseases. Next, we will introduce the global properties and characteristics of large-scale biological networks: scale-free topology, small-worldness, disassortativity, and modular structures. In the third section, we will discuss in brief three common types of network analyses and show how they have been applied: (1) topological analysis to identify proteins/genes of interest, (2) motif analysis to extract and identify small subnetwork motif structures, and (3) modular analysis to cluster large molecular networks to self-contained subnetworks as functional units or modules. Finally, examples of analyses will be provided, including: (1) human-*Pseudomonas* interaction networks and network-based identification of drug targets for treatment of *Pseudomonas* infections and,

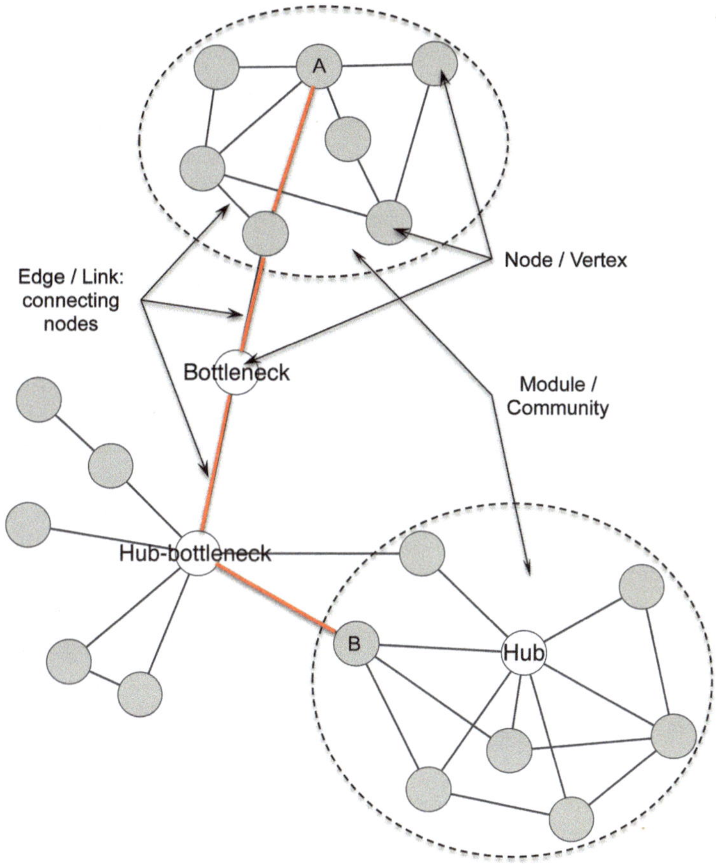

Fig. 15.1 Basic features of networks. Nodes or vertices are entities in a network, and edges or links connect pairs of nodes. The *red* edges form a path that connects nodes *A* and *B*. A hub node is defined as a node with a large number of neighbors, and a bottleneck is defined as a node that many shortest paths go through. In this sample network, the node in the *lower left* is both a hub and a bottleneck. Fifteen nodes in a network with 22 nodes form two different densely intra-connected and loosely inter-connected modular subnetwork structures called modules or communities (in *dashed ellipses*)

(2) analysis of proteins and protein-protein interactions in subspecies of high density lipoprotein particles that play a role in cardiovascular disease , and (3) construction of brain functional connectivity maps from neuroimaging data in children with a focus on attention-deficit/hyperactivity disorder (ADHD). Chapters 16 (Bioinformatics and Orphan Diseases) and 17 (Functional Genomics – Transcriptional Regulatory Networks) provide additional examples of the application of network analysis to studies of childhood diseases.

15.2 Biological Networks: Properties and Characteristics

Large-scale biological networks of various types are similar in a number of topological characteristics, namely scale-free, small-worldness, disassortativity, and modular structures. Similar to many other large-scale networks such as the World Wide Web and social networks, most large-scale biological networks of model organisms are observed to approximate a scale-free structure. Scale-free means that distributions of nodes within these networks follow a power-law distribution, denoted as $P(k) \sim k^{-\gamma}$, where $P(k)$ is the fraction of nodes that have k connections to other nodes (degree k), and the degree exponential constant γ is a constant usually smaller than 3 (Barabasi and Oltvai 2004). Figure 15.2 shows the scale-free topology of a large-scale human protein-protein interaction (PPI) network, using PPI data from the current release of the Human Protein Reference Database (Goel et al. 2012). Scale-free topology indicates that in a network, the higher the degree of the node, the greater the number of connections it has. A small fraction of nodes have a very high degree of connections (large number of connections) and are called hubs. Hubs are critically important to the overall functioning of the network. The scale-free topology contributes to the robustness of the network. For example, one

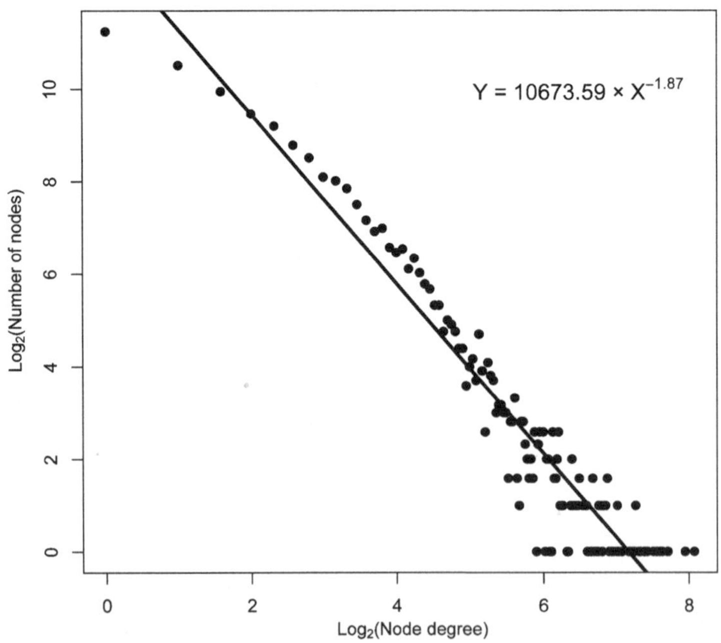

Fig. 15.2 The scale-free topology of a large-scale Protein-Protein Interaction (PPI) network. The node degree of a large-scale PPI network based on the ninth release of the Human Protein Reference Database (HPRD) follows a power-law distribution. A total of 9,517 proteins and 37,004 pairwise PPIs exist in the network

study showed that removing up to 80 % of randomly selected nodes in a network would not disconnect the rest of the network, due to the scale-free topology and sparseness of connections of most of the nodes (Albert et al. 2000). On the other hand, networks are more vulnerable, if there is an attack on the highly influential hub nodes (non-random attack).

Most large-scale biological networks have small-world properties as a consequence of their scale-free topology. Small-worldness is often defined and depicted as high clustering coefficient and low characteristic path length of a network, which means that nodes in a biological network are involved in self-contained community-like subnetworks and most node pairs are connected via short paths (Fig. 15.1). For example, in a *Escherichia coli* metabolic network of 287 nodes and 317 edges, where a node is a metabolite and an edge represents a reaction, a typical path between the most distant metabolites contains only four reactions (Wagner and Fell 2001). The small-world property also contributes to the robustness of a network, because when any perturbation occurs, most of the network components can conduct a timely response because of the small-worldness.

Disassortativity, which means hub and non-hub nodes are more likely to be connected than randomly expected, was discovered in large-scale PPI as well as transcriptional regulatory networks (Maslov and Sneppen 2002). The disassortativity, in accord with scale-free and small-world characteristics, strengthens the robustness of the network. The property also helps to separate subnetwork structures in the network, which may correspond to functional units that perform particular functions in the biological processes.

Large-scale networks are composed of functional modules that perform distinct, but relevant, functions (Hartwell et al. 1999). In biological networks, such functional modules correspond to modular subnetwork structures, which are highly intra-connected within themselves and loosely inter-connected with each other. The molecules within each module are likely annotated with the same or similar function. The modular structures are hierarchically organized, as observed in most large-scale biological networks, because of both modular and scale-free properties (Ravasz 2009).

15.3 Network Analysis and Applications: Topology, Motifs and Modules

The most common types of network analysis include topological, motif, and modular analysis. In this section, we introduce these methods of network analysis and their applications to knowledge discovery. Table 15.1 lists some tools commonly used in network analysis, their function, and how to access them. Typical outputs of these analyses are shown in Figs. 15.1, 15.4, 15.5, and 15.6.

Given a biological network, it is of interest to characterize its topology. In order to quantitatively measure the topological structure of a network, a number of classic

Table 15.1 Examples of tools (software applications) commonly used in network analysis

Tool	Function	How to access
Cytoscape (Shannon et al. 2003)	Cytoscape is an open source software platform for visualizing complex networks and integrating these with any type of attribute data. Cytoscape supports many use cases in molecular and systems biology, genomics, and proteomics. Various plug-ins for analysis and visualization have been developed in addition to its core functionalities.	www.cytoscape.org/
Gephi (Bastian et al. 2009)	Gephi is an open source standalone software platform for visualizing networks and complex systems. Gephi provides topological and modular analysis for biological and social networks. A variety of plug-ins are also available for analysis, visualization layout, and applications of specific purposes.	www.gephi.org
VisANT (Hu et al. 2009)	VisANT provides a webstart application, a standalone Java application, and a batch mode for visualization and analysis. It has multiple layout options and is scalable to visualize genome-scale biological networks. Users can perform statistical, motif, and modular enrichment analysis using VisANT.	visant.bu.edu
Pajek (Batagelj and Mrvar 2004)	Pajek is a standalone application for network visualization. Pajek supports 3D layout and can perform modular analysis of networks to decompose them into modules.	vlado.fmf.uni-lj.si/pub/networks/pajek/
tYNA (Yip et al. 2006)	tYNA is a Web server that for biological network visualization. tYNA supports basic network analysis of major topological characteristics and motifs.	tyna.gersteinlab.org/
JUNG (O'Madadhain et al. 2003)	JUNG, short for Java universal network/graph framework, is a Java-based open-source software library package. JUNG provides various implemented algorithms for network modeling, visualization and analysis. It is a general library and is not specific to biological network applications.	jung.sourceforge.net/
PINA (Wu et al. 2009)	PINA is a Web server primarily for protein-protein interaction network visualization and statistical analysis. Important nodes such as hubs and bottlenecks can be identified. PINA can perform GO term enrichment analysis. It has good scalability.	cbg.garvan.unsw.edu.au/pina/

(continued)

Table 15.1 (continued)

Tool	Function	How to access
NeAT (Brohee et al. 2008)	NeAT is suite of Web-server tools for network analysis and visualization. NeAT supports topological analysis, pathway extraction, and modular analysis (multiple network clustering algorithms such as clique enumeration and MCL clustering). It also provides functionalities like network comparison, cluster comparison, and heat map generation.	rsat.ulb.ac.be/neat/
N-Browse (Kao and Gunsalus 2008)	N-Browse is both a webstart application and a Web-server for visualizing molecular networks. It is connected to and integrated with multiple databases such as modMine database for biological data mining.	http://aquila.bio.nyu.edu/NBrowse2/

network statistics from graph theory have been adopted. Commonly used network statistics include: degree, centrality, clustering coefficient, shortest path length, and eccentricity. These network statistics not only can describe single nodes in the network, but also characterize the network as a whole.

In an undirected network such as a PPI network, node degree is defined as the number of connections linked to a node. In a directed network such as a transcriptional regulatory network, the number of out-going and in-coming links of a node can be measured separately as out-degree and in-degree. The average degree of all nodes in a network is an indication of how dense or sparse the connections are. Similar to other real-life networks, biological networks are sparse, which is defined as having far fewer than the possible maximum links $\frac{n\times(n-1)}{2}$, where n is the number of nodes. The clustering coefficient of a node ranges from 0 to 1 and can be defined as the fraction of links compared with all possible links for its neighboring nodes. The average clustering coefficient for all nodes in a network is an indication of whether these nodes are involved in densely connected subnetwork clusters. The higher the average clustering coefficient of a network, the more likely that its nodes form self-contained subnetwork modules. The shortest path length between two nodes is the minimum number of links needed to connect two nodes. Characteristic path length for a network is defined as the median of the means of the shortest path lengths of all nodes taken together. Small-worldness is defined as having high clustering coefficient and low characteristic path length.

In biological networks, hubs and bottlenecks, are of special interest. A hub is an influential local connector, while a bottleneck is a bridge-like connector between different communities (Fig. 15.1). A number of studies have reported enriched functional essentiality in hub or bottleneck nodes of protein-protein interaction networks of model organisms, or high correlation between topological connectivity and functional essentiality (Jeong et al. 2001; Yu et al. 2007). Based on these observations, one application of retrieving hubs and bottlenecks is to identify

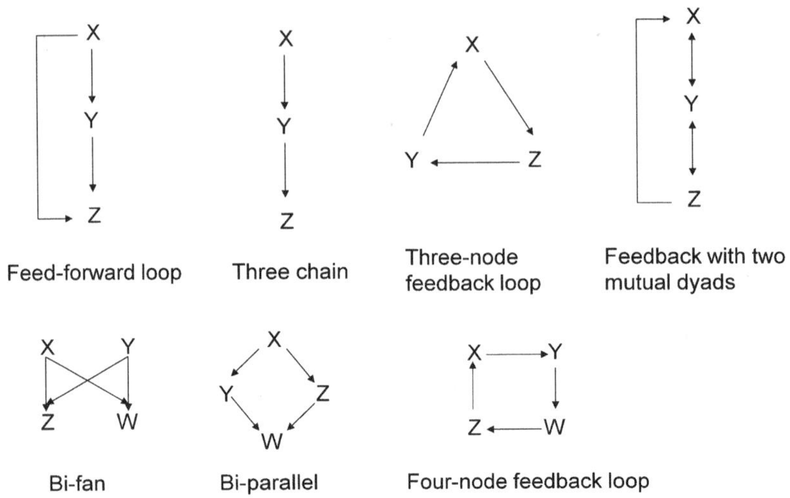

Fig. 15.3 Commonly identified motifs in networks. Each letter "x", "y", "z", or "w" indicates a node. Each motif consists of 3–4 nodes in these commonly identified motifs

important proteins or genes in biological networks. For example, hubs or bottlenecks in the PPI networks of human pathogens may help identify essential proteins for the survival of these pathogens. Such proteins are potential targets for development of drugs to treat infections caused by the pathogens.

A network motif is defined as an interconnected subnetwork structure of several nodes that occurs much more frequently than random expectation. Motifs have been identified in a wide range of networks, such as transcriptional regulatory networks, social networks, electrical circuits, the World Wide Web, and ecological food webs. Figure 15.3 shows several types of commonly identified motifs. Despite different sizes, types, and complexity of networks, the approaches to identify network motifs are generally similar. A common method involves scanning a network for all patterns with several interconnected nodes, recording frequencies of their occurrence, and comparing pattern frequencies with those in networks that have randomly rewired links. A subnetwork pattern is considered a motif, if it has a much higher frequency than randomly expected, often defined as two or more standard deviations greater than the average number of occurrences in random networks. Different types of motifs are over-represented in networks with different types of functions. For example, in transcriptional regulatory networks, the "feed-forward loop" motif is common.

Biological networks are highly modular. Structurally, they are composed of densely interrelated nodes of proteins or genes. These subnetworks have high connectivity density and are loosely interconnected (Fig. 15.1). Functionally, modules in biological networks are those interconnected molecules that together perform specific functions. Structural modules are believed to correspond to functional modules. In the past decade, the research community has made efforts to uncover functional modules through the extraction of structural modules from large-scale biological networks. Various graph clustering algorithms and methods have been developed or adopted for modular analysis. These methods can be roughly categorized into four types based on the underlying methodologies, including density-based, partition-based, centrality-based, and hierarchical clustering approaches (Zhang et al. 2010). Many of the algorithms are implemented in the tools listed in Table 15.1.

A density-based clustering method seeks to identify densely connected subnetwork structures in a network, for example fully connected or near-fully connected subgraphs. Clique enumeration is a simple and straightforward density-based method, which can identify all cliques of up to n nodes (Spirin and Mirny 2003). Other density-based methods may seek to find densely connected subnetworks with less stringent connectivity criteria than fully connected cliques. For example, Palla et al. developed a clustering method to identify k-clique subnetwork structures, which are defined as unions of all cliques of k nodes that share $k - 1$ nodes (Palla et al. 2005). In a biological network such as a PPI network, proteins in a clique or clique-like module likely correspond to those belonging to the same protein complex. Modules identified by density-based methods can overlap with each other. Such a property is favored in biological networks because proteins and genes may participate in different functional units under different contexts. One limitation of density-based methods is that a module may never cover loosely connected nodes in a network.

Unlike density-based clustering methods, partition-based methods seek to identify an optimal partition of all nodes in a network based on a certain cost function. Cost function is an important concept in the theory of networks. It is a measure of how far away a particular solution is from an optimal solution to the problem to be solved. The cost function is dependent on what is being modeled and the *a priori* assumptions about the optimal solution. Such a method often starts with a random partition of a network, followed by iterative reassignment of nodes from different previous partitions until an optimal cost is achieved. In each iteration, a cost function is calculated, and the method outputs a partition when an optimal cost is achieved. One example of such cost function is "modularity", a score between 0 and 1, where a higher score indicates many more within-module links and many less between-module links than expected. A limitation of such methods is that network partitioning assigns nodes to distinct modules and does not allow overlapping modules.

Centrality-based clustering methods are based on network centrality statistics. One commonly used measurement of centrality-based clustering is edge betweenness, which is defined as the number of paths (edges that are connected by nodes) of the shortest lengths between all possible pairings of node pairs that pass through an edge. Edges with high betweenness correspond to bridge-like connectors in a network, the disconnection of which would result in self-contained node clusters. A classic edge betweenness based clustering method takes all the links in a network as input, and iteratively removes an edge with the highest edge betweenness (Girvan and Newman 2002). A network can be clustered into subnetwork modules after a certain percentage of edge removal. In a biological network, the resulting modules from a centrality-based clustering method often form a hierarchy because of the hierarchical modularity properties of biological networks.

Classic hierarchical clustering methods have also been applied to clustering biological networks. In a PPI network, a distance measure between a pair of proteins in a network can be as simple as their pairwise distance (shortest path length d) or a measurement based on the shortest path length, e.g. d^2. The proteins are grouped together from the closest in the network to the farthest. The grouping forms a hierarchy, and given a cutoff of the distance, proteins can be put into different clusters. One limitation of hierarchical clustering is that different linkage methods and different cutoffs would result in different output modules. In addition, classic hierarchical clustering algorithms do not consider overlapping clusters.

Prediction of the function of proteins is also a major application of network modular analysis in PPI networks. In PPI networks, subnetworks identified via network clustering likely correspond to functional modules. Despite the variety of clustering methods and algorithms, protein members in each subnetwork tend to have the same or similar functional annotations, e.g., Gene Ontology annotations (Berardini et al. 2010). Based on grouping of proteins from clusters, those proteins with unknown functions are predicted to have functional annotations that are highly enriched for the group (Brun et al. 2004; King et al. 2004).

15.4 Applications of Network Analysis

15.4.1 Network Pharmacology – Identification of Targets for Development of New Drugs to Treat Pseudomonas Infections

Network analysis can be applied to the identification of potential drug targets for treating *Pseudomonas* infections. *Pseudomonas aeruginosa* (*PA*), a common bacterium, is a ubiquitous opportunistic pathogen that can cause chronic infections in patients with damaged tissues or reduced immunity. For example, *PA* often infects the lungs of children with cystic fibrosis. *PA* possesses a remarkable capacity to resist multiple front-line antibiotics, and more effective drugs to treat *PA* infection are urgently needed.

The philosophy in traditional drug design, i.e., the "one gene, one drug, one disease" paradigm, focuses on the individual properties of a protein, for example, whether the deletion of a gene is lethal to an organism and if it is, then inhibition of the product of the gene by a drug should also be lethal. On the other hand, many effective drugs have been found to affect a handful of targets instead of a single protein (Huang 2002). Recently, understanding of biological networks and molecule functionality has given rise to a new discipline, network pharmacology, that provides the opportunity to identify new drug targets in an infectious organism, while avoiding targets that cause toxicity in the host (Zhang et al. 2012).

In order to utilize the approach advocated by network pharmacology for drug target identification to treat *PA* infection, the establishment of a genome-scale PPI network is a prerequisite. Currently, the knowledge of *PA* PPI networks is limited. Experimentally verified PPIs in *PA* are limited to a handful of protein pairs from individual small-scale studies (Goll et al. 2008). In a recent study, we computationally predicted a genome-scale PPI network in *PA* using machine learning (Zhang et al. 2012). Specifically, we first constructed a positive reference set of *PA* PPIs by homology mapping based on PPI data from three closely related species of bacteria: *C. jejuni, E. coli, and H. pylori.* Experimentally verified PPIs of the three organisms were retrieved from DIP, BIND, and literature (Xenarios et al. 2002; Peregrin-Alvarez et al. 2009; Hu et al. 2009). We then collected and compiled a set of eight genomic features of *PA* genes or their protein products, each of which may be highly relevant to protein physical interactions and thus have predictive power of PPI. We performed training and testing by a random forest classifier, and conducted PPI prediction using the trained classifier. As the output, a confidence score is associated with each predicted PPI representing the probability of the physical interaction or the co-involvement in a protein complex.

To verify these predicted PPIs, we performed three types of computational validations and confirmations. We first used an independent testing set of 35 experimentally verified *PA* PPIs from MPIDB (Goll et al. 2008), the majority of the PPIs were predicted positive in our results, while none of these PPIs can be identified when using homology mapping. In addition, we compared the network structure of the predicted network with established networks and found that they were largely the same. For example, the node degree distribution of the predicted network follows a power-law with a degree exponent of 1.69. Furthermore, we extracted 20 known *PA* drug target proteins in the network from DrugBank (Wishart et al. 2006) and observed that 12 of them hubs. This is consistent with the hypothesis that drug targets tend to be topologically important hubs and bottlenecks and functionally annotated with essentiality.

After extensive computational validation of the predicted PPI network, we performed subsequent analysis to identify putative drug target proteins based on their topological importance and functional essentiality. We also identified a set of essential functional modules that are highly enriched with these essential proteins (Zhang et al. 2012). The essential hubs and modules that are not targets of presently available drugs provide potential targets for development of new drugs that are likely to be effective in treatment of *PA* infections.

The infectious process of bacterial pathogenesis often involves PPIs between bacterial and host proteins. A map of human-*PA* protein interactions should help elucidate the disease mechanisms of *PA* infections in cystic fibrosis patients. We extracted 12 human-*PA* protein interactions between 11 human proteins and three *PA* proteins from pathogen interaction gateway (PIG) (Driscoll et al. 2009). Combining these human-*PA* interactions with the human PPIs from human protein reference database (HPRD) (Keshava Prasad et al. 2009) and *PA* PPIs from the high-confidence PPI network that contain proteins involved in human-*PA* protein interactions, we constructed a map of human-*PA* interactions (Fig. 15.4). Based

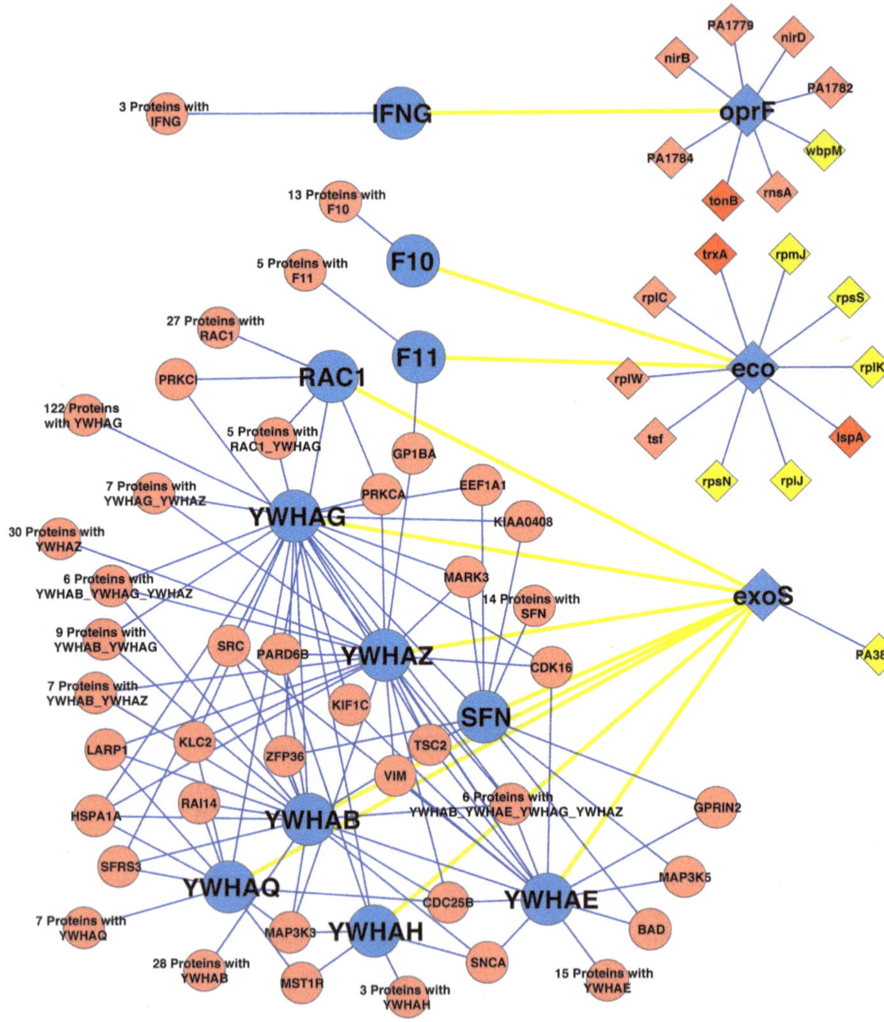

Fig. 15.4 A map of human-*Pseudomonas aeruginosa*(*PA*)PPIs. A *round node* is a human protein, and a *diamond node* is a *PA* protein. *Blue nodes* are proteins involved in human-*PA* PPIs that are denoted by *yellow edges*. *Yellow* and *red PA* proteins are essential proteins, and *red* ones are predicted to be potential drug targets. *Blue* edges denote corresponding PPIs in a human interactome and the high-confidence *PA* PPI network

on this human-*PA* interaction network, we found that *Cellular protein metabolic process* (GO:0044267) was enriched among *PA* proteins (p-value = 7.9e-6). In addition, top significantly enriched GO annotations by human proteins were identified, such as *ATP binding* (GO:0005524) of molecular functions, and *nerve growth factor receptor signaling pathway* (GO:0048011), *blood coagulation* (GO:0007596), *intracellular signaling pathway* (GO:0023034), and *platelet activation* (GO:0030168) (p-value < 1e-12). Interestingly, 10 of 22 *PA* proteins in the map of human-*PA* interactions are essential proteins, and three proteins (TonB, thioredoxin TrxA, and lipoprotein signal peptidase LspA) were predicted as potential drug targets by this study, both numbers being significantly higher than randomly expected (p-value < 1e-3). Such intensified relevance of the proteins in the map supports the validity of their being candidate drug targets for further investigations.

15.4.2 Protein-Protein Interaction Networks in High Density Lipoprotein Particles

In this section, we will illustrate the application of network analysis to characterization of protein-protein interactions (PPIs) in complex mixtures of proteins within subcellular particles using as an example our recent study on high-density lipoprotein (HDL) subspecies. Databases and tools used include Gene Ontology (GO) (Berardini et al. 2010), Human Protein Reference Database (HPRD) (Keshava Prasad et al. 2009) and The Database for Annotation, Visualization and Integrated Discovery (DAVID) v6.7 (Huang da et al. 2009).

HDLs are blood-borne complexes consisting of proteins and lipids that play critical roles in cardiovascular disease (CVD) (Boden 2000; Lewis and Rader 2005; Cuchel and Rader 2006). There is growing controversy about whether and/or how HDL prevents CVD. A widely accepted mechanism is called reverse cholesterol transport, where HDL promotes cholesterol efflux from peripheral cells such as macrophage-derived foam cells in the vessel wall to transport excess cholesterol and other lipids back to the liver for catabolism (Franceschini et al. 1991). HDL is also involved in other CVD-protective functions, including anti-oxidation, anti-inflammation and endothelial relaxation (Watson et al. 1995; Naqvi et al. 1999; Nofer et al. 2002; Barter et al. 2004; Negre-Salvayre et al. 2006; Mineo et al. 2006).

There is evidence that HDL is composed of numerous distinct particle subpopulations, each containing a unique protein makeup that plays critical physiological roles (Gordon et al. 2010a, b). The major activities of HDLs rely on the cooperative interactions among its protein components. Recent proteomic studies on HDL have identified upwards of 75 distinct HDL associated protein components (Gordon et al. 2010a; Heller et al. 2005; Karlsson et al. 2005; Rezaee et al. 2006; Vaisar et al. 2007; Davidson et al. 2009). We fractionated HDL particles by gel filtrations and identified the proteins within subpopulations using mass spectrometry (Gordon et al. 2010a). In total, 106 HDL-associated proteins were identified, 17 of which have not been reported previously to be associated with HDL particles (Gordon et al. 2010a). To elucidate the potential roles of these newly identified HDL proteins, we first constructed a PPI network among these HDL associated proteins using data retrieved from the HPRD (Keshava Prasad et al. 2009) (Fig. 15.5). Eight out of the 17 new HDL proteins have a direct interaction with known HDL proteins, and an additional three can be connected to known HDL proteins by one intermediate

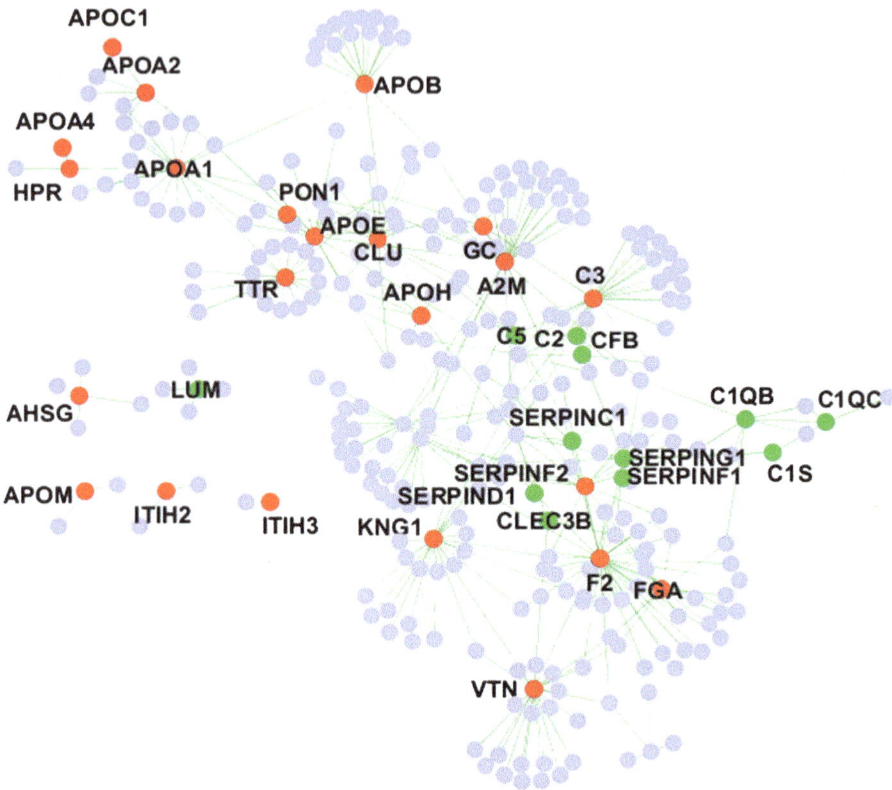

Fig. 15.5 Protein interaction networks between newly identified lipid associated proteins and known HDL associated proteins. In this PPI network, *red nodes* denote known HDL proteins; *green nodes* denote new phospholipid associated proteins

protein. This supports the possibility that these new proteins play a role in HDL function. We also examined the functions of the proteins in this network based on Gene Ontology (GO) (Berardini et al. 2010). GO annotations provide information about the distribution of proteins in different biological processes and molecular functions. We were able to assign new functional roles to several of these newly identified proteins, such as complement function and serine protease inhibitors (Gordon et al. 2010a).

In addition to discovering new HDL-associated proteins, the proteomic data generated in gel filtration has also allowed us to identify distinct HDL particles using a network approach. We measured two proteins' likelihood to co-occur in the same particle by quantifying the similarity between the migration patterns exhibited in gel filtration. The result is a similarity network that contains 90 nodes and 278 edges (Fig. 15.6). In the network, each node is an HDL-associated protein and each edge is an interaction between two proteins that have highly similar co-migration patterns. Given the similarity network, next we identified cliques that may correspond to the

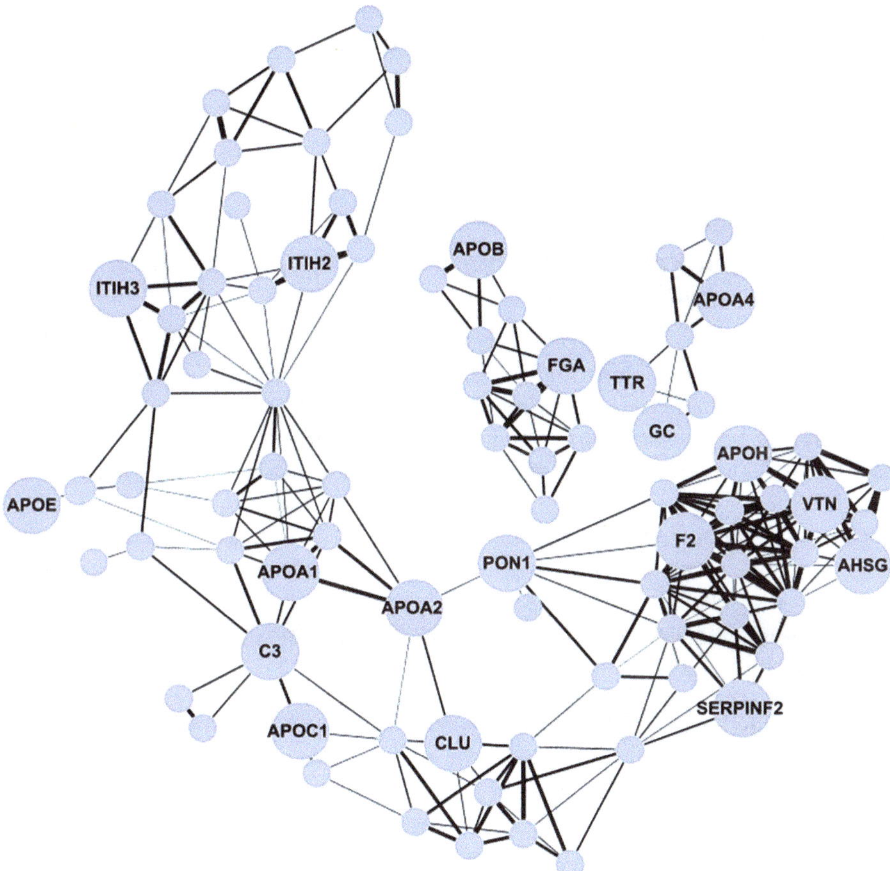

Fig. 15.6 The co-migration similarity network of HDL proteins, where each node is a HDL protein and the width of the edge is proportional to the co-migration similarity score between the connected two proteins

distinct HDL subspecies. A clique in a network is a subset of its nodes, where every two nodes in this subset are connected by an edge; a maximal clique is a clique that cannot be extended by including one more adjacent node. Thus, a maximal clique in the similarity network may correspond to the entirety or partial of a HDL subspecies, where its components have quite similar co-migration patterns. We identified all the maximal cliques from the similarity network, which generated 52 maximal cliques sizing from 3 to 10 nodes (proteins).

We elucidated the functions of each clique by performing functional enrichment analysis on these cliques based on GO annotations. We found 35 out of the 52 maximal cliques have significantly enriched functions. The most enriched functions are regulation of cholesterol esterification, complement activation, coagulation, innate immune response, and regulation of acute inflammatory response. These results are consistent with previously known HDL functions such as reverse cholesterol transport, anti-oxidation, anti-inflammation and endothelial relaxation, but suggest that subpopulations of the HDL particles differ in functional profiles. These findings are guiding the design of mouse gene knockout experiments to validate the existence of these cliques and their proposed functions.

15.4.3 Detection of Brain Functional Connectivity Map in Children with Attention-Deficit/Hyperactivity Disorder (ADHD)

Imaging techniques, such as magnetoencephalography (MEG), positron emission tomography (PET), and magnetic resonance imaging (MRI), provide opportunities for analyzing both structural and functional organization of human brains. Traditional analysis of brain images focuses on the study of individual brain regions. The differences between the brains of two groups of humans, e.g., normal versus diseased, are compared on a region-by-region basis. There are limitations to this approach. First, some disorders may affect a large number of brain regions, and the significance of changes in any individual region is difficult to establish. Second, the functional impairment may be caused by abnormalities in the interconnections among several brain regions. To overcome these limitations, the different anatomic regions of the brain must be analyzed as an interconnected whole.

Network analysis takes into account patterns of connections among brain regions or voxels in digital images. Different from molecular networks presented in the previous sections, the nodes in brain networks represent voxels or brain regions, depending on the expected granularity of the analysis, and the edges represent the structural or functional connections among the different voxels or brain regions.

There are three main types of brain networks: structural networks, functional networks, and effective networks. The edges in these three brain networks represent physical connections, functional correlations, and causal functional relations, respectively. Brain networks can be represented as weight matrices or graphs. Figure 15.7 shows an example of a functional brain network. The forms of representation for the structural, functional, and effective networks are similar to one another.

Topological analysis can be performed on the brain networks to reveal the connection patterns of different groups of brains, e.g., healthy versus diseased. The results of analysis can reveal the topological differences that cannot be revealed by traditional non-network based approaches. The topological differences can provide new insights into the mechanism of the diseases.

Small-world property is one of the topological features found in healthy brains. It is reported that a variety of psychiatric disorders, such as Alzheimer's disease in adults and ADHD in children, are associated with disrupted small-worldness. In the following, we use a study of brain network of ADHD as an example to illustrate how the brain network approach can be applied to study the mechanisms of the diseases.

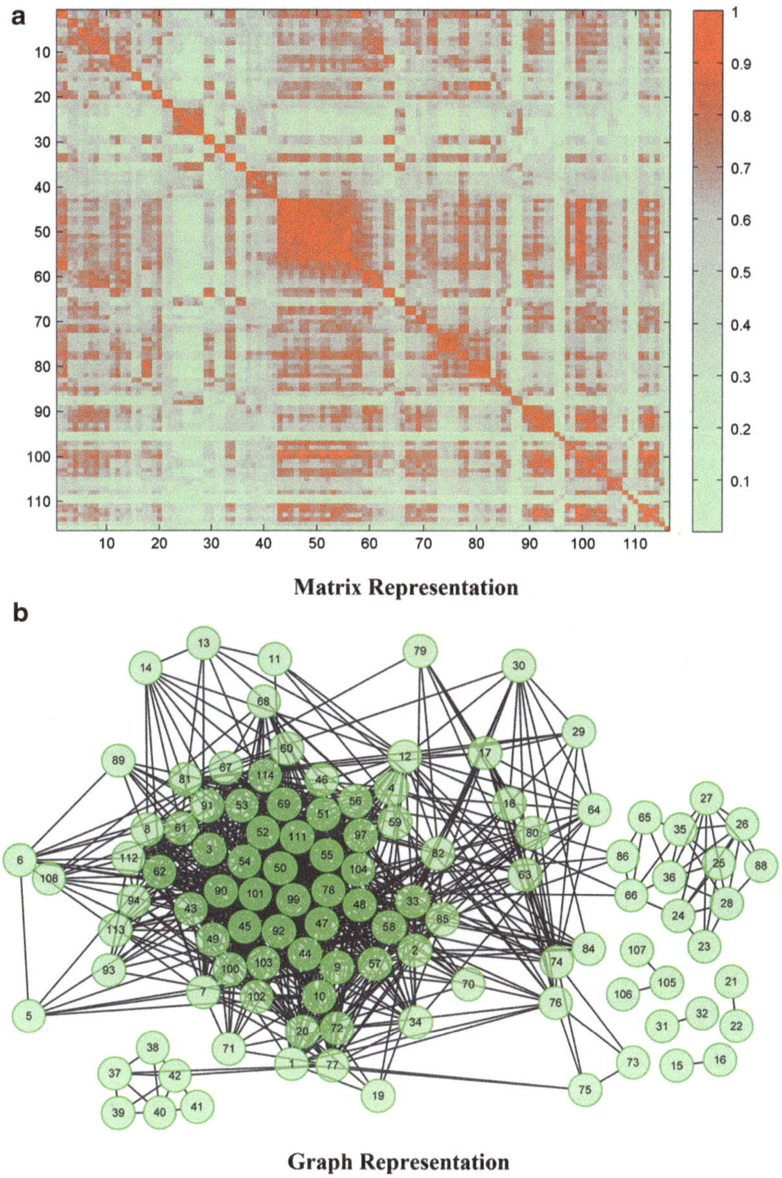

Fig. 15.7 An example of a functional brain network generated from the fMRI data reported in (Wang et al. 2009) and reanalyzed using tools described in the text of this chapter. The whole brain is parcellated into 116 brain regions based on an automated anatomical labeling (AAL) template. The correlation between every pair of brain regions is calculated and shown in the correlation matrix in (**a**). The correlations are thresholded at 0.8 and the resulting graph is shown in (**b**)

ADHD is a psychiatric disorder that is widely studied in children. Wang et al. (2009) investigated the topological changes of the functional brain networks in children with ADHD. The whole brain was parcellated into 90 regions using the anatomical automatic labeling (AAL) template, and a mean time course was calculated for each region. Pair-wised Pearson's correlation between the regional time courses was calculated to construct the 90×90 connectivity matrix, which was subsequently thresholded at different levels of cost to generate the binary graph (network G), with nodes representing brain regions and edges representing undirected connections. Unlike other investigators, they used the global and local efficiency instead of clustering coefficient and characteristic path length, to evaluate the small-worldness of the networks. Then, they generated populations of regular networks (G_{reg}) and random networks (G_{rand}) that have the same number of nodes and edges as network G, respectively. The network G was considered to be small-world, if it has a global efficiency larger than G_{reg} while smaller than G_{rand}, and a local efficiency larger than G_{rand} while smaller than G_{reg}.

For the topological analysis, they first compared the global efficiencies and local efficiencies of brain networks with those of the random networks, in which nodes are randomly connected, and regular networks, in which nodes are connected to a fixed number of other nodes. Results demonstrated that brain networks in both ADHD group and control group exhibited small-world topology. Comparing the ADHD group with the control group, no significant difference in global efficiency was found, whereas the local efficiency was significantly different between the two groups at a range of cost ($0.12 < \text{cost} < 0.16$). At the cost of 0.15, when the between-group difference in local efficiency was most significant, the ADHD group exhibited an increased local efficiency (p-value = 0.016) and a decreased global efficiency (p-value = 0.144). The patterns of the changes in the network efficiency supported the notion that ADHD promoted a shift from small-world networks toward regular networks. Furthermore, it has been suggested that the small-world configuration of brain networks maximizes the efficiency of information processing at a relatively low wiring cost. Disease-related shift toward either random or regular networks may reflect disrupted local/global communication in the brain, although the biological causes underlying the shift remain unclear.

Despite several recent successes, there are still many challenges to applying network approaches to study brain diseases. Further studies are necessary to address questions such as why there is a network topological alteration, when do changes in network topology first occur, how changes in topology progress, and how differences among groups of brain networks can be measured more accurately? There are also specific challenges in applying network approaches to the study of psychiatric disorders in the brains of children. Children's brains have a larger functional reorganization capability than adults' brains. The plasticity of the children's brains could cause difficulty in identifying the abnormality of the brain networks.

15.5 Summary

In this chapter, we discussed general biological network analysis with a focus on network topology and structures of subnetworks. We also provided three examples of applications of network analysis in translational research. In general, biological network analysis has proved useful in applications such as predicting the function of proteins, guiding the design of wet-lab experiments, and identifying markers of various stages of disease. Applications of network analysis have gone beyond molecular entities to characterization of many other complex relationships among entities, e.g., regions of the brain. However, they have not been fully exploited in studies of pediatric diseases. Chapters 16 (Bioinformatics and Orphan Diseases) and 17 (Functional Genomics – Transcriptional Regulatory Networks) provide additional examples of how tools of network analysis can be applied to identification of potential new therapies of relatively rare diseases of childhood and to research on development and maturation of human organ systems.

References

Albert R, Jeong H, Barabasi AL. Error and attack tolerance of complex networks. Nature. 2000;406(6794):378–82.

Barabasi AL, Oltvai ZN. Network biology: understanding the cell's functional organization. Nat Rev Genet. 2004;5(2):101–13.

Barter PJ, et al. Antiinflammatory properties of HDL. Circ Res. 2004;95(8):764–72.

Bastian M, Heymann S, Jacomy M. Gephi: an open source software for exploring and manipulating networks. In: International AAAI conference on weblogs and social media. 2009;361–362.

Batagelj V, Mrvar A. Pajek – analysis and visualization of large networks. In: Graph drawing software. Berlin/New York: Springer; 2004. p. 77–103.

Berardini TZ, et al. The gene ontology in 2010: extensions and refinements. Nucleic Acids Res. 2010;38(Database issue):D331–5.

Boden WE. High-density lipoprotein cholesterol as an independent risk factor in cardiovascular disease: assessing the data from Framingham to the Veterans Affairs High–Density Lipoprotein Intervention Trial. Am J Cardiol. 2000;86(12A):19L–22.

Brohee S, et al. Network analysis tools: from biological networks to clusters and pathways. Nat Protoc. 2008;3(10):1616–29.

Brun C, Herrmann C, Guenoche A. Clustering proteins from interaction networks for the prediction of cellular functions. BMC Bioinformatics. 2004;5:95.

Cuchel M, Rader DJ. Macrophage reverse cholesterol transport: key to the regression of atherosclerosis? Circulation. 2006;113(21):2548–55.

Davidson WS, et al. Proteomic analysis of defined HDL subpopulations reveals particle-specific protein clusters: relevance to antioxidative function. Arterioscler Thromb Vasc Biol. 2009;29(6):870–6.

Driscoll T, et al. PIG–the pathogen interaction gateway. Nucleic Acids Res. 2009;37(Database issue):D647–50.

Franceschini G, Maderna P, Sirtori CR. Reverse cholesterol transport: physiology and pharmacology. Atherosclerosis. 1991;88(2–3):99–107.

Girvan M, Newman ME. Community structure in social and biological networks. Proc Natl Acad Sci U S A. 2002;99(12):7821–6.
Goel R, et al. Human Protein Reference Database and Human Proteinpedia as resources for phosphoproteome analysis. Mol Biosyst. 2012;8(2):453–63.
Goll J, et al. MPIDB: the microbial protein interaction database. Bioinformatics. 2008;24(15):1743–4.
Gordon SM, et al. Proteomic characterization of human plasma high density lipoprotein fractionated by gel filtration chromatography. J Proteome Res. 2010a;9(10):5239–49.
Gordon S, et al. High-density lipoprotein proteomics: identifying new drug targets and biomarkers by understanding functionality. Curr Cardiovasc Risk Rep. 2010b;4(1):1–8.
Guelzim N, et al. Topological and causal structure of the yeast transcriptional regulatory network. Nat Genet. 2002;31(1):60–3.
Hartwell LH, et al. From molecular to modular cell biology. Nature. 1999;402(6761 Suppl): C47–52.
Heller M, et al. Mass spectrometry-based analytical tools for the molecular protein characterization of human plasma lipoproteins. Proteomics. 2005;5(10):2619–30.
Hu Z, et al. VisANT 3.5: multi-scale network visualization, analysis and inference based on the gene ontology. Nucleic Acids Res. 2009;37(Web Server issue):W115–21.
Huang S. Rational drug discovery: what can we learn from regulatory networks? Drug Discov Today. 2002;7(20 Suppl):S163–9.
Huang da W, Sherman BT, Lempicki RA. Systematic and integrative analysis of large gene lists using DAVID bioinformatics resources. Nat Protoc. 2009;4(1):44–57.
Jeong H, et al. Lethality and centrality in protein networks. Nature. 2001;411(6833):41–2.
Kao HL, Gunsalus KC. Browsing multidimensional molecular networks with the generic network browser (N-Browse). Curr Protoc Bioinformatics. 2008;Chapter 9:Unit 9 11.
Karlsson H, et al. Lipoproteomics II: mapping of proteins in high-density lipoprotein using two-dimensional gel electrophoresis and mass spectrometry. Proteomics. 2005;5(5):1431–45.
Keshava Prasad TS, et al. Human protein reference database–2009 update. Nucleic Acids Res. 2009;37(Database issue):D767–72.
King AD, Przulj N, Jurisica I. Protein complex prediction via cost-based clustering. Bioinformatics. 2004;20(17):3013–20.
Lewis GF, Rader DJ. New insights into the regulation of HDL metabolism and reverse cholesterol transport. Circ Res. 2005;96(12):1221–32.
Maslov S, Sneppen K. Specificity and stability in topology of protein networks. Science. 2002;296(5569):910–13.
Mineo C, et al. Endothelial and antithrombotic actions of HDL. Circ Res. 2006;98(11):1352–64.
Naqvi TZ, et al. Evidence that high-density lipoprotein cholesterol is an independent predictor of acute platelet-dependent thrombus formation. Am J Cardiol. 1999;84(9):1011–17.
Negre-Salvayre A, et al. Antioxidant and cytoprotective properties of high-density lipoproteins in vascular cells. Free Radic Biol Med. 2006;41(7):1031–40.
Nofer JR, et al. HDL and arteriosclerosis: beyond reverse cholesterol transport. Atherosclerosis. 2002;161(1):1–16.
O'Madadhain J, et al. The JUNG (Java Universal Network/Graph) framework. Technical Report UCI-ICS 03-17. Irvine: UC Irvine; 2003.
Palla G, et al. Uncovering the overlapping community structure of complex networks in nature and society. Nature. 2005;435(7043):814–18.
Peregrin-Alvarez JM, et al. The modular organization of protein interactions in *Escherichia coli*. PLoS Comput Biol. 2009;5(10):e1000523.
Ravasz E. Detecting hierarchical modularity in biological networks. Methods Mol Biol. 2009;541:145–60.
Rezaee F, et al. Proteomic analysis of high-density lipoprotein. Proteomics. 2006;6(2):721–30.
Rual JF, et al. Towards a proteome-scale map of the human protein-protein interaction network. Nature. 2005;437(7062):1173–8.

Shannon P, et al. Cytoscape: a software environment for integrated models of biomolecular interaction networks. Genome Res. 2003;13(11):2498–504.
Spirin V, Mirny LA. Protein complexes and functional modules in molecular networks. Proc Natl Acad Sci U S A. 2003;100(21):12123–8.
Stuart JM, et al. A gene-coexpression network for global discovery of conserved genetic modules. Science. 2003;302(5643):249–55.
Vaisar T, et al. Shotgun proteomics implicates protease inhibition and complement activation in the antiinflammatory properties of HDL. J Clin Invest. 2007;117(3):746–56.
Wagner A, Fell DA. The small world inside large metabolic networks. Proc Biol Sci. 2001;268(1478):1803–10.
Wang L, et al. Altered small-world brain functional networks in children with attention-deficit/hyperactivity disorder. Hum Brain Mapp. 2009;30(2):638–49.
Watson AD, et al. Protective effect of high density lipoprotein associated paraoxonase. Inhibition of the biological activity of minimally oxidized low density lipoprotein. J Clin Invest. 1995;96(6):2882–91.
Wishart DS, et al. DrugBank: a comprehensive resource for in silico drug discovery and exploration. Nucleic Acids Res. 2006;34(Database issue):D668–72.
Wu J, et al. Integrated network analysis platform for protein-protein interactions. Nat Methods. 2009;6(1):75–7.
Xenarios I, et al. DIP, the Database of Interacting Proteins: a research tool for studying cellular networks of protein interactions. Nucleic Acids Res. 2002;30(1):303–5.
Yip KY, et al. The tYNA platform for comparative interactomics: a web tool for managing, comparing and mining multiple networks. Bioinformatics. 2006;22(23):2968–70.
Yu H, et al. The importance of bottlenecks in protein networks: correlation with gene essentiality and expression dynamics. PLoS Comput Biol. 2007;3(4):e59.
Zhang M, et al. Molecular network analysis and applications. In: Alterovitz G, Ramoni M, editors. Knowledge-based bioinformatics: from analysis to interpretation. Chichester: Wiley; 2010.
Zhang M, et al. Prediction and analysis of the protein interactome in *Pseudomonas aeruginosa* to enable network-based drug target selection. PLoS One. 2012;7(7):E41202.

Chapter 16
Orphan Diseases, Bioinformatics and Drug Discovery

Anil G. Jegga, Cheng Zhu, and Bruce J. Aronow

Abstract In general, a rare or orphan disease is any disease that affects a small percentage of the population. Since a majority of the known orphan diseases are genetic, they are present throughout the life of affected individuals. Many of the orphan diseases appear early in life and approximately 30% of children with orphan diseases die before the age of 5. The bulk of genes and pathways underlying these diseases remain unknown and pose a major gap in orphan disease research. In spite of technological advances and opportunities available to understand the causes of orphan diseases and for developing innovative medical approaches, most of the current efforts are focused either on a single or related group of orphan diseases. Relatively few studies have attempted global analysis of all orphan diseases. Constructing networks that underlie biological processes and pathways associated with orphan diseases and orphan drugs facilitate identification of the functional units that respond to genetic perturbations and potentially affect disease risk or therapeutics. Analysis of these biological networks can also identify common pathways or processes for multiple orphan diseases that are biologically related. Comprehensive understanding of such molecular bases may provide opportunities for novel interventions that are beneficial for an array of related orphan diseases.

A.G. Jegga, DVM, M.S. (✉)
Department of Pediatrics, University of Cincinnati College of Medicine, Cincinnati, OH, USA

Division of Biomedical Informatics, Cincinnati Children's Hospital Medical Center, 3333 Burnet Ave, MLC 7024, Cincinnati, OH 45229-3039, USA
e-mail: Anil.Jegga@cchmc.org

C. Zhu, M.S.
School of Computing Sciences and Informatics, University of Cincinnati College of Engineering and Applied Science, Cincinnati, OH, USA

B.J. Aronow, Ph.D.
Department of Pediatrics, University of Cincinnati College of Medicine, Cincinnati, OH, USA

Genome Informatics Core Laboratory, Division of Biomedical Informatics, Cincinnati Children's Hospital Medical Center, 3333 Burnet Ave, MLC 7024, Cincinnati, OH 45229-3039, USA

J.J. Hutton (ed.), *Pediatric Biomedical Informatics: Computer Applications in Pediatric Research*, Translational Bioinformatics 2, DOI 10.1007/978-94-007-5149-1_16,

In this chapter, we review some of the current bioinformatic analytical options available for orphan disease and drug research including computational approaches for candidate gene prioritization. We also discuss strategies and present examples and case studies of common drugs being repositioned for treatment of orphan diseases.

16.1 Introduction

A rare or orphan disease (OD) is any disease that affects a small percentage of the population. Most of the known ODs are genetic, and therefore are present throughout the life of an affected individual. Many appear early in life and approximately 30% of children with ODs die before the age of 5. In the United States, the Rare Disease Act of 2002 defines an OD as any disease or condition that affects fewer than 200,000 persons in the United States, while the European Commission on Public Health, defines ODs as those that are life-threatening or chronically debilitating and are of such low prevalence (1 in 2,000 people) that special combined efforts are needed to address them. On the other hand, in Japan an OD is defined as one that affects fewer than 50,000 patients. While the incidence of an individual OD may be small, cumulatively, in the US itself, the 8,000 known ODs affect about 25 million Americans, or nearly 10% of the US population (Rados 2003). While some of the listed ODs are well-known and well-studied (e.g., cystic fibrosis), very little is known about a majority of ODs primarily because several ODs affect patient populations of fewer than a hundred. Additionally, about 250 new ODs and conditions are described each year (Wastfelt et al. 2006).

Even though opportunities now exist to accelerate progress toward understanding the basis for many more ODs and for developing innovative medical approaches, relatively few efforts have successfully addressed scientific or technical questions across a spectrum of ODs (Zhang et al. 2011; Field and Boat 2010). Constructing networks that underlie biological processes and pathways associated with ODs and identifying the functional units that respond to genetic perturbations and potentially affect disease risk or therapeutic response may facilitate systematic advancement of OD research and health care in a favorable direction (Zhang et al. 2011). For instance, studies of biological networks can identify common pathways or processes for multiple ODs that are biologically related. A comprehensive understanding of these networks may provide opportunities for interventions that are beneficial for an array of related ODs. Furthermore, this capability may open the door for the discovery of single therapies that can not only treat multiple ODs, but, potentially, also more common diseases (Field and Boat 2010).

In this chapter, we review some of the current bioinformatics and genomics-based approaches to study the ODs and drug discovery process including opportunities to reposition known drugs, i.e. show efficacy of the drugs as treatment of diseases where they were not previously known to be effective. We present several case studies and examples to illustrate these approaches. In the first section we outline some of the fundamental concepts and introduce organizations and resources related to orphan disease and drugs. In the second and third sections we specifically focus on bioinformatics and genomics centered approaches used to investigate ODs and discover drug targets and drug repositioning candidates for ODs.

16.2 The Orphan Drug Act, 1983

The United States' Orphan Drug Act (ODA; 1983) includes both rare diseases and any non-rare diseases for which there is no reasonable expectation that the cost of developing and making available a drug for such a disease in the US can potentially be recovered from sales of that drug in the US. Since the definitions of rare diseases refer to treatment availability, resource scarcity and disease severity, rare diseases are now commonly referred to as orphan diseases (ODs), especially after the orphan drug movement that began in the US in 1983. The ODA was intended to encourage the development and marketing of drugs ("orphan drugs") to treat ODs and conditions.

The ODA evolved in response to the small number of orphan drugs that were approved in the US in the years prior to the approval of the ODA (United States Food and Drug Administration 1992). About 8,000 ODs have been identified, and a list is maintained by the Office of Rare Diseases (ORD) at the NIH. The Food and Drug Administration (FDA) administers the ODA and reviews applications for orphan designations. Some of the incentives provided to the drug companies involved in production of orphan drugs include 7-year marketing exclusivity to sponsors of approved orphan products, a tax credit of 50% of the cost of conducting human clinical testing, and research grants for clinical testing of new therapies to treat ODs (The Orphan Drug Act – implementation and impact 2001). The ORD designates diseases that qualify and administers the small grants program, while the Center for Drug Evaluation and Research and the Center for Biologics Evaluation and Research review applications for marketing approval (The Orphan Drug Act – implementation and impact 2001).

16.3 Orphan Diseases and Drugs – Organizations and Resources

In the following sections, we briefly outline some of the organizations and resources focused on orphan diseases and drugs (see Table 16.1).

16.3.1 Genetic and Rare Diseases Information Center (GARD)

The Genetic and Rare Diseases Information Center was created in 2002 by the National Human Genome Research Institute (NHGRI) and the Office of Rare

Table 16.1 List of resources relevant to orphan diseases and drugs

Resource name	Description and URL
EURODIS	European organization for rare diseases; http://www.eurordis.org
List of marketed orphan drugs in Europe	http://www.fda.gov/orphan/designat/Approvals.htm
List of orphan drugs in Australia	http://www.tga.gov.au/docs/html/orphand2.htm
List of the American orphan diseases	http://rarediseases.info.nih.gov/RareDiseaseList.aspx
NIH rare diseases (GARD)	Genetic and Rare Diseases; http://rarediseases.info.nih.gov/GARD/
NORD	National Organization for Rare Disorders; http://www.rarediseases.org
OMIM	Online Mendelian Inheritance in Man; http://omim.org
OOPD at FDA	Office of Orphan Product Development; http://www.fda.gov/orphan
Orphan Disease Networks	Orphan disease and gene networks; http://research.cchmc.org/od
Orphan drugs at FDA	Orphan drug designations and approvals; http://www.accessdata.fda.gov/scripts/opdlisting/oopd/index.cfm
Orphanet	Portal for rare diseases and orphan drugs; http://www.orpha.net
RDRD from FDA	Rare Disease Repurposing Database; http://www.fda.gov/ForIndustry/DevelopingProductsforRareDiseasesConditions/

Diseases Research (ORDR) – two agencies at the National Institutes of Health (NIH) – to help people find useful information about genetic and orphan diseases. In December 2011, NIH established the National Center for Advancing Translational Sciences (NCATS) to speed movement of discoveries from the laboratory to patients. NCATS now includes the ORDR, which coordinates and supports OD research, and TRND, the Therapeutics for Rare and Neglected Diseases, a program to encourage and speed the development of new drugs for rare and neglected diseases. With an aim to work closely with partners in the regulatory, academic, nonprofit, and private sectors, NCATS goals are to identify and overcome hurdles that slow the development of effective treatments and cures.

16.3.2 European Organization for Rare Disorders (EURORDIS)

The European Organization for Rare Diseases, a European equivalent of the USA's GARD, is a patient-driven alliance of patient organizations and individuals active in the field of rare diseases. Its mission is to build a strong pan-European community of patient organizations and people living with rare diseases, to be their voice at the European level, and – directly or indirectly – to fight against the impact of rare diseases on their lives.

16.3.3 Orphanet

Orphanet (Ayme 2003) is the reference portal for information on rare diseases and orphan drugs. It is perhaps the best known and most comprehensive database for ODs and drugs. It is led by a European consortium of around 40 countries, coordinated by the French team. National teams are responsible for the collection of information on specialized clinics, medical laboratories, ongoing research and patient organizations in their country. The French coordinating team is responsible for the infrastructure of Orphanet, management tools, quality control, rare disease inventory, classifications and production of the encyclopedia. Orphanet is governed by various committees, which independently supervise the project in order to ensure its coherence, evolution and viability. However, because it predominantly follows European Union (EU) regulations and definitions of OD, there is a considerable difference between the US-based NIH list of rare diseases (http://rarediseases.info.nih.gov/RareDiseaseList.aspx) and the EU prevalence-based Orphanet list of rare diseases.

16.3.4 OMIM Database

The OMIM (Online Mendelian Inheritance in Man) database (Hamosh et al. 2000) is a comprehensive and authoritative compendium of human genes and genetic phenotypes. The full-text, referenced overviews in OMIM contain information on all known Mendelian disorders and over 12,000 genes. OMIM focuses on the relationship between phenotype and genotype. It is updated daily, and the entries contain copious links to other genetics resources. OMIM, while useful, is not limited to rare conditions. Neither Orphanet nor OMIM databases have analytical servers that can facilitate further analysis (e.g., gene set enrichment analysis of all genes associated with a particular OD).

16.4 Mutant Gene Network Analyses of Orphan Diseases

Most of the OD-related research efforts focus on either a single OD or a small set of related ODs. Additionally, previous studies on disease networks did not separate out the ODs from the common diseases. To address these issues, we

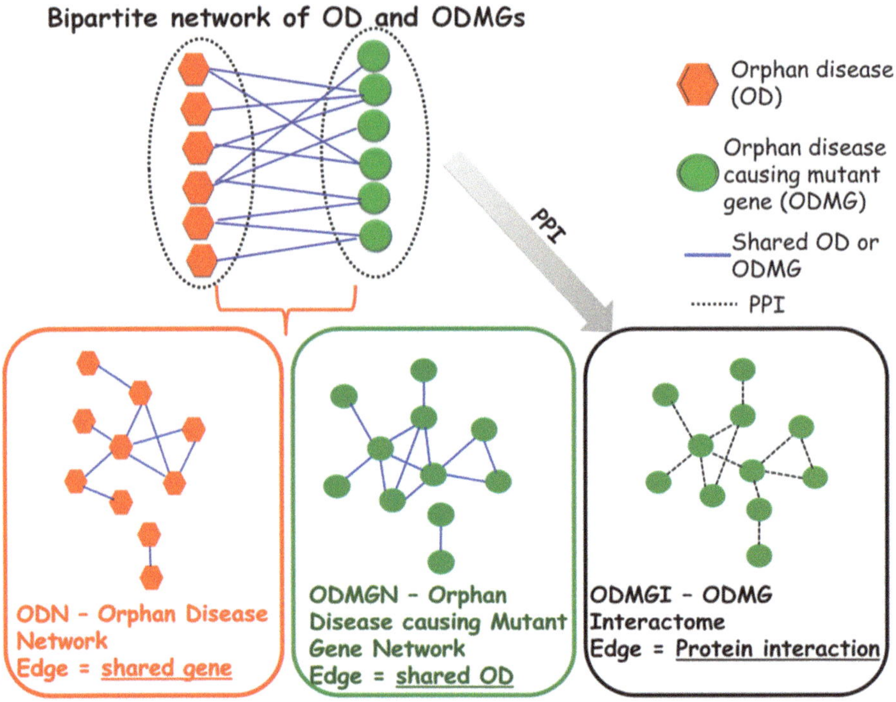

Fig. 16.1 Schematic representation of various networks related to ODs and ODMGs. Starting with a bipartite network (*top panel*) of orphan diseases (*ODs*; *red hexagons*) and their known causal genes (orphan disease-causing mutant genes or *ODMGs*; *green circles*), three types of networks are generated. The ODN (*lower panel left-hand side*) is an OD network where a node represents OD while the edge represents shared ODMG. The ODMGN (*lower panel, central panel*) represents an ODMG network with nodes being ODMGs while the edges represent shared ODs. The third network, ODMG interactome is built using the human protein-protein interactions and here the nodes are ODMGs while the edge represents a protein interaction between the ODMG-encoded proteins

have recently conducted a bioinformatic-based global analysis of all ODs with known OD-causing mutant genes (ODMGs) to identify and investigate relationships based on shared genes or shared functional features (Zhang et al. 2011). Most importantly, using bioinformatics- and network analyses-based approaches we tackled an interesting question of potential differences in properties between ODMGs and causal genes of common diseases. The orphan diseases and mutant gene information was downloaded from the Orphanet database (Ayme 2003). Starting with a bipartite network of known OD and OD-causing mutant genes (1,772 ODs and 2,124 ODMGs) and using the human protein interactome, we first constructed and topologically analyzed three networks (see Fig. 16.1): orphan disease network (ODN), orphan disease-causing mutant gene network (ODMGN) and orphan disease-causing mutant gene interactome (ODMGI). In the ODN, nodes are ODs while edges are shared genes; the ODMGN comprised ODMGs as nodes and shared ODs as edges. The orphan disease gene interactome is the protein interactions network of ODMGs. We also compared these networks with earlier studies focusing on all diseases (Goh et al. 2007; Feldman et al. 2008) primarily to answer the question as to whether there is any difference in the network topology or functional properties of orphan disease genes and common disease genes. Network-based approaches helped us in finding that ODMGs are predominantly essential, a finding that is in contrast to the previous reports of disease genes being non-essential (Goh et al. 2007; Feldman et al. 2008).

By integrating data from mouse knock-out models with the network-based approaches, we also found that ortholog gene knock-out models of orphan disease genes in the mouse in most cases either result in premature death or are embryonic lethals. Thus, informatics-based approaches integrating heterogeneous data sources (e.g., human orphan disease-causing mutant genes, protein interactome, and data from mouse knock-out models) and applying network analyses helped to better understanding the orphan diseases. For example, the two findings – ODMGs are predominantly essential and are associated with premature deaths – probably explain why most cases of orphan diseases are life-threatening. Most of the previous studies elucidating relationships between diseases are genecentric and therefore are limited in their discovery of new and unknown disease relationships (Suthram et al. 2010).

To address this, we (Zhang et al. 2011) and others (Suthram et al. 2010; Li and Agarwal 2009; Linghu et al. 2009) recommend using functional linkage maps. As part of this study, we therefore built functional linkage networks (Fig. 16.2) of ODs based on shared significant pathways or biological processes rather than just shared genes. Briefly, we first extracted all those ODs with four or more mutant genes from our original data set. Starting with this filtered sub-bipartite network of 196 ODs and 1,087 genes (1,283 total nodes and 1,395 total edges), we built OD-OD networks based on shared genes and shared functions. The enriched functions ($p < 0.05$) for each of the 196 ODs were determined with the ToppFun application (http://toppgene.cchmc.org) (Chen et al. 2009a). Using the enriched features for each of the orphan diseases, we rebuilt the orphan disease networks. However, this time the edge between two ODs represents an enriched shared function (BP, CC, Pathway, or MP) and not necessarily a shared gene (Fig. 16.2). After generating these function-based OD networks, we compared them with the gene-based orphan disease networks to find the overlapping nodes and edges. The results were surprising. Although the node agreement between the gene-based and function-based ODNs was relatively higher, the edge agreement was much lower and indicates that their wiring is significantly different. This suggests that the relationship between the ODs cannot be fully captured by the gene-based networks alone. Thus, by considering functional connectivity between causative genes involved in different orphan diseases, relationships between orphan diseases that are based on underlying molecular mechanisms can be revealed. Such associations can potentially be used

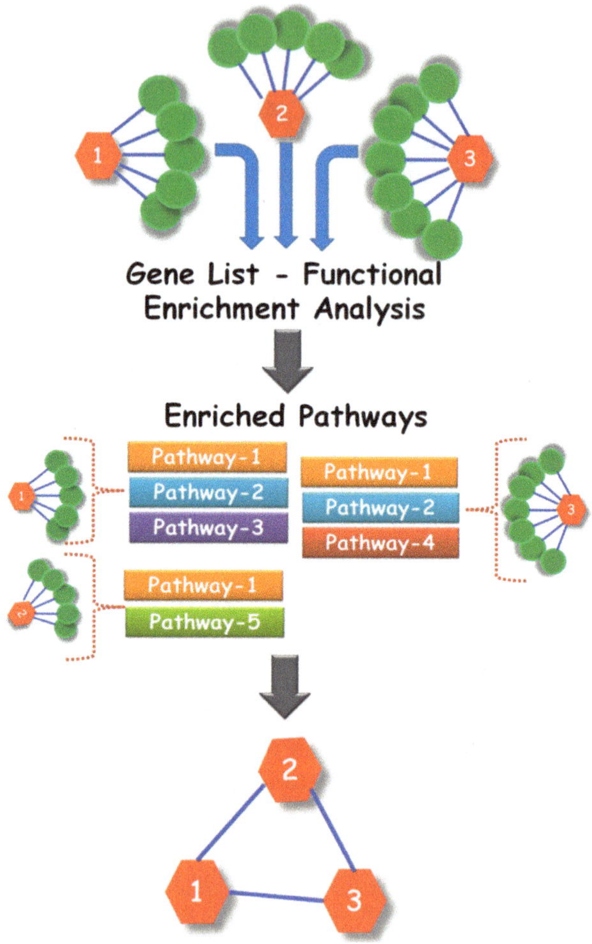

Fig. 16.2 Schematic representation of steps involved in building an orphan disease functional linkage network. ODs (e.g., 1, 2, 3; *red hexagons*) with known causal genes (*green circles*) are subjected to enrichment analyses (using ToppFun; http://toppgene.cchmc.org) to identify enriched features (e.g., pathways; *colored rectangles*). Two ODs are then connected if they share an enriched feature (pathway, in this case; *blue edges* represent shared enriched pathways). Thus, although ODs 1, 2 and 3 do not share any genes (*top panel*), they can still be connected via shared enriched pathways (*lower panel*)

to generate novel hypotheses on the molecular mechanisms of diseases, and can in turn guide the development of relevant therapy (Linghu et al. 2009) or potential drug repositioning opportunities (Zhang et al. 2011).

Apart from leading to new insights into the biological underpinnings of various orphan diseases, a global analysis of the *orphan diseasome* will encourage the development of new and innovative research on these rare conditions, which have been hitherto understudied. Additionally, the global analysis of all ODs can help in analyzing co-morbidities and the underlying molecular basis apart from establishing potential networking opportunities. The functional linkage networks of ODs apart from the conventional gene-based connectivity maps of diseases have direct implications to the drug discovery process. For more details the readers are referred to the original publication (Zhang et al. 2011).

16.5 Orphan Disease Gene Identification and Prioritization – Computational Approaches

Despite the advances in genomewide techniques such as linkage analysis and association studies, the selected disease loci are actually a region of the chromosome rather than location of a "gene". A "locus" identified in this way may contain several hundreds of candidate genes. For instance, in the OMIM database, over 900 ODs are described that have been mapped to one or more such gene map loci and are classified as having an 'unknown molecular basis' (OMIM IDs prefixed with "#"). The prioritization of the positional candidate genes in these OD loci is therefore an important step to facilitate OD-gene identification for further experimental analysis. As shown in Table 16.2, several candidate gene prioritization methods have been developed to overcome the limitations of high-throughput, genome-wide studies like linkage analysis and gene expression profiling, both of which typically result in the identification of hundreds of potential candidate genes (Chen et al. 2007, 2009a; Freudenberg and Propping 2002; Turner et al. 2003; Tiffin et al. 2005, 2006; Adie et al. 2005, 2006; Aerts et al. 2006; Thornblad et al. 2007; Zhu and Zhao 2007) (for additional details see (Piro and Di Cunto 2012)). Most of these computational approaches are based on the assumption that similar phenotypes are caused by genes with similar or related functions (Goh et al. 2007; Chen et al. 2009a; Turner et al. 2003; Jimenez-Sanchez et al. 2001; Smith and Eyre-Walker 2003). They, however, differ by the data sources utilized and by the strategy adopted in calculating similarity (Tranchevent et al. 2008). Except for ENDEAVOUR (Aerts et al. 2006; Tranchevent et al. 2008) and ToppGene (Chen et al. 2007, 2009a), most of the existing approaches utilize a very limited number of data sources.

Because biological networks have been found to be comparable to communication and social networks (Junker et al. 2006) through commonalities such as scale-freeness and small-world properties (further discussed in Chap. 15), the algorithms used to analyze social and Web networks have been successfully used for disease gene identification and ranking. These network-based approaches predominantly use protein-protein interaction networks (PPIN) and the candidate genes are typically ranked based on their connectivity to known disease genes (seed or training set). While PPINs have been used widely to identify novel disease candidate genes (George et al. 2006; Kann 2007; Kohler et al. 2008; Wu et al. 2008; Xu and Li 2006), several recent studies (Kohler et al. 2008; Wu et al. 2008; Chen et al. 2006, 2009b; Ortutay and Vihinen 2009) report also using them for candidate gene prioritization. For additional details about network-based approaches see the Chap. 15.

A principal limitation of almost all of the current disease gene identification and prioritization approaches is that they are gene-centric. However, it has been speculated that complex traits result more often from noncoding regulatory variants than from coding sequence variants (King and Wilson 1975; Korstanje and Paigen 2002; Mackay 2001), yet analyzing noncoding regulatory variants is replete with

Table 16.2 List of current bioinformatics approaches and tools to rank human disease candidate genes

Approach	Online availability	Data types used	Training set (input)
Approaches based on disease gene properties			
DGP (Lopez-Bigas and Ouzounis 2004)	http://cgg.ebi.ac.uk/services/dgp/	Sequence	Not applicable (N/A)
PROSPECTR (Adie et al. 2005)	http://www.genetics.med.ed.ac.uk/prospectr/	Sequence	N/A
Approaches using links between genes and phenotypes			
Genes2Diseases (Perez-Iratxeta et al. 2002, 2005)	http://www.ogic.ca/projects/g2d_2/	Sequence, Gene Ontology (GO), literature mining	Phenotype GO terms Known genes
BITOLA (Hristovski et al. 2005)	http://www.mf.uni-lj.si/bitola/	Literature mining	Concept
GeneSeeker (van Driel et al. 2003, 2005)	http://www.cmbi.ru.nl/GeneSeeker/	Expression, phenotype, literature mining	N/A
GFINDer (Masseroli et al. 2004, 2005)	http://www.bioinformatics.polimi.it/GFINDer/	Expression, phenotype	N/A
TOM (Rossi et al. 2006)	http://www-micrel.deis.unibo.it/~tom/	Expression, GO	Known genes and/or disease loci
Approaches using functional relatedness between candidate genes			
OMIM phenome map (van Driel et al. 2006)	http://www.cmbi.ru.nl/MimMiner/	Phenotype, sequence, GO, protein interactions	N/A
SUSPECTS (Adie et al. 2006)	http://www.genetics.med.ed.ac.uk/suspects/	Sequence, expression, GO	Known genes
Prioritizer (Franke et al. 2006)	http://www.prioritizer.nl/	Expression, GO, protein interactions	Disease loci
Endeavour (Aerts et al. 2006)	http://www.esat.kuleuven.be/endeavour/	Sequence, expression, GO, pathways, literature mining	Known genes
ToppGene (Chen et al. 2009a)	http://toppgene.cchmc.org	Mouse phenotype, expression, GO, pathways, literature mining	Known genes
ToppNet (Chen et al. 2009b)	http://toppgene.cchmc.org	Protein interactions	Known genes

The first column has the source or the name of the tool (including reference, if available) while the second column has the URL of the corresponding web application. The third column shows the list of genomic annotation types/features used by each of the methods for candidate gene ranking. The last column has details of the training or the input data, if used (*Note: modified from Kaimal et al.* (2011)*, this list is extensive, but not exhaustive; reference* (Piro and Di Cunto 2012) *provides an additional list of tools*)

problems. For instance, functional consequences of coding region variants are relatively easy to assess (e.g., missense, nonsense, splicing, etc.) while interpreting the consequences of noncoding sequence variants is more complicated and the relationships between noncoding sequence variation and gene expression level or phenotypes are relatively less well understood (Kaimal et al. 2011).

16.6 Exome Sequencing to Decipher Orphan Diseases

NextGen whole genome or exome sequencing can be used to search for Mendelian disease genes in an unbiased manner (Gilissen et al. 2012). Of the ODs with a known causal gene mutation, about 70% are monogenic (Zhang et al. 2011) and as per the current version of OMIM, there are about 5,000 monogenic ODs and for half of these the underlying genes remain unknown. The OD causal gene identification thus represents the first step to a better understanding of the pathophysiological mechanisms underlying ODs, which in turn can lead to developing effective therapeutic interventions. While massively parallel DNA sequencing technologies have rendered the whole genome resequencing of individual humans increasingly practical, the associated expenses continue to be a major hurdle. However, about 85% of the known genetic causes for Mendelian disorders affect the protein coding exonic regions (Botstein and Risch 2003), which accounts for approximately 2% of the genome. An alternative approach involves the targeted re-sequencing of all protein-coding subsequences (exome sequencing), potentially overcoming the financial hurdle to routine comprehensive sequencing for diagnostic purposes in the near future (Bainbridge et al. 2011; Kingsmore and Saunders 2011). Thus, whole exome sequencing using the next generation technologies provides a new and transformational approach for identifying causative mutations in ODs (see Table 16.3) and is likely to become the most commonly used tool for disease gene identification (Gilissen et al. 2012).

While the technology and software for sequencing are widely and readily available, there is an increasing need for better and novel computational approaches to annotate the variants, integrate and mine the variant data, prioritize the variants and candidate genes, and analyze pathways for potential genetic heterogeneity. Although there is no doubt about the promise of exome sequencing in OD research, it has some limitations. First, the cause of an OD could be a noncoding variation or a large indel or structural genomic variant, all of which are missed by exomic sequencing. Second, some of the variants may not be identified because of lack of sequence coverage of the variant or could be related to technical errors (e.g. bioinformatics variant calling issues) (Gilissen et al. 2012).

Table 16.3 Orphan disease gene identification by exome sequencing

Orphan disease	Identified OD-gene	References
Amelogenesis imperfecta	*FAM20A*	O'Sullivan et al. (2011)
Amyotrophic lateral sclerosis	*VCP*	Johnson et al. (2010)
Autoimmune lymphoproliferative syndrome	*FADD*	Bolze et al. (2010)
Cerebral cortical malformations	*WDR62*	Bilguvar et al. (2010)
Chondrodysplasia and abnormal joint development	*IMPAD1*	Vissers et al. (2011)
Combined hypolipidemia	*ANGPTL3*	Musunuru et al. (2010)
Congenital chloride diarrhea	*SLC26A3*	Choi et al. (2009)
Fowler syndrome	*FLVCR2*	Lalonde et al. (2010)
Hajdu-Cheney syndrome	*NOTCH2*	Simpson et al. (2011), Majewski et al. (2011), and Isidor et al. (2011)
Hereditary spastic paresis	*KIF1A*	Erlich et al. (2011)
Hyperphosphatasia mental retardation syndrome	*PIGV*	Krawitz et al. (2010)
Infantile mitochondrial cardiomyopathy	*AARS2*	Götz et al. (2011)
Kabuki syndrome	*MLL2*	Ng et al. (2010a)
Kaposi sarcoma	*STIM1*	Byun et al.(2010b)
Miller syndrome	*DHODH*	Ng et al. (2010)
Osteogenesis imperfecta	*SERPINF1*	Becker et al. (2011)
Perrault syndrome	*HSD17B4*	Pierce et al. (2010)
Progeroid syndrome	*BANF1*	Puente et al. (2011)
Retinitis pigmentosa	*DHDDS*	Züchner et al. (2011)
Schinzel-Giedion syndrome	*SETBP1*	Hoischen et al. (2010)
Sensenbrenner syndrome	*WDR35*	Gilissen et al. (2010)
Spinocerebellar ataxia	*TGM6*	Wang et al. (2010)

16.7 Exome Sequencing and Bioinformatics Applications to Identify Novel Orphan Disease Causal Variants – Two Case Studies

In the following sections we present two examples from recently published studies (Benitez et al. 2011; Erlich et al. 2011) where disease-network analysis approach is used along with exome sequencing to identify novel orphan disease causal variants.

In a recent study, Erlich et al. (2011) demonstrated how gene prediction tools when used in combination with traditional mapping approaches can be successfully applied to prioritization of OD candidate genes from exome resequencing experiments. This study illustrates the potential of combining genomic variant and gene level information to identify and rank novel causal variants of orphan diseases. The authors used three different candidate gene prioritization tools (Endeavour (Aerts et al. 2006), ToppGene (Chen et al. 2009a), and SUSPECTS (Adie et al. 2006)) to prioritize *KIF1A* as the most likely candidate gene for hereditary spastic paraparesis (HSP). In this study, a familial case of hereditary spastic paraparesis (HSP) was analyzed through whole-exome sequencing and the four largest homozygous regions (containing 44 genes) were identified as potential HSP loci. The authors then applied several filters to narrow the list further. For example, a gene was considered as potentially causative if it contains at least one variant that is either not inherited from the parents or absent in dbSNP or the 1,000 Genomes Project data. Because the majority of the known orphan disease variants affect coding sequences, they also checked whether the variant is non-synonymous. After this filtering step, 15 candidate genes were identified and this list was further prioritized using three computational methods (Endeavour (Aerts et al. 2006), ToppGene (Chen et al. 2009a), and SUSPECTS (Adie et al. 2006)). As a training set, a list of 11 seed genes associated with a pure type of HSP was compiled through literature mining. Interestingly, the top-ranking gene from all the three bioinformatics approaches (each of which uses different types of data and algorithms for prioritization) was *KIF1A*. Subsequent Sanger sequencing confirmed that *KIF1A* indeed is the causative variant.

In a second study, Benitez et al. (2011) used disease-network analysis as supporting *in silico* evidence of the role of the adult neuronal ceroid-lipofuscinosis (NCL) candidate genes identified by exome sequencing. The authors used Endeavour (Aerts et al. 2006) and ToppGene (Chen et al. 2009a) to rank the NCL candidate variant genes identified by exome sequencing. Known causal genes of other NCLs along with genes that are associated with phenotypically close disorders were used as the training set for both ToppGene and Endeavour. Interestingly, the three variants identified by exome sequencing (*PDCD6IP, DNAJC5* and *LIPJ*) were in the top five genes in the combined analysis using ToppGene and Endeavour suggesting that they may be functionally or structurally related with NCLs encoded genes and constituting true causative variants for adult NCL.

16.8 Drug Repositioning Strategies

Drug development in general is complicated, time-consuming, and expensive with extremely low success rates. To overcome or by-pass this productivity gap, more and more companies are resorting to "*Drug Repositioning*" or "*Drug Repurposing,*" or simply identifying and developing new uses for existing or abandoned pharmacotherapies (Ashburn and Thor 2004). This approach can significantly reduce the risks associated with drug development. Repositioned drugs can enter clinical phases more rapidly and at a lower cost than novel compounds because the starting point is usually approved compounds with known bioavailability and safety profiles, proven formulation and manufacturing routes, and well-characterized pharmacology (Boguski et al. 2009). It is therefore no surprise that in recent years, of the new medicines that reach their first markets, repurposed drugs account for ~30%!

Following the FDA Orphan Drug Act in 1983, there has been a dramatic rise in new treatment options for ODs (~325 orphan are drugs now available in the market). However, these drugs cover only about 5% of known ODs. Bypassing the traditional drug discovery process and discovering OD indications for already approved compounds could therefore be a viable strategy to jump-start the orphan drug discovery process. The recent release of a 235-drug database (Xu and Cote 2011) of approved compounds and products that show promise in ODs by the FDA further supports the need for a more systematized analysis of approved drugs for novel OD indications.

Drug repositioning is primarily based on two key principles: (a) the "promiscuous" nature of the drug and (b) targets relevant to a specific disease or pathway may also be critical for other diseases or pathways (Pujol et al. 2010; Sardana et al. 2011). While understanding the potential off-target interactions of existing drugs is of major interest in pharmaceutical research to understand the molecular basis of adverse drug reactions and for discovering novel indications of drugs, identifying these targets in a context that also provides basic information about medical exploitability is a major challenge. In this context, recent rapid advances in genomic, proteomic, functional, and systems studies of the known drug targets and disease proteins have enabled the discovery of drugs, multitarget agents, combination therapies, and analysis of on-target and off-target toxicity and pharmacogenetic responses (Padhy and Gupta 2011). Such information is now publicly available in the form of different databases (e.g., DrugBank (Wishart et al. 2006), PharmGKB (Altman 2007), STITCH (Kuhn et al. 2012), Therapeutic Target Database (Zhu et al. 2012)) which provide target and drug profiles. Druggable targets on the other hand can be identified by integrating biochemistry and cell biology with genetics and physiology, as well as bioinformatics and network analysis-based approaches. Potential candidates can also be identified using FDA's "Disc" (Discontinued Drug Products) list, which contains thousands of drugs that made it through Phase I testing but were withdrawn subsequently for reasons other than safety.

Although there are several advantages, rational drug repositioning poses formidable challenges, especially in the case of ODs because the molecular basis and the underlying mechanism of most ODs and drug actions are unknown, intricate, or are not amenable to human or computational data mining techniques. In the following sections we will discuss some of these challenges and present an overview of current strategies to overcome them to translate pharmacological and biomedical discoveries into safe and effective OD therapeutics (Sardana et al. 2011).

With the value of drug repositioning becoming increasingly evident, a number of companies have developed systematic approaches which can be broadly categorized into three strategies (see recent review by Sardana et al. 2011) which are presented below along with case studies

16.8.1 Strategy 1: Knowledge-Based Drug Repositioning

This strategy involves integration of the accumulating heterogeneous pharmacological, genomic, biomedical, and chemical data and mining it using novel analytical algorithms and approaches. The "virtual screening" is performed to discover unrecognized or non-explicit connections between a drug, target, and disease. For example, Iorio et al. (2010a) developed MANTRA (Mode of Action by Network Analysis), a bioinformatics tool to analyze novel drugs and potential drug repositioning candidates of known and FDA approved drugs by the assignment of previously unrecognized putative therapeutic applications. The analysis is based on a novel similarity measure among the cellular responses elicited by a large set of compounds in humans. For each compound, this response is summarized by a prototype genome-wide ranked list (PRL) of genes. This PRL is built *in silico* by combining genome-wide profiles of differential expression, following treatments with the compound on a number of different human cell lines. The PRLs are then compared to each other with a novel approach based on Gene Set Enrichment Analysis (GSEA) ending up with an exhaustive set of drug pair-wise distance values. Using these distance measures, a drug-drug network is built and analyzed to identify modules or highly interconnected subnetworks or communities. Communities enriched for drugs with similar mechanism of action are then identified using known knowledge and literature searches. These communities are used to make novel hypotheses on known drugs and to classify novel drugs when they are integrated in the network. Using this framework, the authors discovered that fasudil (a Rho-kinase inhibitor) could potentially be repositioned as treatment for neurodegenerative disorders.

In another study, Suthram et al. (2010) integrated molecular profiles of diseases with protein–protein interaction data, to infer protein functional modules and networks that were shared among many diseases. The molecular profiles of diseases were obtained from the NCBI Gene Expression Omnibus (GEO) (Barrett et al. 2007) while the protein-protein interaction (PPI) data for human was obtained from the Human Protein Reference Database (HPRD) (Goel et al. 2012). Using this human PPI data that was organized into 'modules' of functionally interacting proteins, a statistical approach was used to evaluate the molecular signatures of diseases for gene functional module activity, whereby the module activity was determined as the mean normalized transcriptional activity of its component genes in the disease molecular profile. A disease-disease network was then formed on the basis of functional module activity shared between diseases and mined to identify subnetworks with multiple known drug targets.

In another interesting study, Chiang and Butte (2009) used a "guilt by association" approach to discover alternative uses for drugs by systematically evaluating a drug treatment-based view of diseases. The authors used the DRUGDEX System, as a gold-standard comprehensive pharmacopeia resource. The DRUGDEX system provides literature-backed drug information on both (FDA) approved indications and off-label uses and is an U.S. government approved source for medical insurance reimbursable off-label uses. Using this system, the authors systematically captured the FDA-approved (FDA approved drug and indications) and practiced (FDA-approved drugs, but off-label drug uses) views into a Drug-Disease Knowledge Base consisting of 726 diseases and 2,022 drugs. Given that the treatment profiles between a small subset of disease pairs can be quite different between the

FDA-approved and practiced views, the authors reasoned that these discrepancies can be mined to suggest novel drug uses. This was based on the hypothesis that if two diseases share similar therapies, then drugs that are currently used for only one of the two diseases may also be extrapolated to the other.

16.8.2 Strategy 2: Rescreening the Pharmacopeia Against New Targets

Taking advantage of the recent and rapid advances in high throughput screening technology, this unbiased strategy takes the semi-blind approach of re-screening existing compounds against a multitude of targets to identify possible therapeutic benefits or side-effects (O'Connor and Roth 2005). For instance, focusing on neglected parasitic diseases, researchers at the University of Dundee carried out a large scale screening to cover a broad range of potential targets (Brenk et al. 2008). Out of 2.3 million commercially available compounds about 222,000 compounds were selected for an *in silico* library, ~57,000 for a diverse general screening library, and ~1,700 compounds for a focused kinase set. To compile these libraries, the authors set several rules to define unwanted groups and also to identify "lead-like" compounds, which facilitate straightforward exploration of structure–activity relationships. To assemble the focused library for hit discovery for kinases, literature and patents were mined and reviewed. Screening the known drugs (common and orphan) against OD causative and druggable proteins (Russ and Lampel 2005) can not only identify potentially new therapeutic agents for validated targets, but also identify and validate new orphan disease-relevant targets and potential drug combinatorials (Sardana et al. 2011). Apart from jump-starting orphan disease therapeutic programs, if a novel target is discovered for an existing drug (common or orphan), its chemical scaffold can be used as a good starting point to identify potential new chemical entities with the premise of developing them as drugs for the novel target (Grau and Serbedzija 2005).

16.8.3 Strategy 3: End-Point Screening

The starting point in this strategy is a phenotype of interest and the existing drugs are screened to discover drugs that produce an unanticipated (either as on- and/or off-target effects of compounds), yet desired, phenotypic result. For example, increasing autophagy may provide clinical benefit in the treatment of various diseases, and therefore there is a great effort in developing drugs enhancing this function (Iorio et al. 2010b). Using autophagy as a phenotype of interest, Iorio et al. (2010b) used a known inducer of autophagy (2-deoxy-D-glucose) and generated a list of drugs that were predicted to share a similar mode of action and identified a novel autophagy enhancer drug (fasudil) along with other previously known inducers of autophagy. The authors used the Connectivity Map (Lamb et al. 2006) dataset which is a large compendia of publicly available gene expression data obtained by treatments of several human cell lines with a large collection of small molecules. A drug network is then built considering similarities in the consensual transcriptional responses and groups of highly interconnected nodes (drugs or small molecules) are then identified. These communities of highly interconnected nodes are subjected to further computational analyses to identify their enriched modes of action.

16.9 FDA's Rare Disease Repurposing Database (RDRD)

Recently, the U.S. FDA's Office of Orphan Products Development has established a new resource: Rare Disease Repurposing Database (RDRD) (Xu and Cote 2011) for drug developers. It is a compilation of drugs that have shown promise for treating orphan diseases and already have FDA approval or designation. It is a database of products that (a) have received orphan status designation (i.e., they have been found "promising" for treating a rare disease); and (b) are already market-approved for the treatment of some other diseases. Since these compounds already have FDA approval, it is anticipated that repositioning these drugs for a new orphan disease indication will be quick and inexpensive for the developer, and will therefore help patients by getting to market quicker. The data are provided as three downloadable excel files: (a) Orphan-designated products with at least one marketing approval for a common disease indication; (b) Orphan-designated products with at least one marketing approval for an OD indication; and (c) Orphan-designated products with marketing approvals for both common and OD indications. These lists offer sponsors a new tool and a "shorter path" for finding special opportunities to develop niche therapies instead of beginning with an untested new therapy compound and obtaining an FDA approval. Additionally, these data sets can be used as a "positive control" or "gold standard" for testing or validating high-throughput and computational approaches for drug repositioning for orphan diseases.

References

Adie EA, et al. Speeding disease gene discovery by sequence based candidate prioritization. BMC Bioinformatics. 2005;6:55.

Adie EA, et al. SUSPECTS: enabling fast and effective prioritization of positional candidates. Bioinformatics. 2006;22(6):773–4.

Aerts S, et al. Gene prioritization through genomic data fusion. Nat Biotechnol. 2006;24(5): 537–44.

Altman RB. PharmGKB: a logical home for knowledge relating genotype to drug response phenotype. Nat Genet. 2007;39(4):426.

Ashburn TT, Thor KB. Drug repositioning: identifying and developing new uses for existing drugs. Nat Rev Drug Discov. 2004;3(8):673–83.

Ayme S. [Orphanet, an information site on rare diseases]. Soins. 2003; (672):46–7.

Bainbridge MN, et al. Whole-genome sequencing for optimized patient management. Sci Transl Med. 2011;3(87):87re3.

Barrett T, et al. NCBI GEO: mining tens of millions of expression profiles–database and tools update. Nucleic Acids Res. 2007;35(Database issue):D760–5.

Becker J, et al. Exome sequencing identifies truncating mutations in human SERPINF1 in autosomal-recessive osteogenesis imperfecta. Am J Hum Genet. 2011;88(3):362–71.

Benitez BA, et al. Exome-sequencing confirms DNAJC5 mutations as cause of adult neuronal ceroid-lipofuscinosis. PLoS One. 2011;6(11):e26741.

Bilguvar K, et al. Whole-exome sequencing identifies recessive WDR62 mutations in severe brain malformations. Nature. 2010;467(7312):207–10.

Boguski MS, Mandl KD, Sukhatme VP. Drug discovery. Repurposing with a difference. Science. 2009;324(5933):1394–5.

Bolze A, et al. Whole-exome-sequencing-based discovery of human FADD deficiency. Am J Hum Genet. 2010;87(6):873–81.

Botstein D, Risch N. Discovering genotypes underlying human phenotypes: past successes for mendelian disease, future approaches for complex disease. Nat Genet. 2003;33(Suppl): 228–37.

Brenk R, et al. Lessons learnt from assembling screening libraries for drug discovery for neglected diseases. ChemMedChem. 2008;3(3):435–44.

Byun M, et al. Whole-exome sequencing-based discovery of STIM1 deficiency in a child with fatal classic Kaposi sarcoma. J Exp Med. 2010;207(11):2307–12.

Chen JY, Shen C, Sivachenko AY. Mining Alzheimer disease relevant proteins from integrated protein interactome data. Pac Symp Biocomput; 2006:367–78.

Chen J, et al. Improved human disease candidate gene prioritization using mouse phenotype. BMC Bioinformatics. 2007;8:392.

Chen J, et al. ToppGene Suite for gene list enrichment analysis and candidate gene prioritization. Nucleic Acids Res. 2009a;37(Web Server issue):W305–11.

Chen J, Aronow BJ, Jegga AG. Disease candidate gene identification and prioritization using protein interaction networks. BMC Bioinformatics. 2009b;10:73.

Chiang AP, Butte AJ. Systematic evaluation of drug-disease relationships to identify leads for novel drug uses. Clin Pharmacol Ther. 2009;86(5):507–10.

Choi M, et al. Genetic diagnosis by whole exome capture and massively parallel DNA sequencing. Proc Natl Acad Sci U S A. 2009;106(45):19096–101.

Erlich Y, et al. Exome sequencing and disease-network analysis of a single family implicate a mutation in KIF1A in hereditary spastic paraparesis. Genome Res. 2011;21(5):658–64.

Feldman I, Rzhetsky A, Vitkup D. Network properties of genes harboring inherited disease mutations. Proc Natl Acad Sci U S A. 2008;105(11):4323–8.

Field MJ, Boat TF. Rare diseases and orphan products: accelerating research and development. In: Field MJ, Boat TF, Institute of Medicine Committee on Accelerating Rare Diseases Research and Orphan Product Development, editors. Rare diseases and orphan products: accelerating research and development. Washington, DC: National Academies Press; 2010.

Franke L, et al. Reconstruction of a functional human gene network, with an application for prioritizing positional candidate genes. Am J Hum Genet. 2006;78(6):1011–25.

Freudenberg J, Propping P. A similarity-based method for genome-wide prediction of disease-relevant human genes. Bioinformatics. 2002;18 Suppl 2:S110–15.

George RA, et al. Analysis of protein sequence and interaction data for candidate disease gene prediction. Nucleic Acids Res. 2006;34(19):e130.

Gilissen C, et al. Exome sequencing identifies WDR35 variants involved in Sensenbrenner syndrome. Am J Hum Genet. 2010;87(3):418–23.

Gilissen C, et al. Disease gene identification strategies for exome sequencing. Eur J Hum Genet. 2012;20(5):490–7.

Goel R, et al. Human Protein Reference Database and Human Proteinpedia as resources for phosphoproteome analysis. Mol Biosyst. 2012;8(2):453–63.
Goh KI, et al. The human disease network. Proc Natl Acad Sci U S A. 2007;104(21):8685–90.
Gotz A, et al. Exome sequencing identifies mitochondrial alanyl-tRNA synthetase mutations in infantile mitochondrial cardiomyopathy. Am J Hum Genet. 2011;88(5):635–42.
Grau D, Serbedzija G. Innovative strategies for drug repurposing. Drug Discov Dev. 2007. http://www.dddmag.com/articles/2007/09/innovative-strategies-drug-repurposing
Hamosh A, et al. Online Mendelian Inheritance in Man (OMIM). Hum Mutat. 2000;15(1):57–61.
Hoischen A, et al. De novo mutations of SETBP1 cause Schinzel-Giedion syndrome. Nat Genet. 2010;42(6):483–5.
Hristovski D, et al. Using literature-based discovery to identify disease candidate genes. Int J Med Inform. 2005;74(2–4):289–98.
Iorio F, et al. Discovery of drug mode of action and drug repositioning from transcriptional responses. Proc Natl Acad Sci U S A. 2010a;107(33):14621–6.
Iorio F, et al. Identification of small molecules enhancing autophagic function from drug network analysis. Autophagy. 2010b;6(8):1204–5.
Isidor B, et al. Truncating mutations in the last exon of NOTCH2 cause a rare skeletal disorder with osteoporosis. Nat Genet. 2011;43(4):306–8.
Jimenez-Sanchez G, Childs B, Valle D. Human disease genes. Nature. 2001;409(6822):853–5.
Johnson JO, et al. Exome sequencing reveals VCP mutations as a cause of familial ALS. Neuron. 2010;68(5):857–64.
Junker BH, Koschutzki D, Schreiber F. Exploration of biological network centralities with CentiBiN. BMC Bioinform. 2006;7:219.
Kaimal V, et al. Integrative systems biology approaches to identify and prioritize disease and drug candidate genes. Methods Mol Biol. 2011;700:241–59.
Kann MG. Protein interactions and disease: computational approaches to uncover the etiology of diseases. Brief Bioinform. 2007;8(5):333–46.
King MC, Wilson AC. Evolution at two levels in humans and chimpanzees. Science. 1975;188(4184):107–16.
Kingsmore SF, Saunders CJ. Deep sequencing of patient genomes for disease diagnosis: when will it become routine? Sci Transl Med. 2011;3(87):87ps23.
Kohler S, et al. Walking the interactome for prioritization of candidate disease genes. Am J Hum Genet. 2008;82(4):949–58.
Korstanje R, Paigen B. From QTL to gene: the harvest begins. Nat Genet. 2002;31(3):235–6.
Krawitz PM, et al. Identity-by-descent filtering of exome sequence data identifies PIGV mutations in hyperphosphatasia mental retardation syndrome. Nat Genet. 2010;42(10):827–9.
Kuhn M, et al. STITCH 3: zooming in on protein-chemical interactions. Nucleic Acids Res. 2012;40(Database issue):D876–80.
Lalonde E, et al. Unexpected allelic heterogeneity and spectrum of mutations in Fowler syndrome revealed by next-generation exome sequencing. Hum Mutat. 2010;31(8):918–23.
Lamb J, et al. The connectivity map: using gene-expression signatures to connect small molecules, genes, and disease. Science. 2006;313(5795):1929–35.
Li Y, Agarwal P. A pathway-based view of human diseases and disease relationships. PLoS One. 2009;4(2):e4346.
Linghu B, et al. Genome-wide prioritization of disease genes and identification of disease-disease associations from an integrated human functional linkage network. Genome Biol. 2009;10(9):R91.
Lopez-Bigas N, Ouzounis CA. Genome-wide identification of genes likely to be involved in human genetic disease. Nucleic Acids Res. 2004;32(10):3108–14.
Mackay TF. Quantitative trait loci in Drosophila. Nat Rev Genet. 2001;2(1):11–20.
Majewski J, et al. Mutations in NOTCH2 in families with Hajdu-Cheney syndrome. Hum Mutat. 2011;32(10):1114–17.

Masseroli M, Martucci D, Pinciroli F. GFINDer: Genome Function INtegrated Discoverer through dynamic annotation, statistical analysis, and mining. Nucleic Acids Res. 2004;32(Web Server issue):W293–300.

Masseroli M, Galati O, Pinciroli F. GFINDer: genetic disease and phenotype location statistical analysis and mining of dynamically annotated gene lists. Nucleic Acids Res. 2005;33(Web Server issue):W717–23.

Musunuru K, et al. Exome sequencing, ANGPTL3 mutations, and familial combined hypolipidemia. N Engl J Med. 2010;363(23):2220–7.

Ng SB, et al. Exome sequencing identifies MLL2 mutations as a cause of Kabuki syndrome. Nat Genet. 2010a;42(9):790–3.

Ng SB, et al. Exome sequencing identifies the cause of a mendelian disorder. Nat Genet. 2010b;42(1):30–5.

O'Connor KA, Roth BL. Finding new tricks for old drugs: an efficient route for public-sector drug discovery. Nat Rev Drug Discov. 2005;4(12):1005–14.

O'Sullivan J, et al. Whole-exome sequencing identifies FAM20A mutations as a cause of amelogenesis imperfecta and gingival hyperplasia syndrome. Am J Hum Genet. 2011;88(5):616–20.

Ortutay C, Vihinen M. Identification of candidate disease genes by integrating gene ontologies and protein-interaction networks: case study of primary immunodeficiencies. Nucleic Acids Res. 2009;37(2):622–8.

Padhy BM, Gupta YK. Drug repositioning: re-investigating existing drugs for new therapeutic indications. J Postgrad Med. 2011;57(2):153–60.

Perez-Iratxeta C, Bork P, Andrade MA. Association of genes to genetically inherited diseases using data mining. Nat Genet. 2002;31(3):316–19.

Perez-Iratxeta C, et al. G2D: a tool for mining genes associated with disease. BMC Genet. 2005;6:45.

Pierce SB, et al. Mutations in the DBP-deficiency protein HSD17B4 cause ovarian dysgenesis, hearing loss, and ataxia of Perrault syndrome. Am J Hum Genet. 2010;87(2):282–8.

Piro RM, Di Cunto F. Computational approaches to disease-gene prediction: rationale, classification and successes. FEBS J. 2012;279(5):678–96.

Puente XS, et al. Exome sequencing and functional analysis identifies BANF1 mutation as the cause of a hereditary progeroid syndrome. Am J Hum Genet. 2011;88(5):650–6.

Pujol A, et al. Unveiling the role of network and systems biology in drug discovery. Trends Pharmacol Sci. 2010;31(3):115–23.

Rados C. Orphan products: hope for people with rare diseases. FDA Consum. 2003;37(6):10–5.

Rossi S, et al. TOM: a web-based integrated approach for identification of candidate disease genes. Nucleic Acids Res. 2006;34(Web Server issue):W285–92.

Russ AP, Lampel S. The druggable genome: an update. Drug Discov Today. 2005;10(23–24):1607–10.

Sardana D, et al. Drug repositioning for orphan diseases. Brief Bioinform. 2011;12(4):346–56.

Simpson MA, et al. Mutations in NOTCH2 cause Hajdu-Cheney syndrome, a disorder of severe and progressive bone loss. Nat Genet. 2011;43(4):303–5.

Smith NG, Eyre-Walker A. Human disease genes: patterns and predictions. Gene. 2003;318:169–75.

Suthram S, et al. Network-based elucidation of human disease similarities reveals common functional modules enriched for pluripotent drug targets. PLoS Comput Biol. 2010;6(2):e1000662.

The Orphan Drug Act – implementation and impact. 2001, Department of Health and Human Services, Office of Inspector Journal.

Thornblad TA, et al. Prioritization of positional candidate genes using multiple web-based software tools. Twin Res Hum Genet. 2007;10(6):861–70.

Tiffin N, et al. Integration of text- and data-mining using ontologies successfully selects disease gene candidates. Nucleic Acids Res. 2005;33(5):1544–52.

Tiffin N, et al. Computational disease gene identification: a concert of methods prioritizes type 2 diabetes and obesity candidate genes. Nucleic Acids Res. 2006;34(10):3067–81.

Tranchevent LC, et al. ENDEAVOUR update: a web resource for gene prioritization in multiple species. Nucleic Acids Res. 2008;36(Web Server issue):W377–84.

Turner FS, Clutterbuck DR, Semple CA. POCUS: mining genomic sequence annotation to predict disease genes. Genome Biol. 2003;4(11):R75.

US Food and Drug Administration. Orphan Drug Act, Pub L. No. 97-144, 96 Stat. 2049. 1982.

van Driel MA, et al. A new web-based data mining tool for the identification of candidate genes for human genetic disorders. Eur J Hum Genet. 2003;11(1):57–63.

van Driel MA, et al. GeneSeeker: extraction and integration of human disease-related information from web-based genetic databases. Nucleic Acids Res. 2005;33(Web Server issue):W758–61.

van Driel MA, et al. A text-mining analysis of the human phenome. Eur J Hum Genet. 2006;14(5):535–42.

Vissers LE, et al. Chondrodysplasia and abnormal joint development associated with mutations in IMPAD1, encoding the Golgi-resident nucleotide phosphatase, gPAPP. Am J Hum Genet. 2011;88(5):608–15.

Wang JL, et al. TGM6 identified as a novel causative gene of spinocerebellar ataxias using exome sequencing. Brain. 2010;133(Pt 12):3510–18.

Wastfelt M, Fadeel B, Henter JI. A journey of hope: lessons learned from studies on rare diseases and orphan drugs. J Intern Med. 2006;260(1):1–10.

Wishart DS, et al. DrugBank: a comprehensive resource for in silico drug discovery and exploration. Nucleic Acids Res. 2006;34(Database issue):D668–72.

Wu X, et al. Network-based global inference of human disease genes. Mol Syst Biol. 2008;4:189.

Xu K, Cote TR. Database identifies FDA-approved drugs with potential to be repurposed for treatment of orphan diseases. Brief Bioinform. 2011;12(4):341–5.

Xu J, Li Y. Discovering disease-genes by topological features in human protein-protein interaction network. Bioinformatics. 2006;22(22):2800–5.

Zhang M, et al. The orphan disease networks. Am J Hum Genet. 2011;88(6):755–66.

Zhu M, Zhao S. Candidate gene identification approach: progress and challenges. Int J Biol Sci. 2007;3(7):420–7.

Zhu F, et al. Therapeutic target database update 2012: a resource for facilitating target-oriented drug discovery. Nucleic Acids Res. 2012;40(Database issue):D1128–36.

Zuchner S, et al. Whole-exome sequencing links a variant in DHDDS to retinitis pigmentosa. Am J Hum Genet. 2011;88(2):201–6.

Chapter 17
Transcriptional Networks – Control of Lung Maturation

Yan Xu and Jeffrey A. Whitsett

Abstract This chapter provides examples of the application of tools of bioinformatics and functional genomics to integration of large-scale gene expression data with array independent genomic information to reveal transcriptional regulatory networks. The focus is on networks that control lung maturation and surfactant homeostasis, which can serve as a prototype for study of other complex biological processes. Prenatal maturation of the respiratory system is fundamentally important for the transition to airbreathing at birth. Lung immaturity is a major cause of morbidity and mortality in newborn infants and underlies the pathogenesis of acute respiratory failure (respiratory distress syndrome) and chronic respiratory dysfunction (bronchopulmonary dysplasia), associated with preterm birth. The immaturity of type II alveolar epithelial cells and the lack of pulmonary surfactant lipids and proteins that are needed to reduce surface tension in the alveolar saccules cause atelectasis and respiratory insufficiency after preterm birth. Lung maturation includes diverse structural, cellular, and biochemical changes in lung architecture and function that are precisely coordinated by genetic and environmental factors that synchronize the length of gestation with the process of lung maturation. Analysis of lung specific gene deletion and mutation mouse models, using tools of functional genomics, permits the identification of new genes and pathways controlling surfactant lipid homeostasis and lung maturation. The signaling and

Y. Xu, Ph.D. (✉)
Department of Pediatrics, University of Cincinnati College of Medicine, Cincinnati, OH, USA

Divisions of Pulmonary Biology and Biomedical Informatics, Cincinnati Children's Hospital Medical Center, 3333 Burnet Avenue, MLC7009, Cincinnati, OH 45229-3039, USA
e-mail: Yan.Xu@cchmc.org

J.A. Whitsett, M.D.
Department of Pediatrics, University of Cincinnati College of Medicine, Cincinnati, OH, USA

Perinatal Institute, Cincinnati Children's Hospital Medical Center, 3333 Burnet Avenue, MLC7009, Cincinnati, OH 45229-3039, USA

J.J. Hutton (ed.), *Pediatric Biomedical Informatics: Computer Applications in Pediatric Research*, Translational Bioinformatics 2, DOI 10.1007/978-94-007-5149-1_17,

transcriptional mechanisms that influence lung growth and maturation needed to support the abrupt adaptation to airbreathing at birth are of considerable interest, and will provide a rational basis for the design of new treatment strategies for neonatal pulmonary disease.

17.1 Introduction

The lung is a complex organ consisting of more than 40 distinct cell types derived from ectodermal, mesenchymal and endodermal precursors that serve diverse specialized functions related to gas exchange, host defense, and ion transport (Whitsett and Matsuzaki 2006). Perinatal lung "maturation" is necessary for the transition from intrauterine to extra-uterine life. The maturation process includes structural, biochemical and physiologic changes that in the mouse begin approximately at embryonic day 15 and increase dramatically prior to birth. During late gestation, epithelial cells lining the airways and lung saccules differentiate in association with increased production of surfactant, a complex mixture of lipids and proteins that is essential for reducing surface tension created at the air-liquid interface in the alveoli required for air breathing at birth (McMurtry 2002; Burri 1984). Epithelial cell differentiation is accompanied by dilation of peripheral lung saccules, the thinning of mesenchyme, and extensive growth of the pulmonary vasculature. Lung maturation is a carefully timed process. Preterm infants with immature lungs are not able to make sufficient surfactant, causing Respiratory Distress Syndrome (RDS), the major cause of mortality and morbidity in preterm infants (Dubin 1990; Grenache and Gronowski 2006). Since the discovery that RDS is caused by the lack of pulmonary surfactant (Avery and Mead 1959), the structure, function and clinical relevance of surfactant lipid and proteins have been studied extensively in both human and animals (Weaver and Beck 1999; Weaver and Whitsett 1991; Whitsett 2006; Whitsett et al. 1995; Whitsett and Weaver 2002). Although the introduction of surfactant replacement therapy and antenatal glucocorticoid treatment for prevention of RDS have significantly reduced perinatal morbidity and mortality, preterm birth rates and associated pulmonary morbidities have increased progressively during recent decades in North America (Goldenberg et al. 2008; Gravett et al. 2010). New and alternative treatments for the clinical problems related to lung immaturity are needed. Progress in prevention and therapy of lung disease in preterm infants will rest on understanding of genetic, environmental, and hormonal inputs that control lung maturation and surfactant homeostasis.

The post-genomic era has provided unanticipated tools enabling the elucidation of the complex biological processes critical for organ formation and function. The availability of complete sequences of the human, mouse and other genomes and the introduction of mRNA microarray and other high throughput technologies for study of gene and protein expression enable simultaneous quantification of expression levels of thousands of genes/proteins. These technologies bring new perspectives to the study of gene networks and their regulation, providing access to molecular mechanisms underlying various diseases and phenotypes. While phenotypic outcomes of experiments that are designed to manipulate the expression or function of genes of interest in animal models of disease have been highly successful in the study of disease pathogenesis, the global impact of gene perturbations remains to be fully explored. Systems biology, supported by bioinformatic, proteomic, and genomic data, is providing increasing insight into the pathogenesis of human disease, including those affecting the lung. The density of bioinformatics data relevant to lung biology is increasing exponentially, the thorough analysis of which will provide insights into the processes underlying lung maturation and function. In this chapter, we share experiences applying mRNA microarray analyses and functional genomic applications in conjugation with transgenic mouse models to identify key genes and pathways regulating lung maturation. We provide examples regarding the application of systems biology to integrating lung specific gene expression data with array independent data to generate transcriptional networks regulating surfactant homeostasis during lung development. These approaches can serve as models for study of transcriptional networks that regulate development in other organ systems, especially the use of informatics tools to analyze large data sets.

17.2 Lung Maturation and Surfactant Homeostasis

Lung development can be divided into five morphologically distinct stages that begin near mid-gestation and continue through the early postnatal period. The embryonic stage is distinguished by the formation of the lung buds and division of the trachea and esophagus (E9–11.5) (days 9–11.5 of embryonic development of the mouse). The pseudoglandular stage (E11.5–15.5) is characterized by branching of the conducting airways, formation of the peripheral acinar tubules and buds, and vasculogenesis; the canalicular stage (E15.5–17.5), by expansion of the acinar tubules and buds, angiogenesis, and differentiation of alveolar epithelial type I and II cells; the saccular stage (E17.5 – PN5), by dilation of the terminal respiratory sacs, thinning of the mesenchyme, and deposition of elastin; and the alveolar stage (PN5-30), by maturation of the alveolar-capillary bed, alveolar ducts and alveoli (Perl and Whitsett 1999; Maeda et al. 2007). "Perinatal lung maturation" occurs within the canalicular and saccular stages, at which time differentiation of peripheral distal respiratory epithelial cells mediates increased production of pulmonary surfactant lipids and proteins that are essential for lung function and host defense after birth.

Lung surfactant is a complex mixture of lipid and protein, composed of approximately 80% phospholipids, 10% neutral lipids (particularly cholesterol) and 10% surfactant-associated proteins (SP-A, -B, -C and -D). The most abundant phospholipid is phosphatidylcholine with saturated dipalmitoyl-PC (DPPC) as the predominant form. DPPC is responsible for the surface tension–lowering properties of surfactant (Goerke 1998). The lipid components are synthesized in the endoplasmic reticulum of the type II cell, modified in Golgi apparatus, transported to the lamellar bodies where surfactant is assembled, stored and secreted by exocytosis into the alveolar airspace (Perez-Gil and Weaver 2010). In late gestation, alveolar type II cells dramatically increase production of surfactant lipid and protein to reduce surface tension created at the air-liquid interface and prepares the lung for extrauterine life (Johansson and Curstedt 1997).

17.3 Gene Expression Profiling and Transcriptional Regulatory Network

Advances in genomic and proteomic technologies have been heralded as the new biological revolution after completion of the human genome project. Gene profiling is considered to be the heart of 'omics' because it not only enables a global view regarding expression of genes as cells contribute to specific biological processes, but also infers regulatory relationships between individual genes (Bonner et al. 2003; Bullinger and Valk 2005; Liang et al. 2004; Margalit et al. 2005).

Lung development is a highly coordinated and precisely timed process. During lung maturation, distinct groups of genes are induced or suppressed at specific developmental stages to guide cell proliferation, morphogenesis, and differentiation. An important aspect of gene regulation is mediated by transcription factors (TFs), which bind to cis-elements of the target genes (TGs). Signaling molecules (SM) activate TF responses to biological signals that change the transcription rates of TGs, allowing cells to make needed proteins at the right times and in the right amounts. The interactions between TF and TG can be graphically represented by Transcriptional Regulatory Networks (TRN), in which TF/SMs and TGs are represented as nodes and the interaction relationships are represented as edges. TFs do not function alone in higher organisms, instead, TFs form complexes and bind to *cis* regulatory modules (CRMs), DNA sequences (typically 50–1,000 bp in size) that contain multiple transcription factor binding sites (TFBSs) clustered into modular structures (Jeziorska et al. 2009). CRMs are responsive to specific combinations of TFs, the precise combinatorial interactions of transcription factors providing transcriptional activation appropriate to cell conduct and function (Jeziorska et al. 2009).

A fundamental challenge in the "post genomic era" is to decode transcriptional networks that direct intricate patterns of gene expression typical of complex organs. Although many key regulators have been identified in the lung, how TFs interact with each other and with SMs to regulate groups of target genes mediating perinatal lung development remains unclear. Here, we summarize recent work seeking to decipher transcriptional networks regulating lung maturation and surfactant homeostasis and consider current efforts directed toward the more challenging problem of generating predictive models that account for the dynamic and context dependent TRNs of lung function and disease.

17.4 Functional Genomics to Study Lung Maturation and Surfactant Homeostasis

Microarray technology has been widely applied to many aspects of pulmonary biology (Mayburd et al. 2006; Minn et al. 2005; Wang et al. 2000; Zuo et al. 2002). Our own studies have contributed to the contemporary body of knowledge applying functional genomics approaches to study the transcriptional regulatory programs controlling lung development, function, and disease. In the lung, distinct sets of signaling molecules and transcription factors interact to implement the structural maturation and cell type specific differentiation of the lung. Through the manipulation of a number of key transcription factors and signaling molecules in transgenic mouse models and the application of genome-wide transcriptional profiling analysis to these models, we have identified target genes, pathways, and physiologic consequences in response to the deletion or mutation of many lung transcription factors (Nkx2-1, Foxa1/a2, C/EBPα, Hif1α, STAT3, NFAT, SREBP, SPDEF, SOX17, PTEN, NF1, Nkx2-9, KLF5, FOXM1, and CATNB) and signaling molecules (MIA, SHH, CSF2R, FGF, RSPO1) (Bridges et al. 2006; Dave et al. 2006; DeFelice et al. 2003; Lian et al. 2004; Martis et al. 2006; Matsuzaki et al. 2006; Metzger et al. 2007; Miller et al. 2004; Mucenski et al. 2005; Wan et al. 2004, 2005, 2008; Xu et al. 2003, 2006, 2007, 2009; Maeda et al. 2011). Many of the key regulators of lung maturation are also critical for early embryogenesis, their disruption often causing embryonic lethality preceding lung formation that begins at approximately E9 in the mouse. The application of conditional mutagenesis in specific lung compartments has been useful in identifying the role of factors critical to early embryogenesis, lung morphogenesis and differentiation. Through these studies, multiple TFs and signaling pathways have been implicated in the structural and functional adaption of the lung at birth (Maeda et al. 2007). Factors important for lung development and function are not exclusively lung-specific; lung specificity is derived from the unique combinations and interactions of TFs. An understanding of the individual TFs and their interactions in the context of lung development requires systemic approaches to connect distinct but interrelated components to define the transcription networks governing lung maturation and differentiation.

Substantial gaps remain in causally linking patterns of gene expression to the transcriptional mechanisms regulating cell behavior. Microarray data alone cannot distinguish direct and indirect cellular responses. While using genome expression data to elucidate TRNS is far from transparent, additional layers of information and integrative approaches are needed to link gene expression, cell biology, and lung function.

17.4.1 Clustering

Genes belonging to the same co-expression cluster are likely to be co-regulated by similar TFs or belong to the same transcriptional regulatory network. Clustering provides important insights into regulatory networks by grouping genes on the basis of similarity of their expression patterns under various experimental conditions. Genes selected from microarray analyses are often grouped into distinct clusters. Genes in each cluster are further classified according to Gene Ontology (GO) and shared transcription factor binding sites in the regulatory regions of genes within the cluster to identify the potential biological themes and common regulatory mechanism represented by these unique gene sets. Classical clustering algorithms including K-means, SOM and Hierarchical clustering generally emphasize clear group separations; any given entity will be assigned to only one cluster. In real biology, however, many proteins have multiple roles in cellular responses to various conditions. Fuzzy Heuristic Partition (Fu and Medico 2007; Gasch and Eisen 2002) considers each gene to be a member of every cluster with a variable degree of membership and enables the assignment of genes to more than one cluster with different degrees of membership. Using stringent membership cutoffs, most of the genes in each cluster are highly correlated across all experimental conditions. As the degree of membership decreases, additional genes join the cluster based on their expression similarity under various experimental conditions, enabling the identification of context-dependent regulation. We prefer to evaluate clustering performance based on its ability to produce biologically meaningful clusters using the GO database as a common reference (Datta and Satten 2008; Pihur et al. 2007) rather than to emphasize cluster separation.

Clustering is most commonly used to identify co-expressed genes. Genes also may be clustered based on their functional annotations, shared promoter/regulatory cis-elements, and biochemical and morphological measurements. Figure 17.1 is an example of using multivariate correlation analysis of mRNA expression with lung physiology, biochemical, and morphological measurements. As depicted in Fig. 17.1a, qPCR data analysis of 53 mRNAs previously associated with lung function and structure revealed three major clusters. Cluster 1 genes (including surfactant, Abca3 and Slc34a2) are induced dramatically before birth; cluster 2 genes (Nkx2-1, Pdgfa, Lpcat, etc.) are moderately induced, while cluster 3 genes do not significantly change during perinatal lung maturation. Using mouse embryonic day 15 (E15) mRNA expression as baseline and 1-day before birth as peak maturation, the rate of gene induction prior to birth can be assessed. Clusters 1 and 2 genes are induced earlier and faster in B6 mice (born after 19.5 days gestation) than in A/J mice (born after 20.5 days gestation), indicating the dynamic mRNA changes in cluster 1 are required for the "shortened" lung maturation process. In Fig. 17.1b, multivariate correlation analysis of mRNA expression with dynamic changes of body weight, lung weight, SatPC and morphometric measurements, we identified a subset of mRNAs, including *Sftpa*, *Sftpb*, *Sftpc*, *Sftpd*, *Slc34a2*, *Scgb1a1*, *Cebpa* and *Aqp5*, that were highly correlated with SatPC, body weight, lung weight, and the fractional area of airspace, biochemical and morphological features of lung maturation. Likewise, mRNAs associated with lipid homeostasis, including *Scd1*, *Abca3*, *Fabp5*, and *Lpcat1*, were correlated with lung weight and fractional area of airspace; while another distinct subset of mRNAs, including *Tubb3*, *Pygb*, and

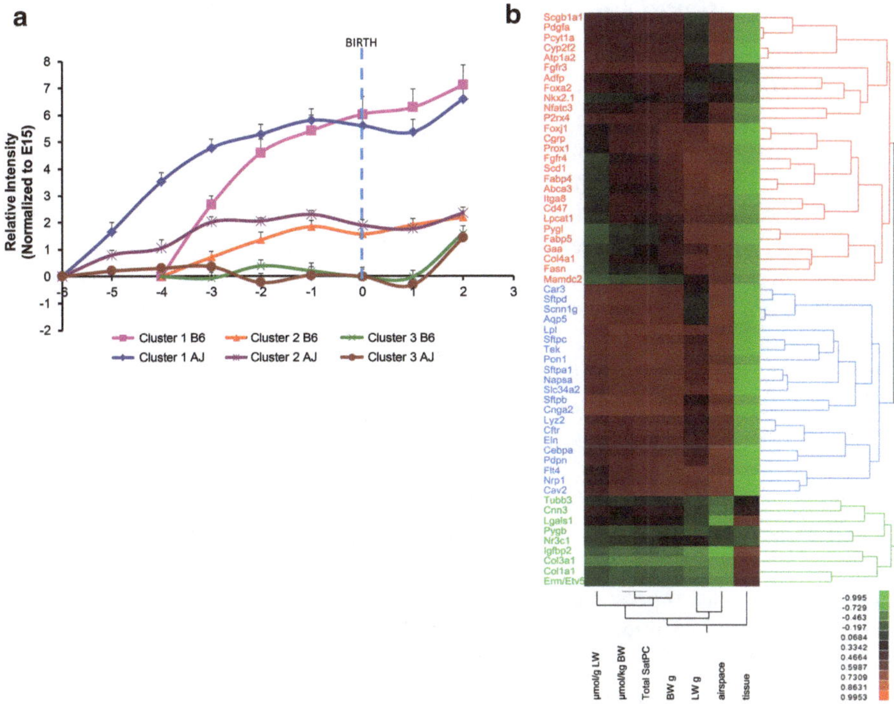

Fig. 17.1 Clustering analysis on selective lung maturation markers: (**a**) Dynamic mRNA expression levels of 53 lung maturation markers were measured by qPCR in both A/J and B6 mouse strains. qPCR data analysis revealed three major clusters. Profile chart is generated based on the normalized cluster mean and standard error of each gene cluster. The X-axis represents days before and after birth (E15-P2): 0 is the day of birth. The Y-axis represents relative expression normalized to E15. (**b**) Dynamic mRNA expression profiles of 53 genes at different gestation ages for the A/J mice were correlated with body weight (*BW*), lung weight (*LW*), SatPC (µmol/gLW, µmol/gBW, and total), and morphometric measurements (airspace, tissue) at corresponding gestational ages using multivariate correction function from JMP 9 (SAS Institute Inc, NC). The heat map was generated based on data from A/J mice using Ward's minimum variance method to estimate cluster similarity. Gradients in the red and green color range indicate positive and negative correlation, respectively. The levels of mRNAs in red clusters were highly correlated with ontogenic changes in lung SatPC and fractional area of airspace mRNAs in blue clusters were moderately correlated with SatPC, but closely correlated with fractional area of airspace mRNAs in green clusters correlated well with the fractional area of the tissue compartment (Adapted with permission from Figure S5 in Besnard et al. 2011)

Igfbp2, were best correlated with fractional area of the tissue compartment (i.e., mesenchyme). Thus, expression of a subset of mRNAs that encode proteins involved in surfactant homeostasis was highly correlated with increasing SatPC (surfactant lipid), body weight, lung weight, and structural maturation of the lung as gestation proceeded.

17.4.2 Functional Classification

After identifying the major co-expressed gene groups, one general question to address is: "What is unique about this gene set?" There are two common approaches to this question; a reductionism or "cherry-picking" approach and a systems approach. A widely used hypothesis driven approach is gene centric, identifying mRNAs of your interest and choosing them for further study. The systems approach is unbiased, seeking to understand the general themes, trends, and biological meanings buried in the data, rather than to identify a single gene or gene network of interest. Biological knowledge and concepts integration represent an unbiased way to identify the potential biological themes represented by distinct gene data sets. Such processes also help in assigning putative roles to previously uncharacterized genes. As each gene is associated with multiple biological annotations from various resources (Gene Ontology terms, Medical Subject Headings and keywords, pathways, protein–protein interactions, protein functional domains, phenotypes, literature/abstract etc.), enrichment of genes in certain functional categories can be determined using Fisher's exact test to compare the occurrence of the term in the experimental gene set of interest, with annotations in the rest of the genome as reference. Thus overrepresented functional categories can be identified in your gene list. Multiple pre-compiled web-based functional annotation tools including Onto-Express (Khatri et al. 2002), GoMiner (Zeeberg et al. 2003), DAVID (Dennis et al. 2003), GSEA (Subramanian et al. 2005), and ToppGene (Chen et al. 2009) have been developed that release the user from the burden of compiling and updating the vast and increasing abundance of annotations. Most of these methods are capable of implementing corrections by comparing functional representations within random gene lists and then generating adjusted p-values that represent the probability of observing a given categorical enrichment in experimental data sets. For genes within a cluster, Kappa similarity can be measured to estimate functional similarity between genes based on the number of shared annotation terms (McGinn et al. 2004). Kappa similarity values range from 0 to 1, the higher the value of Kappa, the stronger the overall agreement in annotation terms.

17.4.3 Identification of Common TFBS Motifs and Modules

It is reasonable to predict that co-expressed genes are likely to be co-regulated through common regulatory mechanisms, via the presence and function of common TFBS and binding modules at regulatory sites within the genome. Because "motif" searches are associated with many false positive predictions due to the short and degenerate nature of many TFBS motifs, several approaches can be used to reduce false positives and improve predictive accuracy.

- ***Apply comparative genomics:*** Programs such as Genome RVista (http://genome.lbl.gov/vista/) and DiRE (http://dire.dcode.org) are used to identify evolutionarily conserved regulatory elements in co-expressed gene clusters. These programs define precompiled evolutionary conserved regions (ECR) via human and mouse whole genome alignment. The locations of putative TFBSs are precomputed for each genome using vertebrate position weighted matrices from TRANSFAC matrix library. Evolutionarily conserved TFBSs are identified at specific genome locations (promoter, intron, UTR or intergenic regions, etc.) at defined strengths.
- ***Search for over-represented TFBSs in proximal promoter regions***: cis-Element over-representation (Clover) (Frith et al. 2004) can be used to identify conserved TFBSs that are over- or under-represented in the given promoter set.
- ***Search for motif cluster and CRM***: Since it is known that TFBS are not evenly distributed, finding motif peaks within the promoter region is likely to indicate functional regulatory regions. Cluster-Buster, a Hidden Markov Model based method (Frith et al. 2003) can be used to identify clusters of motifs in a given gene sequence. Matbase (Genomatix) contain well documented, experimentally confirmed promoter modules with synergistic, antagonistic or additive functions. Comparison of predicted CRMs with the known TF modules is used to identify and cross validate meaningful TFBS combinations.

17.4.4 Concept Integration

The combination of approaches that include unsupervised clustering analysis, gene set enrichment analysis, promoter and literature mining with microarray analysis is useful in identifying general modes of action and in forming initial hypotheses regarding the potential targets and regulatory mechanisms underlying experimental data. Bioinformatics data mining and wet-bench confirmation is then integrated to study selected genes and associated pathways. Figure 17.2 provides an example of an integrative analysis of a gene cluster containing 45 genes derived from microarray analysis of 194 samples under 27 conditions. GO classification indicates a significant enrichment of genes involved in lipid metabolism ($P = 0.000472$) and phosphate transport ($P = 0.0055$). Motif and module search were carried out to search for phylogenetically conserved common regulatory elements. EBOX, SP1F, SREB and EGRF are the most significantly enriched TFBS for this gene cluster. Among these, EBOX was detected in 43 out of 45 gene promoters. SREB shares high matrix similarity with EBOX. SREBP binds to both EBOX and SREB *in vivo* (Bennett et al. 1995). Further, SREB/SP1F and EGRF/ZBPF likely form a regulatory module in the promoters of genes in this cluster. SREBP-SP1 may play important role in controlling lung lipid homeostasis, a concept supported by the demonstration that SREB regulates a number of genes involved in lipogenesis in the lung (Plantier et al. 2012; Besnard et al. 2009; Mason et al. 2003).

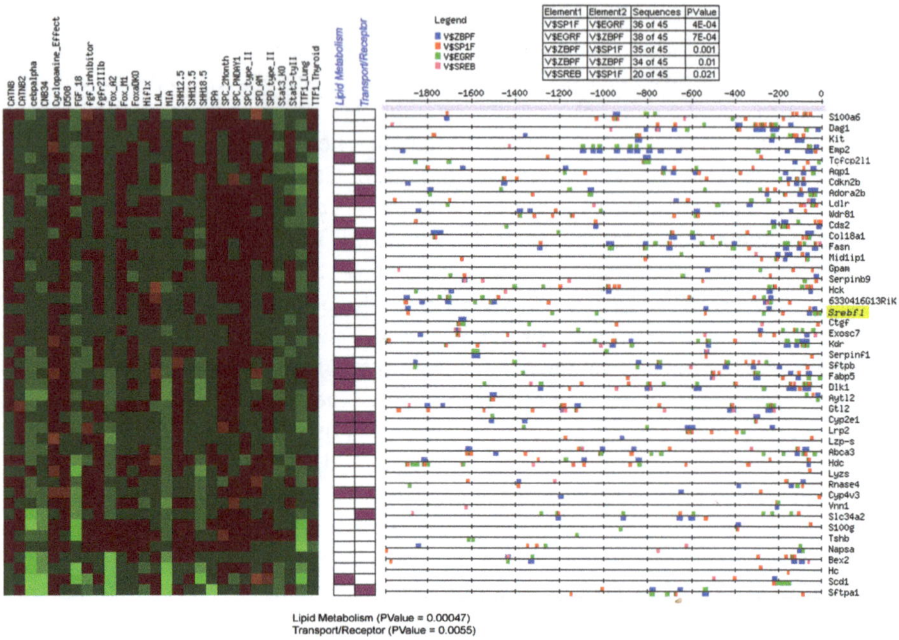

Fig. 17.2 Integrative analysis of a "Lipid" enriched cluster: (**a**) Heatmap represents gene expression profiles across 27 conditions. Intensity in the *red* and *green* color range indicates up-regulated and down-regulated mRNAs, respectively. Each row represents a single gene. Each column represents a particular experimental condition. Each box represents the normalized RNA intensity value. Similarity measures were assessed utilizing Euclidean distance. (**b**) Significant functional classifications. *Colored boxes* indicate genes with the respective classification. The calculated P value for the enriched classifications is shown at the *bottom*. (**c**) Promoter analysis. SREB (*pink*), SP1F (*red*), EGRF (*green*) and ZFPF (*blue*) were indicated in the 2 kb upstream promoter sequences of the cluster genes. Those TFBSs are overly represented in this cluster, conserved across human and mouse genomes and tend to form composite modules as shown in the *top Table*

17.4.5 *Meta-Analyses of Microarray Data*

As a result of the wide application of mRNA microarray technology, there are rapidly growing collections of available microarray data sets that can be used for subsequent analysis. "Meta-analysis" of microarray data uses statistical tools to combine and synthesize results from several related, but independent, microarray experiments (Hong and Breitling 2008). The objectives include increasing power to detect an overall treatment effect, assessment of the variability among studies, and maximizing the use of data available. We have carefully studied genome wide responses to experimental perturbations in more than 80 distinct mRNA microarray experiments. These data are a useful resource for mapping transcriptional networks regulating various aspects of lung development, function, and disease. Applying meta-analysis to combinations of existing lung microarray datasets has been useful in testing hypotheses and revealing insights into potential regulatory mechanisms. Using this strategy, we have compared mRNA microarray experiments from distinct mouse models that share delayed lung maturation phenotypes (details see below).

17.4.6 Analysis of Genetic Models Influencing Lung Maturation

Lung function at birth is highly dependent on the differentiation and function of the respiratory epithelium that, in turn, produces pulmonary surfactant lipids and proteins. Studies from the conditional deletion or mutation of specific genes led to the identification of several transcription factors and signaling molecules that serve cell-autonomous roles in the respiratory epithelium that are critical for respiratory adaptation at birth, including NKX2-1, FOXA2, C/EBPα, MIA1 and CNB1. Thyroid transcription factor gene 1 (TTF-1/NKX2-1) belongs to the NK2 class of homeobox transcription factors. *Nkx2-1(−/−)* embryos die at birth from respiratory failure due to a profound failure of lung formation (Kimura et al. 1996); absence of *Nkx2-1* activity leads to inhibition of distal lung morphogenesis and epithelial cell differentiation (Yuan et al. 2000). Mice bearing a mutation in *Nkx2-1* (i.e., serine phosphorylation sites are mutated to alanine), substantially rescued lung formation, but impaired lung maturation, causing respiratory failure at birth (DeFelice et al. 2003). Deletion of *Foxa2* (forkhead box protein A2, a winged-helix transcription factor), *Cebpa* (CCAAT enhancer binding protein α) and *Cnb1* (Calcineurin b1) from lung epithelial cells or misexpression of *Mia1* (Melanoma inhibitory activity protein) in lung epithelial cells caused respiratory distress at birth with phenotypic and biochemical changes similar to those observed in NKX2-1 mutant mice, namely: decreased expression of surfactant mRNA and proteins, lack of appropriate differentiation of type I and II cells, and absence of lamellar body formation, indicating delayed peripheral lung maturation (Dave et al. 2006; DeFelice et al. 2003; Martis et al. 2006; Wan et al. 2004; Lin et al. 2008). To better understand the mechanisms underlying the similarity of the perinatal lung maturational defects, we employed genome-wide transcriptional profiling to study genomic responses to phosphorylation mutation of *Nkx2-1*, lung epithelial cell specific deletion of *Foxa2, Cebpa, Cnb1* and misexpression of *Mia1*. Meta-analysis of these microarray datasets showed that although these transcription factors and signaling molecules act through different signaling pathways and bind to distinct cis-elements, they influence the expression of many common targets involved in surfactant protein and lipid biosynthesis (e.g., Abca3, Scd1, Pon1, Sftpa, Sftpb, Sftpc and Sftpd), fluid and solute transport (e.g., Aqp5, Scnn1g, Slc34a2) and innate host defense (e.g., Lys, Sftpa, Sftpd and Scgb1a1), suggesting that Foxa2, CEBPα, Cnb1, Mia1 and TTF-1 may share a common transcriptional network regulating perinatal lung maturation and postnatal respiratory adaptation (Fig. 17.3a).

a

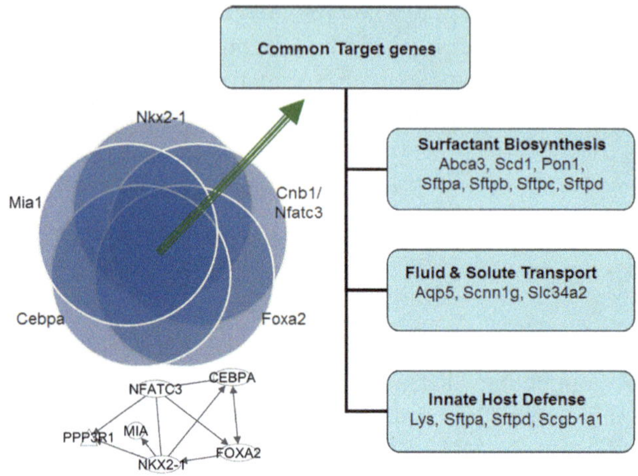

b

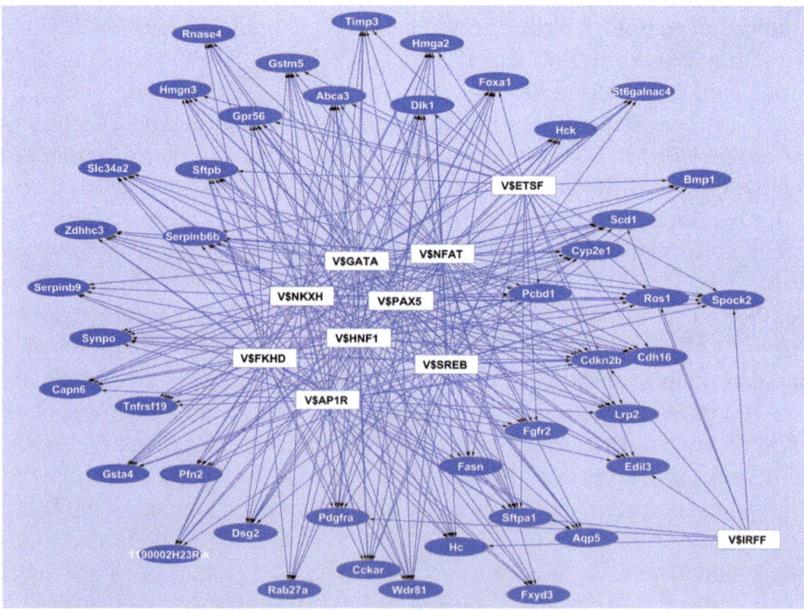

c

Matrix Family	TF Symbol	Binding Frequency
V$GATA	Gata1, Gata5, Gata6	42
V$HNF1	Hnf1a, Hnf1b	40
V$PAX5	Pax9	40
V$NFAT	Ilf3, Nfatc3	39
V$NKXH	Nkx2-1	35
V$AP1R	Bach1, Bach2	33
V$FKHD	Foxa1, Foxa2	32
V$SREB	Srebf1	28
V$ETSF	Gabpa, Gabpb1	24
V$IRFF	Irf1	8

Based on the assumption that co-expressed genes sharing similar phenotypes are likely controlled by the same sets of TFs, the altered genes from the "phenocluster" analysis were subjected to promoter common TFBS search to identify statistically over-represented TFBSs and CRMs as compared to their general occurrence in promoter regions of the mouse genome. Using a p-value of 0.05, TFBSs were ranked based on the total binding frequency in the promoter region of genes in the cluster presence (Fig. 17.3c). Cytoscape v2.8.2 was used to generate transcriptional regulatory networks for visualizing molecular interaction networks between this group of genes and predicted TFs. Nfatc3, Nkx2-1 and Foxal/a2 were among the most over-represented TFs in the lung with high connectivity to this group of genes in the network (Fig. 17.3b). We linked the TFs to their potential target genes, generating data that further supports the concept that these TFs work in concert to control the transcription of genes involved in lung maturation.

17.4.7 mRNA Microarray and ChIP-seq Integration

Chromatin immunoprecipitation (ChIP) followed by genomic tiling microarray (ChIP-chip) or massive parallel sequencing (ChIP-seq) are two of the most widely used approaches for genome-wide identification of physical interactions between TFs and the regulatory DNA sequences to which they bind. Such studies provide direct evidence of regulatory relationships (Ho et al. 2011). The integration of ChIP-seq and mRNA microarray results enables further dissection of the direct vs. indirect effects of selected TFs *in vivo*. Differentially expressed genes identified from mRNA microarray analysis with positive binding sites from ChIP-seq analysis are considered to be direct transcriptional targets. Genes with altered expression in mRNA microarray analysis, but not in ChIP-seq, are likely indirect targets. Promising peaks that are not associated with changes in mRNA level are likely to be nonfunctional binding sites. Because expression data and ChIP-seq data provide complementary information, predictions of TF-TG relationships based on integration of their data are more accurate than predictions based on single data sources. For instance, to more fully understand the NKX2-1 transcriptional activities

Fig. 17.3 Meta-analysis of microarrays from mouse models sharing common perinatal respiratory distress phenotypes at birth: (**a**) Comparative microarray analysis of lung selective deletions of *Foxa2, Cebpa, Cnb1*, mutation of *Nkx2-1* and misexpression of *Mia1* revealed common targets involved in surfactant biosynthesis, fluid and solute transport and innate defense. (**b**) Over-represented TFBS were identified using Clover and TRANSFAC software in the 2 kb promoter regions of the genes altered in all five microarrays. Cytoscape v2.8.2 was used to generate TRN via mapping TF matrix to predicted target genes. Each *white rectangle* represents a TF matrix family and each *blue oval* represents a target gene. (**c**) The summary of binding frequency of each TF to its potential TGs within 2 kb promoter regions. If multiple TFs are associated with the same TF matrix family, only TFs with abundantly expressed in lung were selected as representative TFs in the table

and direct downstream targets in lung, we preformed integrative analysis of NKX2-1 ChIP-seq and NKX2-1 over-expression mRNA microarray data (Secondary analysis of data from (Maeda et al. 2012)) in following steps: (1) identify differentially expressed genes in response to the NKX2-1 overexpression through mRNA microarray analysis, (2) determine high confidence peaks containing a NKX2-1 binding sites through peak calling of ChIP-seq data, (3) annotate the peak regions and find closest gene in the region, (4) map differentially expressed genes to high confidence peak regions, (5) discover novel motifs and mapping known TFBSs by scanning the known TFBS library using position weighted matrix, (6) build an NKX2-1 consensus binding site, and (7) identify NKX2-1 containing cis-regulatory modules by searching for the presence of additional TFBS within NKX2-1 containing peak sequences that represent potential NKX2-1 interacting partners. We show that the AP1, GATA, RXR, and PAX family of TFs are ranked highest to form cis-regulatory modules with NKX2-1 (Fig. 17.4). Based on our previous studies, *Nkx2-1, Cnb1, Cebpa and Foxa2* belong to same "phenoclusters" (i.e., when perturbed, exhibit phenotypes with similar morphological features) and shared a group of common downstream genes that are known to play important roles in lung maturation and function. Taking advantage of next generation sequencing, we assessed the probability that these TFs form cis-regulatory modules. We scanned the presence of NFAT, CEBPA and FOXA2 binding sites within the peak regions containing NKX2-1 binding site as well as their association with changes in expression from corresponding mRNA microarray analysis. We used random sequences of fragments as reference to calculate the binomial probability of their association. Interestingly, the co-binding probability and frequency of NKX2-1/CEBPA, NKX2-1/FOXA2 and NKX2-1/NFAT were significantly enriched in the positive peak regions compared with random DNA fragments, as indicated in Fig. 17.4. These data support the concept that CEBPA, FOXA2 and NFAT act as interaction partners with NKX2-1 to regulate gene expression during lung development and maturation.

Through comparative mRNA microarray data from experiments with related phenotypes, clustering, promoter analysis, gene set functional enrichment analysis, literature mining and correlation with experimental data (physiology, biochemistry and morphometric measurements), we have made substantial progress in mapping the transcription regulation of lung "maturation." A similar approach could be taken to identification of transcriptional regulatory networks underlying the development and maturation of other tissues and organs.

17.5 Toward a Systems Level Understanding of Surfactant Homeostasis

While microarray data can be used for unbiased gene and pathway discovery, identifying direct versus indirect genomic responses remain challenging. Functional enrichment analysis and literature mining enable the association of genes with

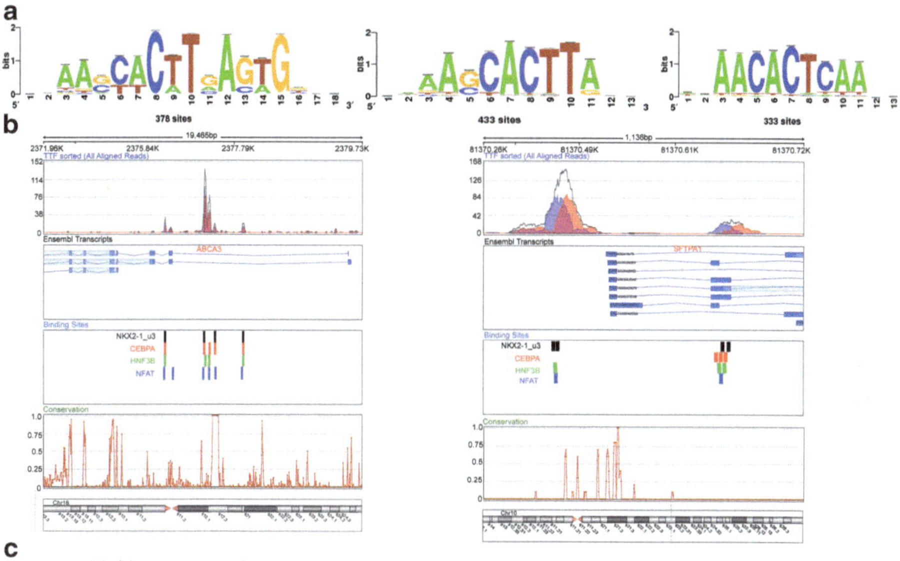

c

Module	pvalue
Nkx2-1_AP1F	0
Nkx2-1_AP1R	0
NKX2-1_RXRF	0
NKX2-1_NR2F	0
Nkx2-1_PAX8	4.8406E-240

Module	pvalue
NKX2-1_NFAT	3.5339E-175
NKX2-1_HNF3B	1.6565E-121
NKX2-1_CEBPA	3.52568E-41

Fig. 17.4 Nkx2-1 Chip-Seq analysis: (**a**) Logos of the motifs discovered by peak-motifs for Nkx2-1. Peaks was selected using MACS peak detection program in Galaxy package with p-value < e-40. Next, we applied 'peak-motifs' in RSAT package to discover binding motifs in the detected peak regions and compare discovered motifs to TRANSFAC and JASPAR databases to predict associated transcription factors. (**b**) Examples of peaks and TFBSs (Nkx2-1: *dark blue*, CEBPA: *red*, FOXA2: *green* and NFAT: *baby blue*) detected in the human SFTPC and ABCA3 gene locus. (**c**) The top enriched TFBS modules that co-occurred within 5–100 bp of NKX2-1-binding sequences were determined by RegionMiner (Genomatix) within all peaks identified from ChIP-seq analysis and listed at *left panel*. The probability of NKX2-1 forming modules with NFAT, FOXA2 and CEBPA were determined within peaks associating with genes altered expression from corresponding NKX2-1 mRNA microarray analysis. The results are compared with random picked sequence fragments with the length similar to the NKX2-1 containing peak regions to determine the binomial probability (*right panel*)

biological processes and pathways, but are limited to current annotated knowledge. Correlations between regulators and potential targets are largely based on their shared expression patterns, taking into account the likelihood that expression patterns of TF and their targets are often correlated and groups of genes sharing highly correlated expression profiles are likely to share TF(s). This approach, however, will miss TFs that regulate their targets via processes that are independent of their own levels of expression, for example, their activity may be mediated primarily by post-transcriptional mechanisms. Analyses seeking conserved or common TFBSs in promoters of co-expressed genes can identify the potential cis-elements, but does not prove the actual binding of the TF; moreover, TFBS prediction is often associated with high numbers of false positive predictions because of the short

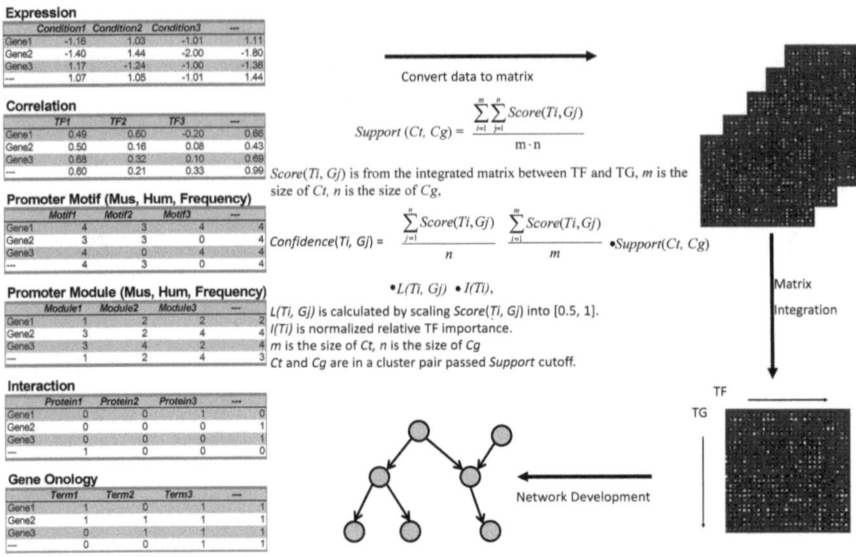

Fig. 17.5 Work flow for construction of a TRN regulating surfactant lipid homeostasis: In order to predict TF-TG interactions using combined evidence, we developed an algorithm to integrate expression profiling with expression-independent data (protein interactions, functional annotation, promoter analyses, and literature mining). TFs were further clustered according to an integrated matrix compiled from data sources including TF-TG functional similarity, TF-TG expression correlation matrix and TF-TG interaction matrix. Each value in the four matrices was scaled from 0 to 1 and summed into the integrated TF-TG matrix. TGs were grouped into sub-clusters based on an integrated matrix, combining and capturing information from four data sources: gene expression, TF-TG correlation, promoter TFBS prediction and GO functional similarity. We calculated the relative confidence score of TF-TG associations by combining the data obtained

(generally encompassing 5–15 base pairs), and degenerate nature of many TFBS motifs. Likewise, the important regulatory roles of non-coding RNAs are not readily identified in the RNA microarray platforms.

To generate a model by which surfactant homeostasis is controlled, we retrieved 194 mRNA microarray samples from 27 distinct mouse models in which TF/SMs modifications were made in mouse models of lung disease. We utilized a systems approach to integrate expression profiling with evidence from other resources, including TF-TG correlation, protein interactions, functional annotation, promoter analyses, and literature mining. Figure 17.5 illustrates the workflow used to build our network model. Detailed methodologies can be found in (Xu et al. 2010). Briefly, we identified 1,498 genes that significantly changed in mRNA microarray analyses in response to the gene perturbations under at least 5 out of 27 experimental conditions. We further clustered differentially expressed genes using Fuzzy clustering by a local approximation of membership algorithm (Fu and Medico 2007). Three co-expressed gene clusters that were highly enriched in genes influencing lipid synthesis and transport were identified. For each cluster, over-represented

TF binding sites, TF-TG expression correlations, TF-TG functional similarity and TF-TG protein-protein interaction were determined. TFs were further clustered according to an integrated matrix compiled from four types of data sources: TFBS-TG scoring, TF-TG functional similarity, TF-TG expression correlations, and TF-TG interaction matrices. Each value in the four matrices was scaled from 0 to 1 and summed into the integrated TF-TG matrix. TGs were grouped into sub-clusters based on an integrated matrix, combining and capturing information from four data sources: gene expression, TF-TG correlation, promoter TFBS prediction and GO functional similarity. In the integrated matrix, each row represents a gene, and each column represents a feature from one of the four matrices. We calculated the relative confidence score of TF-TG associations by combining the resulting data. *Confidence* describes the possibility of a true positive TF-TG relationship according to the integrated information available. The TFBS-TG pairs with the highest *Confidence* scores represent candidates for experimental validation.

Based on predicted, ranked TF-TG relationships, we constructed a "lung lipid regulatory network." In Fig. 17.6, we show a sub-network consisting of the TFs with the highest connectivity (score $\geq$0.6, top 4.5%) among three gene clusters. SREBP, HNF3, ETSF, CEBP, GATA and IRFF are clear regulatory hubs in this network, and are TFs likely to be key regulators controlling surfactant lipid homeostasis via control of genes within three lipid-related clusters. The roles of several key TFs in the proposed network have been partially confirmed by previous studies, including SREBP1, FOXA2, CEBPA, ETV5 and GATA6 (Martis et al. 2006; Wan et al. 2004; Besnard et al. 2009; Bruno et al. 2000; Lin et al. 2006; Liu et al. 2002). IRF1 encodes interferon regulatory factor 1, a member of the interferon regulatory transcription factor family. The finding that IRF may serve as an important regulator in lung lipid homeostasis merits further experimental validation. Transcriptional regulators of surfactant homeostasis that had previously been experimentally validated were identified as key hubs in this unbiased network, supporting the reliability of the proposed model. The TFBS of SREBP, HNF3 and CEBP are commonly enriched in all three lipid related clusters and share many downstream targets, suggesting complex interactions among CEBP, SREBP and HNF3 in the proposed lung lipid network. Many of the network predictions for the targets of *Cebpa, Srebfl and Foxa2* were validated through the combination of promoter reporter assays, transgenic animal models, and literature mining (Xu et al. 2010).

17.6 Transcriptional Programs Controlling Perinatal Lung Maturation

Complex genetic programs that are influenced by multiple environmental and temporal dependent factors likely control the timing of lung maturation. Cross-sectional integrative gene expression profiling analysis does not take into account the dynamic nature of the transcriptional programs accompanying lung maturation. We have recently combined genetic, genomic and bioinformatics methods to

a

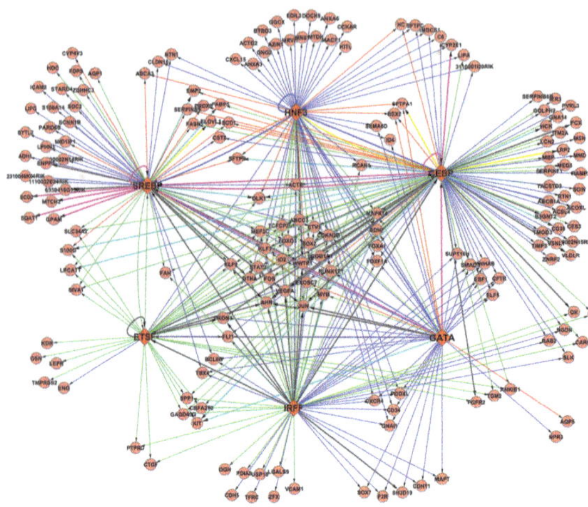

b

TFBS	Total Connectivity	TF in Lung
CEBP	447	Cebpa, Cebpb, Cebpd, Cebpg
IRFF	404	Irf1, Irf2, Irf3, Irf5, Irf7
HNF3	359	Foxa1, Foxa2
GATA	358	Gata6, Gata1
ETSF	344	Ets1, Ets2, Etv5, Nfe2l2, Elf2
SREB	312	Srebf1
FOXO	268	Foxo1, Foxo4, Foxo3a
FKHD	213	Foxf2, Foxc1
HAND	201	Lmo2
STAT	182	Stat6, Stat3
MEF2	176	Mef2a
NFAT	169	Ilf3, Nfatc3
CP2F	168	Atf4, Tcfcp2, Atf3, Atf1
NFKB	166	Nfkb1
EREF	165	Esrra
LEFF	150	Lef1
HFH	134	Foxf1a, Foxi1
PARF	134	Tef, Tead1
AP1R	129	Nfe2
LEFF	121	Tcf4
CIZF	117	Znf384
HAND	113	Tcf12
BARBIE	111	Unknown
NKXH	106	Nkx2-1
SORY	104	Sox5, Sox9, Sox2
NR2F	98	Hnf4a, Nr2f1, Nr2f2
OCT	92	Pou2f1, Pou6f2
CREB	79	Creb1
MYOD	74	Myog
NKX7	62	Nkx6-2
EBOX	59	Tcf4, Max
P53F	57	Trp53
RORA	57	Rora
HAML	54	Runx2, Pebp1
RXRF	54	Nr1h2
GREF	48	Nr3c1
BRN5	46	Pou6f1
HESF	46	Hes1
EGRF	45	Egr1, Wt1

Fig. 17.6 TRN composed of predicted TF-TG pairs with the highest connectivity: (**a**) The graphic representation of a sub-network consisting of predicted TF-TG pairs with confidence cutoff as 0.60 (top 4.5%) and the top 6 TFs with the highest connectivity. The network has 183 nodes and 386 links. *Round nodes* represent TGs, *red diamond nodes* represent TFs. *Blue edges* indicate the TF-TG predictions from C1, *red edges* for C2, *green* for C28, *yellow* for both C1 and C2, *brown* for both C1 and C28, *light blue* for both C2 and C28, and *pink edges* for TF-TG predication from C1, C2, and C28. The thickness of the edge corresponds to the frequency of the TF-TG prediction from all three clusters. (**b**) Confidence score was calculated based on the integrative evidence of TF-TG relationship. The overall connectivity of each TF to its potential TGs within three clusters were calculated and summarized in **b**. The corresponding TFs expressed in lung were also listed (Adapted with permission from Figure 2 in Xu et al. 2010)

elucidate the relationship between the length of gestation and lung function at birth in two inbred mouse strains (C57BL/6 J and A/J), whose gestational length differed by 1 day (Besnard et al. 2011; Murray et al. 2010; Xu et al. 2012). Lung maturation, as indicated by SatPC (surfactant lipid), lung histology, and the expression of surfactant genes, occurs earlier in B6 than A/J mice. Shorter gestation in B6 mice was associated with advanced morphological and biochemical pulmonary development and better perinatal survival when compared to A/J pups born prematurely (Besnard et al. 2011).

17.6.1 Dynamic Profiling

Taking into account the dynamic nature of the transcriptional programs accompanying lung maturation, we designed genome-wide, time-course microarray studies to systemically explore the dynamic regulation of lung maturation in both B6 and A/J mouse strains in order to discover genes, pathways, and associated transcriptional networks underlying lung maturation in the two strains of mice that differ in gestational length (Xu et al. 2012). Briefly, lung samples from each mouse strain were collected daily from E15.5 to PN0 at precise gestational ages (Murray et al. 2010). Lung RNAs isolated from the two mouse strains at different time points in development were hybridized to Mouse Gene 1.0 ST Array ($n = 3$/strain/time). Dynamic lung mRNA expression profiling from B6 and A/J mice were compared to identify: (1) genes and bioprocesses commonly altered in both strains during lung maturation, (2) transcription factors and signaling molecules (TF/SMs) that changed at different stages of lung maturation, (3) pathways and transcriptional networks controlling lung maturation, and (4) strain dependent effects on lung maturation.

To identify temporal-dependent gene expression changes during lung maturation, a functional Bayesian approach (Angelini et al. 2008) was used to analyze time dependent changes in lung mRNAs from both mouse strains. Next, we identified temporal dependent expression patterns and matched dynamic profiles of transcription factors and targets during lung maturation using STEM (Short Time-series Expression Miner), a clustering algorithm specifically designed for the analysis of short time series gene expression datasets (5–8 time points) (Ernst and Bar-Joseph2006). Comprehensive knowledge integration was employed to identify temporal dependent bioprocesses, key TF/SMs and associated TRNs and to reveal the potential biological interrelationships among the matched TF/SMs and their proposed target genes within each cluster.

To identify strain dependent gene expression during lung maturation, we selected genes that were differentially expressed in A/J and B6 mice at E18.5, but unchanged when comparing A/J at E19.5 vs. B6 at E18.5. The selection is based on two sets of evidence: (1) sample correlation analysis suggested that B6 mRNAs at E18.5 were most similar to A/J mRNAs at E19.5 and least similar to A/J mRNAs at E18.5. (2) A/J fetuses delivered 2 days prematurely (at E18.5) failed to expand their lungs and died of respiratory failure soon after birth, while A/J fetuses delivered 1 day prematurely (at E19.5) survived (87.4%), a survival rate similar to that for B6 mice born 1 day prematurely at E18.5 (82.5%) (Besnard et al. 2011). Together these data support the concept that lung mRNAs differ most between the two strains at E18.5 and that mRNAs modulated during "catch up" in A/J mice at E19.5, are likely important for lung maturation and function.

Through comprehensive bioinformatics data mining, both temporal and strain dependent gene expression patterns were identified during lung maturation. As illustrated in Fig. 17.7, bioprocesses and key regulators associated with different stages of lung development were identified. Lung development, cell adhesion and movement, lipid metabolism, and proliferation were induced early in lung maturation (E15–16 in the pseudoglandular stage). *Hopx, Cebpa, Tcf21* and *Klf5* were predicted to be important transcriptional regulators at this stage, a finding consistent with gene deletion studies that support their important roles in prenatal lung maturation (Martis et al. 2006; Wan et al. 2008; Quaggin et al. 1999; Yin et al. 2006). TF/SMs regulating vasculature development and apoptosis were induced at E16–17

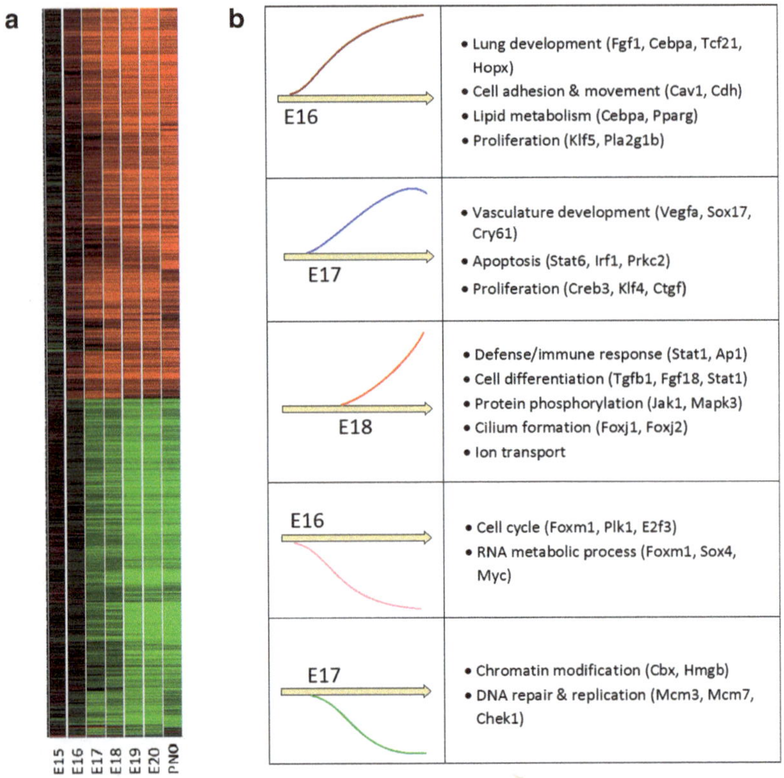

Fig. 17.7 Dynamic regulation of lung maturation: (**a**) Heatmap of temporal dependent gene expression changes during lung maturation (E15-PN0). (**b**) Schematized depiction of bioprocesses and predicted key regulators changed dynamically with advancing gestation

(canalicular stage), *Vegfa, Sox17 and Stat3/6* representing important regulators at this stage. Innate defense/immune responses, cell differentiation, protein phosphorylation, ion transport, and cilium formation were induced at later gestational ages (E18–20, in the saccular stage), *Stat1, Tgfb1 and Foxj1* being important regulators associated with this stage of maturation. Cell cycle and chromatin assembly were repressed during lung maturation. FOXM1, PLK1, chromobox, SWI/SNF and high mobility group families of transcription factors were predicted to play important roles in the negative regulation of lung cell proliferation that occurs in late gestation.

Prior to birth, innate immune responses and surfactant production are critical and connected processes that positively influence lung maturation necessary for respiration and survival after birth. In contrast, epigenetic regulators are likely to play a repressive role by altering chromatin structure and controlling the cell cycle. We hypothesize that precise regulation and balance among the positive and negative gene networks are likely critical determinants coordinating the timing of lung maturation with gestational length that differs in the B6 and A/J mouse strains.

17.6.2 *Sub-networks Control Distinct Biological Processes During Lung Maturation*

Transcriptional regulatory networks can be divided into sub-networks of interconnected genes, each of which represents a functional unit of the entire network. Each unit is driven by tissue or cell type specific TFSM hubs and acts at specific times and in specific cell types. All units work coordinately to control spatiotemporal processes of lung development and maturation. Due to the complexity and modularity of the TRNs, it is more desirable and experimentally feasible to focus on sub-networks. For example, in our recent genome-wide time-course mRNA microarray study in two strains of mice, we identified multiple bioprocesses induced during the saccular stage of lung development, at E16.5–E17.5, including cell adhesion, lipid metabolism/transport and vasculature development. These major bioprocesses were controlled by CAV1/CDH1, CEBPA/PPARG and VEGFA centered sub-networks, respectively. Innate defense/immune responses were induced at later gestational ages (E19.5–20.5). STAT1, AP1, and EGFR are important regulators of these responses in the sub-network. Expression of RNAs associated with the cell cycle was repressed during prenatal lung maturation and was associated with a FOXM1/PLK1 centered sub-network. These sub-networks consist of a small group of effector genes, usually centered around one or several interrelated hubs. Effector genes in the sub-network tend to be co-expressed, transcriptionally co-regulated, and perform a similar cellular function or work in concert to influence a specific developmental process. Perturbations and experimental validation of hubs and effector genes in the sub-networks will help further delineate the complicated biological processes involved in lung maturation.

17.7 Conclusions and Future Directions

In this chapter, we have summarized functional genomic and systems biology approaches that can be applied to the study of transcriptional regulation of tissue and organ development. The example chosen to illustrate uses of tools and databases is perinatal lung maturation and surfactant homeostasis. We have emphasized the importance of using integrative approaches to achieve comprehensive understanding of regulatory mechanisms controlling lung maturation at systems level. To date, none of the transcription factors participating in lung development are exclusively lung-specific. The unique combinations and interactions among TF/SMs are likely to provide the basis for emergence of lung structural and functional specificities that drive lung maturation. A systems level identification of the individual TF/SMs and their unique interactions in the context of different stages of lung development will be needed to further understand the temporal regulation of structural and functional changes involved in lung maturation prior to birth. A thorough understanding of the molecular mechanisms controlling the timing of normal lung maturation will promote the understanding of the pathogeneses of lung diseases associated with preterm birth, providing new therapeutic and diagnostic tools to treat pulmonary diseases in infants. This approach takes full advantage of genomic resources to provide an accurate prediction of lung transcriptional networks. The systems biology strategy discussed in this review is powerful and will be highly relevant to studies of other diseases and developmental processes. Like RDS, most lung diseases are complex, and a single gene or protein approach is less likely to identify the cause of acute and chronic lung diseases. Systems biology strategies will facilitate our understanding the pathogenesis and treatment of diseases of the lung and other organs in the future.

Looking forward, we anticipate that transcriptome analysis, ChIP-seq, expression profiling, and assessment of the important regulatory roles of non-coding RNAs will provide essential building blocks for the construction of TRNs regulating signaling and metabolic pathways underlying the complexity of organ formation and function. The emerging view is that many developmental sub-networks are largely operative at the level of individual cells, each of which expresses a unique combination of proteins. Studies of the transcriptome at the level of single cells level pose major technical challenges that have not yet been solved. Currently, the use of mRNA microarray technology has been widely applied at the organ level. The examples described in this chapter are largely based on studies using whole lung samples. Nevertheless, the lung is a complex organ consisting of more than 40 distinct cell types that may be regulated by cell type-specific TF/SMs combinations that activate additional sub-networks of genes to determine the cell fate and function, for example, those involved in Type I and II cell differentiation and surfactant production. Isolating cell-type-specific mRNA, using Laser Capture Microdissection (LCM) and Fluorescence Activated Cell Sorting (FACS) will be useful in measuring RNA expression patterns at a single cell level (Okaty et al. 2011; Chung et al. 2005; Arlotta et al. 2005). These approaches will significantly improve the sensitivity and resolution of mRNA analyses and enable the construction of TRNs that directly reflect the regulation of expression in a cell autonomous manner.

The validation and reconstruction of TRNs based on experimental data presents a major scientific challenge. Currently, TRNs can be experimentally validated via TF-centered (Protein to DNA) or gene-centered (DNA to protein) methods (Arda and Walhout 2010). TF-centered methods such as chromatin immunoprecipitation and gene-centered methods such as reporter gene assays are commonly used to delineate TF-TG relationships. However, these methods are difficult to apply to larger sets of genes, complex network predictions, and within the context of intact chromatin. The recent introduction of ChIP-seq and high throughput siRNA/shRNA based loss-of-function studies provide powerful tools for functional validation of predictions made by computational analyses. How to translate cell culture findings to *in vivo* responses also remains a major issue. New technologies and experimental approaches may be required to bring the process of experimental validation to the next level that are compatible with the complex computational network predictions and high-throughput, genome-wide technologies.

References

Angelini C, Cutillo L, De Canditiis D, Mutarelli M, Pensky M. BATS: a Bayesian user-friendly software for analyzing time series microarray experiments. BMC Bioinform. 2008;9:415.

Arda HE, Walhout AJ. Gene-centered regulatory networks. Brief Funct Genomics. 2010;9(1):4–12.

Arlotta P, Molyneaux BJ, Chen J, Inoue J, Kominami R, Macklis JD. Neuronal subtype-specific genes that control corticospinal motor neuron development in vivo. Neuron. 2005;45(2):207–21.

Avery ME, Mead J. Surface properties in relation to atelectasis and hyaline membrane disease. AMA J Dis Child. 1959;97(5, Part 1):517–23.

Bennett MK, Lopez JM, Sanchez HB, Osborne TF. Sterol regulation of fatty acid synthase promoter. Coordinate feedback regulation of two major lipid pathways. J Biol Chem. 1995;270(43):25578–83.

Besnard V, Wert SE, Stahlman MT, Postle AD, Xu Y, Ikegami M, Whitsett JA. Deletion of scap in alveolar type II cells influences lung lipid homeostasis and identifies a compensatory role for pulmonary lipofibroblasts. J Biol Chem. 2009;284(6):4018–30.

Besnard V, Wert SE, Ikegami M, Xu Y, Heffner C, Murray SA, Donahue LR, Whitsett JA. Maternal synchronization of gestational length and lung maturation. PLoS One. 2011;6(11):e26682.

Bonner AE, Lemon WJ, You M. Gene expression signatures identify novel regulatory pathways during murine lung development: implications for lung tumorigenesis. J Med Genet. 2003;40(6):408–17.

Bridges JP, Xu Y, Na CL, Wong HR, Weaver TE. Adaptation and increased susceptibility to infection associated with constitutive expression of misfolded SP-C. J Cell Biol. 2006;172(3):395–407.

Bruno MD, Korfhagen TR, Liu C, Morrisey EE, Whitsett JA. GATA-6 activates transcription of surfactant protein A. J Biol Chem. 2000;275(2):1043–9.

Bullinger L, Valk PJ. Gene expression profiling in acute myeloid leukemia. J Clin Oncol. 2005;23(26):6296–305.

Burri PH. Fetal and postnatal development of the lung. Annu Rev Physiol. 1984;46:617–28.

Chen J, Bardes EE, Aronow BJ, Jegga AG. ToppGene Suite for gene list enrichment analysis and candidate gene prioritization. Nucleic Acids Res. 2009;37(Web Server issue):W305–11.

Chung CY, Seo H, Sonntag KC, Brooks A, Lin L, Isacson O. Cell type-specific gene expression of midbrain dopaminergic neurons reveals molecules involved in their vulnerability and protection. Hum Mol Genet. 2005;14(13):1709–25.

Datta S, Satten GA. A signed-rank test for clustered data. Biometrics. 2008;64(2):501–7.

Dave V, Childs T, Xu Y, Ikegami M, Besnard V, Maeda Y, Wert SE, Neilson JR, Crabtree GR, Whitsett JA. Calcineurin/Nfat signaling is required for perinatal lung maturation and function. J Clin Invest. 2006;116(10):2597–609.

DeFelice M, Silberschmidt D, DiLauro R, Xu Y, Wert SE, Weaver TE, Bachurski CJ, Clark JC, Whitsett JA. TTF-1 phosphorylation is required for peripheral lung morphogenesis, perinatal survival, and tissue-specific gene expression. J Biol Chem. 2003;278(37):35574–83.

Dennis Jr G, Sherman BT, Hosack DA, Yang J, Gao W, Lane HC, Lempicki RA. DAVID: database for annotation, visualization, and integrated discovery. Genome Biol. 2003;4(5):P3.

Dubin SB. Assessment of fetal lung maturity: in search of the Holy Grail. Clin Chem. 1990;36(11):1867–9.

Ernst J, Bar-Joseph Z. STEM: a tool for the analysis of short time series gene expression data. BMC Bioinform. 2006;7:191.

Frith MC, Li MC, Weng Z. Cluster-buster: finding dense clusters of motifs in DNA sequences. Nucleic Acids Res. 2003;31(13):3666–8.

Frith MC, Fu Y, Yu L, Chen JF, Hansen U, Weng Z. Detection of functional DNA motifs via statistical over-representation. Nucleic Acids Res. 2004;32(4):1372–81.

Fu L, Medico E. FLAME, a novel fuzzy clustering method for the analysis of DNA microarray data. BMC Bioinform. 2007;8:3.

Gasch AP, Eisen MB. Exploring the conditional coregulation of yeast gene expression through fuzzy k-means clustering. Genome Biol. 2002;3(11):RESEARCH0059.

Goerke J. Pulmonary surfactant: functions and molecular composition. Biochim Biophys Acta. 1998;1408(2–3):79–89.

Goldenberg RL, Culhane JF, Iams JD, Romero R. Epidemiology and causes of preterm birth. Lancet. 2008;371(9606):75–84.

Gravett MG, Rubens CE, Nunes TM. Global report on preterm birth and stillbirth (2 of 7): discovery science. BMC Pregnancy Childbirth. 2010;10 Suppl 1:S2.

Grenache DG, Gronowski AM. Fetal lung maturity. Clin Biochem. 2006;39(1):1–10.

Ho JW, Bishop E, Karchenko PV, Negre N, White KP, Park PJ. ChIP-chip versus ChIP-seq: lessons for experimental design and data analysis. BMC Genomics. 2011;12:134.

Hong F, Breitling R. A comparison of meta-analysis methods for detecting differentially expressed genes in microarray experiments. Bioinformatics. 2008;24(3):374–82.

Jeziorska DM, Jordan KW, Vance KW. A systems biology approach to understanding cis-regulatory module function. Semin Cell Dev Biol. 2009;20(7):856–62.

Johansson J, Curstedt T. Molecular structures and interactions of pulmonary surfactant components. Eur J Biochem. 1997;244(3):675–93.

Khatri P, Draghici S, Ostermeier GC, Krawetz SA. Profiling gene expression using onto-express. Genomics. 2002;79(2):266–70.

Kimura S, Hara Y, Pineau T, Fernandez-Salguero P, Fox CH, Ward JM, Gonzalez FJ. The T/ebp null mouse: thyroid-specific enhancer-binding protein is essential for the organogenesis of the thyroid, lung, ventral forebrain, and pituitary. Genes Dev. 1996;10(1):60–9.

Lian X, Yan C, Yang L, Xu Y, Du H. Lysosomal acid lipase deficiency causes respiratory inflammation and destruction in the lung. Am J Physiol Lung Cell Mol Physiol. 2004;286(4):L801–7.

Liang M, Cowley AW, Greene AS. High throughput gene expression profiling: a molecular approach to integrative physiology. J Physiol. 2004;554(Pt 1):22–30.

Lin S, Perl AK, Shannon JM. Erm/thyroid transcription factor 1 interactions modulate surfactant protein C transcription. J Biol Chem. 2006;281(24):16716–26.

Lin S, Ikegami M, Xu Y, Bosserhoff AK, Malkinson AM, Shannon JM. Misexpression of MIA disrupts lung morphogenesis and causes neonatal death. Dev Biol. 2008;316(2):441–55.

Liu C, Glasser SW, Wan H, Whitsett JA. GATA-6 and thyroid transcription factor-1 directly interact and regulate surfactant protein-C gene expression. J Biol Chem. 2002;277(6):4519–25.

Maeda Y, Dave V, Whitsett JA. Transcriptional control of lung morphogenesis. Physiol Rev. 2007;87(1):219–44.

Maeda Y, Chen G, Xu Y, Haitchi HM, Du L, Keiser AR, Howarth PH, Davies DE, Holgate ST, Whitsett JA. Airway epithelial transcription factor NK2 homeobox 1 inhibits mucous cell metaplasia and Th2 inflammation. Am J Respir Crit Care Med. 2011;184:421–29.

Maeda Y, Tsuchiya T, Hao H, Tompkins DH, Xu Y, Mucenski M, Du L, Keiser AR, Fukazawa T, Naomoto Y, Nagayasu T, Whitsett JA. $KRAS^{G12D}$ and Nkx2-1 haploinsufficiency induce invasive mucinous adenocarcinoma of the lung (submitted to J Clin Invest, 2012).

Margalit O, Somech R, Amariglio N, Rechavi G. Microarray-based gene expression profiling of hematologic malignancies: basic concepts and clinical applications. Blood Rev. 2005;19(4):223–34.

Martis PC, Whitsett JA, Xu Y, Perl AK, Wan H, Ikegami M. C/EBP{alpha} is required for lung maturation at birth. Development. 2006;133(6):1155–64.

Mason RJ, Pan T, Edeen KE, Nielsen LD, Zhang F, Longphre M, Eckart MR, Neben S. Keratinocyte growth factor and the transcription factors C/EBP alpha, C/EBP delta, and SREBP-1c regulate fatty acid synthesis in alveolar type II cells. J Clin Invest. 2003;112(2):244–55.

Matsuzaki Y, Xu Y, Ikegami M, Besnard V, Park KS, Hull WM, Wert SE, Whitsett JA. Stat3 is required for cytoprotection of the respiratory epithelium during adenoviral infection. J Immunol. 2006;177(1):527–37.

Mayburd AL, Martlinez A, Sackett D, Liu H, Shih J, Tauler J, Avis I, Mulshine JL. Ingenuity network-assisted transcription profiling: identification of a new pharmacologic mechanism for MK886. Clin Cancer Res. 2006;12(6):1820–7.

McGinn T, Wyer PC, Newman TB, Keitz S, Leipzig R, For GG. Tips for learners of evidence-based medicine: 3. Measures of observer variability (kappa statistic). CMAJ. 2004;171(11):1369–73.

McMurtry IF. Introduction: pre- and postnatal lung development, maturation, and plasticity. Am J Physiol Lung Cell Mol Physiol. 2002;282(3):L341–4.

Metzger DE, Xu Y, Shannon JM. Elf5 is an epithelium-specific, fibroblast growth factor-sensitive transcription factor in the embryonic lung. Dev Dyn. 2007;236(5):1175–92.

Miller LA, Wert SE, Clark JC, Xu Y, Perl AK, Whitsett JA. Role of Sonic hedgehog in patterning of tracheal-bronchial cartilage and the peripheral lung. Dev Dyn. 2004;231(1):57–71.

Minn AJ, Gupta GP, Siegel PM, Bos PD, Shu W, Giri DD, Viale A, Olshen AB, Gerald WL, Massague J. Genes that mediate breast cancer metastasis to lung. Nature. 2005;436(7050):518–24.

Mucenski ML, Nation JM, Thitoff AR, Besnard V, Xu Y, Wert SE, Harada N, Taketo MM, Stahlman MT, Whitsett JA. Beta-catenin regulates differentiation of respiratory epithelial cells in vivo. Am J Physiol Lung Cell Mol Physiol. 2005;289(6):L971–9.

Murray SA, Morgan JL, Kane C, Sharma Y, Heffner CS, Lake J, Donahue LR. Mouse gestation length is genetically determined. PLoS One. 2010;5(8):e12418.

Okaty BW, Sugino K, Nelson SB. A quantitative comparison of cell-type-specific microarray gene expression profiling methods in the mouse brain. PLoS One. 2011;6(1):e16493.

Perez-Gil J, Weaver TE. Pulmonary surfactant pathophysiology: current models and open questions. Physiology (Bethesda). 2010;25(3):132–41.

Perl AK, Whitsett JA. Molecular mechanisms controlling lung morphogenesis. Clin Genet. 1999;56(1):14–27.

Pihur V, Datta S, Datta S. Weighted rank aggregation of cluster validation measures: a Monte Carlo cross-entropy approach. Bioinformatics. 2007;23(13):1607–15.

Plantier L, Besnard V, Xu Y, Ikegami M, Wert SE, Hunt AN, Postle AD, Whitsett JA. Activation of sterol response element binding proteins (SREBP) in alveolar type II cells enhances lipogenesis causing pulmonary lipotoxicity. J Biol Chem. 2012;287(13):10099–114.

Quaggin SE, Schwartz L, Cui S, Igarashi P, Deimling J, Post M, Rossant J. The basic-helix-loop-helix protein pod1 is critically important for kidney and lung organogenesis. Development. 1999;126(24):5771–83.

Subramanian A, Tamayo P, Mootha VK, Mukherjee S, Ebert BL, Gillette MA, Paulovich A, Pomeroy SL, Golub TR, Lander ES, Mesirov JP. Gene set enrichment analysis: a knowledge-based approach for interpreting genome-wide expression profiles. Proc Natl Acad Sci U S A. 2005;102(43):15545–50.

Wan H, Xu Y, Ikegami M, Stahlman MT, Kaestner KH, Ang SL, Whitsett JA. Foxa2 is required for transition to air breathing at birth. Proc Natl Acad Sci U S A. 2004;101(40):14449–54.

Wan H, Dingle S, Xu Y, Besnard V, Kaestner KH, Ang SL, Wert S, Stahlman MT, Whitsett JA. Compensatory roles of Foxa1 and Foxa2 during lung morphogenesis. J Biol Chem. 2005;2804(14):13809–16.

Wan H, Luo F, Wert SE, Zhang L, Xu Y, Ikegami M, Maeda Y, Bell SM, Whitsett JA. Kruppel-like factor 5 is required for perinatal lung morphogenesis and function. Development. 2008;135(15):2563–72.

Wang T, Hopkins D, Schmidt C, Silva S, Houghton R, Takita H, Repasky E, Reed SG. Identification of genes differentially over-expressed in lung squamous cell carcinoma using combination of cDNA subtraction and microarray analysis. Oncogene. 2000;19(12):1519–28.

Weaver TE, Beck DC. Use of knockout mice to study surfactant protein structure and function. Biol Neonate. 1999;76 Suppl 1:15–8.

Weaver TE, Whitsett JA. Function and regulation of expression of pulmonary surfactant-associated proteins. Biochem J. 1991;273(Pt 2):249–64.

Whitsett JA. Genetic disorders of surfactant homeostasis. Paediatr Respir Rev. 2006;7 Suppl 1:S240–2.

Whitsett JA, Matsuzaki Y. Transcriptional regulation of perinatal lung maturation. Pediatr Clin North Am. 2006;53(5):873–87, viii.

Whitsett JA, Weaver TE. Hydrophobic surfactant proteins in lung function and disease. N Engl J Med. 2002;347(26):2141–8.

Whitsett JA, Nogee LM, Weaver TE, Horowitz AD. Human surfactant protein B: structure, function, regulation, and genetic disease. Physiol Rev. 1995;75(4):749–57.

Xu Y, Clark JC, Aronow BJ, Dey CR, Liu C, Wooldridge JL, Whitsett JA. Transcriptional adaptation to cystic fibrosis transmembrane conductance regulator deficiency. J Biol Chem. 2003;278(9):7674–82.

Xu Y, Liu C, Clark JC, Whitsett JA. Functional genomic responses to cystic fibrosis transmembrane conductance regulator (CFTR) and CFTR(delta508) in the lung. J Biol Chem. 2006;281(16):11279–91.

Xu Y, Ikegami M, Wang Y, Matsuzaki Y, Whitsett JA. Gene expression and biological processes influenced by deletion of Stat3 in pulmonary type II epithelial cells. BMC Genomics. 2007;8(1):455.

Xu Y, Saegusa C, Schehr A, Grant S, Whitsett JA, Ikegami M. C/EBP {alpha} is required for pulmonary cytoprotection during hyperoxia. Am J Physiol Lung Cell Mol Physiol. 2009;297(2):L286–98.

Xu Y, Zhang M, Wang Y, Kadambi P, Dave V, Lu LJ, Whitsett JA. A systems approach to mapping transcriptional networks controlling surfactant homeostasis. BMC Genomics. 2010;11:451.

Xu Y, Wang Y, Besnard V, Ikegami M, Wert SE, Heffner C, Murray SA, Donahue LR, Whitsett JA. Transcriptional programs controlling perinatal lung maturation. PLoS One. 2012;7(8):e37046.

Yin Z, Gonzales L, Kolla V, Rath N, Zhang Y, Lu MM, Kimura S, Ballard PL, Beers MF, Epstein JA, Morrisey EE. Hop functions downstream of Nkx2.1 and GATA6 to mediate HDAC-dependent negative regulation of pulmonary gene expression. Am J Physiol Lung Cell Mol Physiol. 2006;291(2):L191–9.

Yuan B, Li C, Kimura S, Engelhardt RT, Smith BR, Minoo P. Inhibition of distal lung morphogenesis in Nkx2.1(-/-) embryos. Dev Dyn. 2000;217(2):180–90.

Zeeberg BR, Feng W, Wang G, Wang MD, Fojo AT, Sunshine M, Narasimhan S, Kane DW, Reinhold WC, Lababidi S, Bussey KJ, Riss J, Barrett JC, Weinstein JN. GoMiner: a resource for biological interpretation of genomic and proteomic data. Genome Biol. 2003;4(4):R28.

Zuo F, Kaminski N, Eugui E, Allard J, Yakhini Z, Ben-Dor A, Lollini L, Morris D, Kim Y, DeLustro B, Sheppard D, Pardo A, Selman M, Heller RA. Gene expression analysis reveals matrilysin as a key regulator of pulmonary fibrosis in mice and humans. Proc Natl Acad Sci U S A. 2002;99(9):6292–7.

Chapter 18
Defining Genetic Blueprints – Kidney and Craniofacial Development

Eric W. Brunskill, Andrew S. Potter, and S. Steven Potter

Abstract Genomics tools are revolutionizing the way we study developmental processes. Microarrays and RNA-Seq allow the sensitive, quantitative and global analysis of gene expression patterns. Instead of examining a few select genes, it is now routine to generate global gene expression profiles of developing systems. The early experiments examined entire developing organs, thereby identifying, for example, all transcription factors, growth factors, and receptors expressed. Subsequent experiments dramatically improved the genomics analysis by providing higher spatio-temporal resolution. Organ development is remarkably complex, with discrete substructures, a developmental time course of events, and the simultaneous differentiation of discrete cell types. It is not possible, therefore, to homogenize the developing organ, apply genomics tools, and derive deep insight. Laser capture microdissection allows the purification of single developmental compartments. Transgenic mice with gene specific promoters driving the expression of markers, such as green fluorescent protein, can be used to purify specific cell types from developing organs. Even higher resolution analysis can be carried out with tools that allow the gene expression profiling of single cells. The use of these genomics approaches produces enormous volumes of data that need to be captured, annotated in a systematic way, stored, integrated and analyzed, using the resources of biomedical informatics.

E.W. Brunskill, Ph.D. • A.S. Potter, M.S.
Division of Developmental Biology, Cincinnati Children's Hospital Medical Center, 3333 Burnet Ave, MLC-7007, Cincinnati, OH 45229-3039, USA

S.S. Potter, Ph.D. (✉)
Division of Developmental Biology, Cincinnati Children's Hospital Medical Center, 3333 Burnet Ave, MLC-7007, Cincinnati, OH 45229-3039, USA

Department of Pediatrics, University of Cincinnati College of Medicine, Cincinnati, OH, USA
e-mail: steven.potter@cchmc.org

J.J. Hutton (ed.), *Pediatric Biomedical Informatics: Computer Applications in Pediatric Research*, Translational Bioinformatics 2, DOI 10.1007/978-94-007-5149-1_18,

18.1 Introduction

Developmental genetics is a fascinating field. We all begin life as a fertilized egg, a zygote, equipped with a nuclear program having only about 22,000 genes. We are just beginning to understand how all of the remarkable complexities of a human being can be made from this single microscopic cell carrying so few genes.

A productive approach is to break a big problem down into a series of smaller problems that can be more easily solved. So, instead of studying the entire organism, developmental geneticists study individual organs, such as the heart, brain, kidney and liver. However, the development of each organ is still incredibly complex. There are sophisticated substructures, a developmental time course, with different things happening at different times, and the simultaneous differentiation of many distinct cell types. A recurring theme of this chapter is that we cannot simply homogenize a developing organ, apply genomics tools, and derive deep insight. The higher the spatio-temporal resolution of the analysis, the better and more detailed the insights. It is necessary to globally define what is going on in each developmental compartment and each cell type at each developmental time point in order to generate a true genomics level blueprint of organogenesis.

There are now tools available to dissect developing organs into coherent anatomic groups. Laser capture microdissection allows the isolation of specific compartments. In addition there are transgenic mice that can mark specific cell types with fluorescent proteins, such as green fluorescent protein (GFP), thereby facilitating their purification. Indeed, there are rapidly evolving technologies that provide global transcription profiles down to the level of single cells. The purpose of this chapter is to illustrate how these various tools can be used to address important questions in developmental biology that are highly relevant to child health.

The use of genomics tools generates huge amounts of data that need to be captured, annotated in a systematic way, stored, integrated and analyzed. In this chapter we provide guiding principles for data analysis and discuss in detail the use of several widely used programs, such as GeneSpring (www.agilent.com/chem/genespring/) for microarray data analysis, Avadis NGS for RNA-Seq analysis (www.avadis-ngs.com/), and ToppGene (toppgene.cchmc.org/), for functional analysis of resulting gene lists.

As prototypes for the genomics analysis of organogenesis, we describe kidney and craniofacial development. The primary job of kidneys is to remove the end products of cellular metabolism. They also regulate blood pressure by controlling blood volume, controlling blood salt levels, and through the production of renin, which regulates vasoconstriction through the renin-angiotensin system. The kidney is also the source of important hormones such as calcitriol, which regulates blood calcium levels, and erythropoietin, which regulates erythropoiesis. The functional unit of the kidney is the nephron. The glomerulus is the filtration element, with fenestrated capillary endothelial cells providing the blood access to the glomerular basement membrane filter. The filtrate then flows through the proximal tubule, the loop of Henle, the distal tubule, and finally the collecting duct, with selective uptake and concentration occurring along the way.

The kidney provides an excellent of example of how ontogeny, the development of the individual, recapitulates phylogeny, the evolution of the species. Three kidneys successively form during mammalian development, with sequentially increases in complexity, reflecting the increasing requirement for water conservation as animals moved from water to land. In this chapter we focus on the final metanephric kidney, which functions in the adult mammal.

The kidney forms from the intermediate mesoderm. It provides a classic example of mutual inductive interactions (Costantini and Kopan 2010). An outgrowth of the nephric duct, the ureteric bud, invades a flanking region of condensed mesenchyme, the metanephric mesenchyme. There is essential crosstalk between the bud and mesenchyme, with the mesenchyme driving branching morphogenesis of the bud, which eventually forms the collecting duct system, and the bud inducing the mesenchyme to form nephrons. During the process of nephron formation there are multiple stages that are morphologically distinct. The nephron progenitor cells flank the branching ureteric bud and are called capping mesenchyme. Following induction they first form a renal aggregate, which undergoes mesenchyme to epithelial transformation, making the renal vesicle, a spherical structure. Clefts then form on the renal vesicle, sequentially making the comma and S-shaped bodies, which fuse to the ureteric bud. The S-shaped body in turn elongates and differentiates to form the glomerulus and the various tubule segments of the mature nephron (Fig. 18.1).

The development of the mammalian face is also quite remarkable. The face is primarily made from neural crest cells that exit the neural tube early in development. Wilhelm His first described neural crest cells over 100 years ago. His reporting that these neural cells, from the top of the neural tube, could give rise to cartilage and bone was quite controversial at the time. It was dogma that skeletal structures were mesodermal in origin, and the proposed ectodermal origin for some of them was heresy. It is now accepted, however, that the muscle, bone and cartilage of the head and face are made from both the neural crest and the paraxial mesoderm.

The mouse is a powerful model for the study of human development. All mammals have very close to the same number of genes, approximately 22,000. For each human gene there is about a 99% chance that there will be a mouse ortholog, which will vary from the human counterpart only in its precise base sequence. In addition, developmental mechanisms in general appear to be extremely well conserved during evolution. Indeed, in some cases it has been possible to demonstrate developmental functional equivalence of orthologous fruit fly and human genes through gene swap experiments (McGinnis et al. 1990). It is therefore quite likely that what we learn through the study of mice will translate well to the human system.

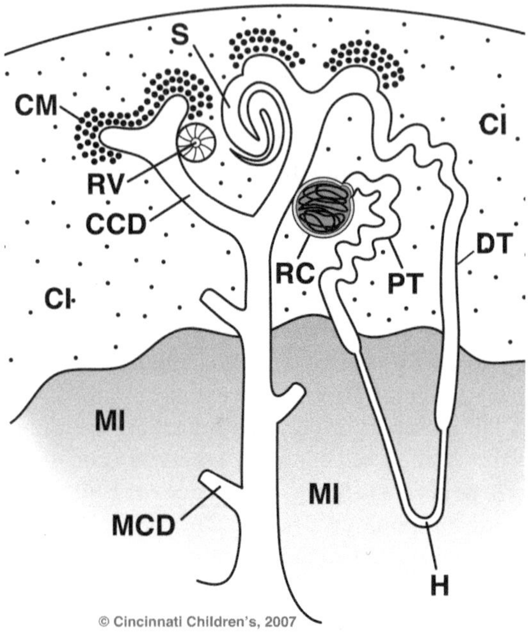

Fig. 18.1 Diagram of kidney development. The branching ureteric bud gives rise to the collecting duct, which can be divided into the medullary (*MCD*) and cortical (*CCD*) regions. The nephron progenitor cells are in the capping mesenchyme (*CM*), which gives rise to the renal vesicle (*RV*), which fuses with the forming collecting duct and gives rise to the S-shaped bodies (*S*), which further grow, segment and differentiate to give the nephron, with renal corpuscle (glomerulus and Bowman's capsule) (*RC*), proximal tubule (*PT*), loop of Henle (*H*) and distal tubule (*DT*). In addition there are insterstitial, or stromal, cells, which can be divided into medullary (*MI*) and cortical (*CI*) regions (From Brunskill et al. 2008)

18.2 Transcription Profiling and Kidney Development: Early Work

The Indian tale of the blind men and the elephant illustrates the fallacy of studies of limited scope. We can reach the wrong conclusions, or at least derive a very incomplete view, by carrying out an investigation with a focus that is too narrow. The great advance of genomics approaches is the novel ability to study, for example, all genes at once. Microarrays and RNA-Seq give global, sensitive and quantitative measures of gene expression. It is no longer necessary to make good or lucky guesses concerning which genes might be involved in a process. Indeed, it is no longer acceptable to restrict an analysis to just a few candidate genes. The challenge of the genomics era, however, is dealing with the veritable flood of data that comes from genomics tools.

Sanjay Nigam and his colleagues were pioneers in applying genomics tools to the study of kidney development (Stuart et al. 2001). They used microarrays to define gene expression profiles of whole kidneys at different developmental stages. These and other early studies described for the first time the complete collection of genes expressed during nephrogenesis, and gave a measure of the changing gene expression profiles as a function of developmental time (Schwab et al. 2003). All of the growth factors, receptors and transcription factors expressed during kidney development were identified. This work was truly ground breaking, but still represented just the first step towards a genomics level understanding of kidney development.

A major limitation of these early studies was a complete lack of spatial definition of gene expression. The analysis was of entire developing kidneys. Genes were identified as expressed, but the localization of the expression remained uncertain. It was difficult to define developmental pathways or cross inductive interactions of specific compartments of the forming kidney when one could not be sure which genes were being expressed where. One cannot homogenize the whole to understand the parts. It would clearly be necessary to generate a higher resolution spatial definition of gene expression patterns in order to generate a useful genomics level understanding of kidney development.

18.3 Divide and Conquer

The early developing kidney is readily manually dissected, following mild treatment with proteases, into metanephric mesenchyme and ureteric bud. Several groups used microarrays to better define the gene expression profiles of these two critical components of the developing kidney, thereby providing significantly deeper insights into the genetic pathways driving early nephrogenesis than do profiles of whole organ homogenates (Challen et al. 2005; Schwab et al. 2006; Stuart et al. 2003; Schmidt-Ott et al. 2005).

This strategy of divide and conquer was carried further by using laser capture microdissection (LCM) to isolate the multiple compartments of the developing kidney (Brunskill et al. 2008). Using a combination of compartment specific lectin stains and distinct morphological structures, it was possible to purify capping mesenchyme, renal vesicles, S-shaped bodies, glomeruli, proximal tubules, distal tubules, the capsule, and several different regions of stromal cells and the developing collecting duct (Fig. 18.1) (Brunskill et al. 2011a). This created the first atlas of gene expression patterns of the developing kidney at microanatomic resolution.

The data were examined with GeneSpring, thereby defining lists of differentially expressed genes, performing clustering and producing heatmaps. The ToppGene web tool, created by bioinformaticians at Cincinnati Children's Hospital, was then particularly useful in the functional analysis of the resulting extensive microarray dataset. It allows one to submit a list of differentially expressed genes and to receive an immediate output of enriched molecular functions, biological processes, cellular components, human phenotypes, mouse phenotypes, domains, pathways, PubMed references, interactions, cytobands, transcription factor binding sites in promoters, candidate regulatory microRNAs, and more.

Several interesting principles emerged. First, there were surprisingly few genes with a strict compartment specific expression pattern at early stages of development. For example, the capping mesenchyme and its developmental derivative, the renal vesicle, and the subsequent S-shaped bodies, all showed a considerable overlap in gene expression profiles. There were fewer than a hundred genes with true compartment specific expression in the entire atlas of early kidney development. This principle broke down, however, during the later stages of development, as compartments began to express large numbers of specific differentiation related genes. For example the proximal tubules showed restricted expression of thousands of genes, many of them transporters involved in selective uptake.

In addition, a phenomenon called anticipatory gene expression was observed. It has also been referred to as lineage priming. It appears that developmental compartments can anticipate their subsequent developmental stage and initiate expression of genes that will then peak in expression in the next compartment. For example the renal vesicle will begin to express genes that will then show much higher expression in the subsequent S-shaped bodies. This makes sense, since the renal vesicle must start to express S-shaped body related genes as it progresses to the S-shaped body stage. Lineage priming contributes significantly to overlapping gene expression patterns.

This microarray based atlas provided the first global definition of compartment level expression patterns of growth factors, receptors, transcription factors and indeed all other genes driving the formation of the kidney. It produced a genomics level description of the complexities of the multiple inductive interactions taking place, and better defined the genetic programs driving the formation of the different elements of the kidney. It represented a useful step forward in the functional genomics analysis of kidney development.

18.4 Cell Type Specific Gene Expression Profiling

Nevertheless, a laser capture microdissection based microarray analysis is still imperfect. For example, a single laser captured structure, such as a developing glomerulus, will include several distinct cell types. The logic that predicts a compartment specific atlas will be superior to a gene expression profile of total kidneys also predicts that a cell type specific atlas will be more valuable still. While a compartment level collection of gene expression profiles will define all of the genes expressed in the glomerulus, for example, it will not determine which of its three cellular constituents, the podocyte, mesangial cell, or endothelial cell, expresses a particular gene of interest.

It is possible to refine the gene expression profiles to specific cell type resolution by using transgenic reporter mice that contain a cell type specific promoter. Gene targeting technology can be used to "knock in" a reporter, such as green fluorescent protein (GFP), into a gene. The regulatory elements of the gene will usually then drive appropriate gene specific expression of the reporter. Another strategy is to

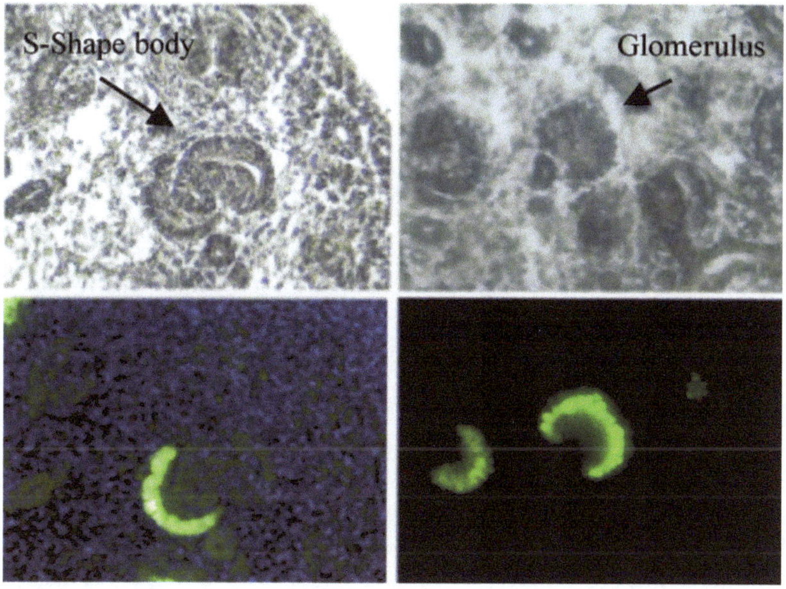

Fig. 18.2 Mafb-GFP specifically marks podocytes in the developing kidney. *Top panels* show brightfield images of embryonic kidneys with S-shaped body (*left*) and glomerulus (*right*) marked. The *bottom panels* show fluorescent images illustrating podocyte specific GFP labeling of early presumptive podocytes in the S-shaped body (*left*), as well as the later glomerulus (*right*) (From Brunskill et al. 2011a)

generate transgenic mice using a bacterial artificial chromosome (BAC) with a reporter inserted into a gene of interest within a BAC. The large size of the BAC often insures that the reporter will then be expressed appropriately. For example the BAC transgenic *Mafb-GFP* mouse specifically marks the podocyte in the kidney (Fig. 18.2) (Brunskill et al. 2011a). It is therefore possible to carry out rapid enzymatic dissociation to single cells, and to use fluorescent activated cell sorting (FACS) to purify the GFP labeled podocytes from the embryonic kidney (Brunskill et al. 2011a).

Similar strategies can be used to isolate other cell type specific populations of cells from the kidney. For example *Tie2-GFP* marks endothelial cells. Using *Tie2-GFP* transgenic mice it is possible to define the gene expression profile of the endothelial cells of the entire kidney (Brunskill and Potter 2010). In addition the kidney can be physically separated into medullary and cortical domains and then the respective endothelial cells are profiled to search for possible differences in gene expression in these physiologically distinct regions. It is also possible to physically separate the glomeruli, using a sieving protocol, and then, following single cell dissociation, the *Tie2-GFP* positive glomerular endothelial cells are FACS purified and profiled. Furthermore, it is possible to use these approaches to define the changing gene expression patterns as a function of disease (Brunskill and Potter 2010).

Similar approaches can be used to examine the gene expression states of other specific cell types of the kidney. *Meis1-GFP* transgenic mice can be used to purify stromal cells and mesangial cells. This transgene marks both cell populations, but by separating out glomeruli, with mesangial cells, using a sieving protocol prior to FACS, it is possible to purify each. *Crym-GFP* transgenic mice allow purification of the capping mesenchyme progenitor cells and renal vesicles. The *Crym-GFP* transgene primarily marks the capping mesenchyme, facilitating their purification. However, following the postnatal burst of nephrogenesis, which consumes the final capping mesenchyme, the specific GFP persists at a lower level to mark the renal vesicles.

18.5 Podocytes

The podocytes are quite remarkable cells. They surround the glomerular capillaries, cooperating with the endothelial cells to synthesize the glomerular basement membrane. Although mesodermal in origin, they exhibit a striking shape, with multiple axonal like projections that reach around the capillaries. From these extend still smaller projections, the foot processes, which precisely interdigitate, leaving between them the slit diaphragms, for the passage of glomerular filtrate.

Many important functions have been associated with podocytes. The slit diaphragm is an extracellular extension of the podocyte and provides the final filtration barrier. Podocytes also function as pericytes, counteracting the hydrostatic pressure within the capillaries. In addition, podocytes are thought to play an important role in cleaning the glomerular basement membrane filter, thereby preventing clogging. Furthermore, podocytes are thought to represent initial sites of injury for a number of kidney diseases, including diabetic nephropathy and focal segmental glomerulosclerosis. Podoctye effacement, or detachment from the glomerular basement membrane, and podocyte loss are among the earliest pathologic events in these diseases.

Podocytes are derived from the capping mesenchyme progenitor cells, and begin to express the *Mafb* podocyte marker in the proximal cleft of the S-shaped bodies, as the podocyte precursors begin to differentiate (Fig. 18.2). These early podocytes also express VEGF, which plays an important role in recruiting endothelial cells to the glomerulus. The gene expression profiling of podocytes during development of the mouse embryo at E13.5 and E15.5, as well as adult, was carried out to better define the unique character of these extraordinary cells (Brunskill et al. 2011a). The data analysis in GeneSpring included filtering for genes with a minimum expression level, for example a raw signal of at least 150 in three samples. Genes with very low levels of expression, near noise, are a major source of artifact when calling a gene as changed or not in expression analyses. A raw signal of 150 is well below the expression level detectable by *in situ* hybridizations. Although it remains possible that some of these very low expression level genes are functionally important, it is nevertheless very useful to remove them to improve the reliability of the resulting gene lists.

Another key requirement for data analysis is an appropriate set of gene expression profiles for comparison. It is rather challenging to interpret gene expression data from a single cell type without comparisons, since around 10,000 genes are typically expressed at a significant level. One can divide the genes into functional groups, such as transcription factors and growth factors. One can also rank genes within groups according to expression levels. In general genes with high expression levels appear to be more functionally significant than those with low expression levels, although there are exceptions. This type of straightforward analysis can begin to define the set of genes that supervise the character of the cells under investigation.

Deeper insight typically requires comparison of the gene expression profiles of one type of cell with those of other types. As for all experiments, there must be the correct controls. The controls should be as similar to the test samples as possible. By comparing gene expression profiles of several different types of samples, it is easier to identify those aspects of the gene expression profile under study that are truly unique. For example to identify the distinguishing aspects of the podocyte gene expression profile, one can compare it to multiple types of cells from kidney, sequentially subtracting out the other profiles to define the distinctive elements of the podocyte. There is a synergistic effect as more and more types of samples are examined. With the accumulation of additional related sets of gene expression profiles, the analysis of each is significantly improved.

Following the selection of comparison sets of data and filtering out of probe sets with low expression levels, the next stage of analysis is typically a statistical test to identify differentially expressed genes. If there are just two sample types being compared, then this is normally an unpaired T-test. For multiple types of samples, analysis of variance (ANOVA) is used, which generalizes the T-test to multiple groups. A P-value of 0.05 or less is generally used to label a result as statistically significant. The next consideration is choice of multiple testing correction methods. The Bonferroni correction, in simple terms, multiplies the uncorrected P value times the number of probe sets in the analysis. For example, if 10,000 probe sets survive the raw signal filter then the T-test P values would all be multiplied by 10,000. This is an extremely stringent correction and generally results in very short gene lists, with many truly differentially expressed genes missed. The Benjamini and Hochberg correction is somewhat less stringent. In this case the best P-value is multiplied by the total number of probe sets, as per Bonferroni, but the next best P-value is multiplied by the number of probe sets minus one, and so on. That is, as the P-values increase the correction decreases. Nevertheless, the Benjamini and Hochberg correction is still quite stringent when dealing with a large number of probesets, and again can give very short gene lists that miss many genuine differences. The Storey multiple testing correction methods are less stringent, and are generally more appropriate unless the number of probesets has been severely trimmed prior to statistical analysis. In practice it is often preferred to use no multiple testing correction, and to appreciate that while this eliminates most false negatives, it results in the inclusion of some false positives. Further filtering based on fold change, normally two or threefold minimum, can effectively remove most of the false positives. We typically observe that over 90% of genes that pass a screen of P value less than 0.05 and fold change greater than two can subsequently be validated using Q-PCR or RNA-Seq. *In situ* hybridizations and Q-PCR are standard methods used to confirm differential gene expression calls made by microarrays. *In situ* hybridizations are not very quantitative, but can detect large fold changes in gene expression present in the embryo or tissue of interest, for example comparing mutant to wild type. RNA-Seq offers an independent global high throughput technology that can also be used to validate microarray data. Q-PCR (quantitative PCR) is a gold standard method used to quantitate gene expression levels, typically applicable to intermediate numbers of genes.

When analyzing the podocyte microarray expression data, two comparisons were carried out. The goal was to define the set of expressed genes uniquely defining the striking multifunctional character of these cells. The first analysis compared the expression profile of the adult podocyte to that of the E13.5 embryo. The early embryonic podocyte is just beginning to express podoctye specific differentiation markers, such as *Mafb*, while the adult podocyte expresses the full complement of these genes. GeneSpring software was used, with summarization algorithm RMA16, probe sets stringently filtered for raw expression of at least 150 in three samples, $P \leq 0.05$, and fold change greater than three, giving a list of 739 probe sets with increased expression in adult podocytes. A second comparison searched for adult podocyte specific expressed genes by subtracting out the gene expression profile of the entire adult kidney cortex. Using a similar screen to that comparing adult and embryonic podocytes, we derived a list of 436 probe sets that showed specific podocyte enriched expression. Of interest, the two strategies gave an overlapping set of 281 probe sets. This suggests that both methods identify similar sets of genes, thereby providing a measure of cross validation.

Each analysis strategy is imperfect. The comparison of adult and embryonic podocytes misses differentiation genes that are podocyte specific, but are active very early in development, such as *Mafb*. The comparison of adult podocytes with adult total kidney cortex misses a set of genes that are very important in podocyte function, but are also expressed elsewhere in the adult kidney. Nevertheless, by combining the two approaches one derives a list of 894 probesets whose expression in large measure specifically defines the adult podocyte. GeneSpring gene ontologies analysis identifies, for example, 54 genes in the transcription factor category, with 34 of these showing strong expression with over 500 raw signal, and 14 genes with very high expression of over 1,000 raw signal. Of interest, three of the highly expressed transcription factor genes, *Wt1*, *Pod1* and *Foxc2*, have been previously shown to play key roles in kidney development. This provides a validation of the analysis, showing that it does identify important regulatory genes. Of particular interest, the transcription factor gene *Foxd1* also showed high expression in podocytes. This was surprising because *Foxd1* had previously been thought to be a specific marker of the stromal lineage in the developing kidney. *In situ* hybridizations confirmed the expression of *Foxd1* in podoctyes (Brunskill et al. 2011a).

The ToppGene web tool was used to further analyze the molecular processes and biological functions the 894 gene list of the podocyte. The results better defined the multifunctional nature of the podocyte, identifying statistically significant sets of genes involved in cytoskeletal organization (39 genes), neurogenesis (79 genes), biological adhesion (59 genes), muscle contraction (16 genes), and phagocytosis (11 genes). Each list of genes, such as the 79 genes involved in neurogenesis, provides deeper molecular insight into the character of the podocyte.

The analysis described above "cast a wide net" to identify a broad set of genes with podocyte elevated expression. There is, however, also considerable value in defining a more strict set of podocyte specific genes. This could identify a new set of podocyte specific molecular markers, aid in the generation of novel transgenic mouse tools for podocyte studies, and better define the most unique aspects of podocyte gene expression. A more stringent screen of the data was therefore carried out, requiring a fivefold enrichment in podocytes, compared to the previous three, deriving a list of 171 probe sets. Their podocyte elevated expression can be illustrated in a heatmap comparing the adult podocyte to a dozen other kidney compartments (Fig. 18.3). These 171 genes provide a distinct podocyte specific gene expression signature.

The podocyte gene expression profile, as well as the profiles of other specialized cell types, allows us to draw some interesting conclusions concerning the gene expression definition of differentiated cells. It is perhaps surprising that out of a total set of about 22,000 genes so few show podocyte cell type specific expression. Of the approximately 10,000 genes expressed in the podocyte very few show digital on/off differences when compared to other cell types. The vast bulk of the podocyte gene expression signature is more quantitative in nature, with differences in gene expression levels, when compared to other cell types. It is therefore clear that the global gene expression codes that delineate cell types are primarily analog and not binary on/off in nature.

18.6 Analyzing Mutants

Genomics tools can be combined with procedures of classical genetic analysis to better understand the genetic programs of development. Gene targeting techniques have been available for over two decades now, allowing the developmental biologist to alter gene function and then to observe the resulting developmental defects in the mouse. Indeed, there is an international effort to target all mouse genes. Targeted ES cells are currently available for over 15,000 genes (http://www.knockoutmouse.org/).

The traditional methods for studies of mutant phenotypes include histological analysis of the altered morphology of the organ of interest, for example the kidney, as well as a time course characterization of the progression of the developmental abnormalities. In addition a series of *in situ* hybridizations and/or immunostains are often carried out to observe changes in the expression of "molecular markers" that might distinguish cell types or compartments, to better define the nature of the developmental defects.

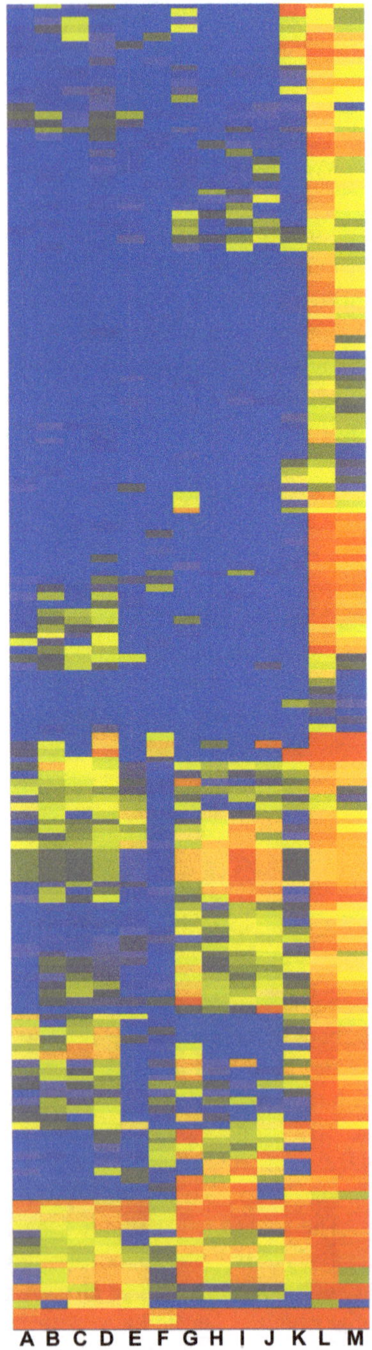

Fig. 18.3 Heatmap showing genes with specifically elevated expression in podocytes. The expression patterns of 171 gene selected for specific podocyte expression are compared across 12 compartments of the kidney. *A*. E13.5 podocytes. *B*. Renal vesicles from P4 mice. *C*. P1 cap mesenchyme. *D*. E15.5 podocytes. *E*. E15.5 endothelial cells. *F*. Adult total kidney cortex. *G*. Adult *Meis1* expressing mesangial cells. *H*. Adult cortex endothelial cells. *I*. Adult glomerular endothelial cells. *J*. Adult medullary endothelial cells. *K*. Adult renal capsule. *L*. Adult podocytes. *M*. Adult total glomerulus, which will include podocytes. *Red* indicates strong expression, *blue* is weak expression, and *yellow* is intermediate expression (From Brunskill et al. 2008)

Genomics tools, such as microarrays and RNA-Seq, allow a much deeper analysis of mutant phenotypes, but these studies must be carried out with care. With *in situ* hybridization or immunostain one typically looks for the altered expression of approximately a dozen genes. These techniques are quite labor intensive and reagent limited. Antibodies are not available for all proteins. Typically, only one or a very few genes can be examined for expression with each experiment. Microarrays and RNA-Seq, however, allow the simultaneous expression analysis of essentially all 22,000 genes. Nevertheless, genomics studies of mutants are often carried out in a manner that gives relatively little insight. A key complication is the potentially changed cellular makeup of the mutant tissue examined. Then the genomics analysis will likely reveal a changed gene expression pattern that reflects the altered cellular composition, which is useful, but might make very little known of the cause of the difference. For example, why are certain cells missing in the mutant?

One key is to perform the microarray analysis at a sufficiently early stage in the progression of the developmental defect to detect the first events, the causative factors, rather than the end results. A second principle is to focus the analysis, as much as possible, on the specific region or compartment showing the developmental perturbation. There are severe dilution effects when the region showing the developmental abnormality represents a very small part of the sample used for genomics analysis. If the genes with differences are expressed at very high levels, and if there are essentially on/off differences in expression between wild type and mutant, then this strategy might still succeed. In general the sensitivity of difference detection will be low when an entire organ, or even an entire embryo, is examined.

For example consider the conditional *Lhx1* mutant. Homozygous mutation of *Lxh1* results in an early block in intermediate mesoderm development, with no kidneys or gonads formed (Shawlot and Behringer 1995). Conditional deletion of *Lhx1* in the progenitor cells that make nephrons reveals later functions in kidney development. Histological analysis reveals that in the conditional mutants there is a developmental arrest at the renal vesicle stage, with the renal vesicle failing to progress to the S-shaped body (Kobayashi et al. 2005). *In situ* hybridizations showed altered expression of two genes, *Dll1* and *Brn1,* in the mutant renal vesicle. Microarray analysis of the entire embryonic kidney in this case would only show the reduced expression of a very large set of genes associated with the later stages of nephrogenesis, including S-shaped bodies, glomeruli, and tubules. The array results, in essence, would only reveal what was already known from the histologic analysis, that the later stages of nephron development failed to take place.

A more useful genomics analysis used laser capture microdissection (LCM) and microarrays to compare the renal vesicle gene expression profiles of wild types and mutants (Potter et al. 2007). In this case the study was focused specifically on the structure of interest, the renal vesicle (RV), which in the mutant failed to progress. What changes in RV gene expression could cause this failure? The array results identified over 100 genes with greater than threefold change in expression in the mutant RV, compared to only two genes that were found *by in situ* hybridization. The microarray results, therefore, greatly expanded our understanding of the perturbed gene expression pattern of the *Lhx1* mutant RV. Some of the observed differences were particularly interesting in terms of the RV developmental arrest in the mutants. For example *Dkk1* expression in the mutant RV was down about tenfold. DKK1 is a WNT inhibitor, binding to the WNT co-receptor LRP5/6 and inhibiting canonical WNT signaling (Bafico et al. 2001; Mao et al. 2001). The microarray results suggest that while WNT9b (Carroll et al. 2005) and WNT4a (Stark et al. 1994) are critically important to drive RV formation, the subsequent down regulation of WNT signaling is essential for the further progression of nephrogenesis.

18.7 RNA-Seq

RNA-Seq technology is rapidly replacing microarrays. For RNA-Seq, the RNA is first reverse transcribed to cDNA, which is then sequenced using high throughput NextGen sequencing. The results provide a digital measure of gene expression levels. More abundant RNAs give rise to more copies of cDNA, which then give more DNA sequence reads.

DNA sequencing technology, the basis of RNA-Seq, is undergoing an astonishing revolution. The Human Genome Project took 13 years, at a total cost of about three billion dollars, to define the first human DNA sequence. As of this writing, it is possible to completely sequence a human genome in a matter of weeks for under 5,000 dollars. So, the price of sequencing the human genome has dropped about a million fold in a decade. Over a dozen entirely new technologies are being developed, and older technologies continue to be refined. The thousand-dollar genome barrier will likely be crossed in the near future. One can confidently predict that the sequencing revolution will continue, making technologies like RNA-Seq increasingly affordable and powerful.

RNA-Seq is unmistakably superior to microarrays. The advantages are many. First, the results are not restricted to the sequences selected for the microarray. Probe sets used for microarrays are sometimes faulty, coming from introns instead of exons, for example, or perhaps from the wrong strand. In addition, no microarray is truly universal. Transcripts detected by RNA-Seq are sometimes quite abundant in the specific cell type or tissue examined, yet are very rare or not present in the EST database, and therefore were not selected for representation on microarrays. RNA-Seq generates sequence from any RNA present, with no bias and no selection required. Second, RNA-Seq has an extremely low background. All sequences that can unambiguously be aligned with the genomic sequence, including known and novel exons, and any splice combinations, are used. Sequences that cannot be aligned are discarded. With microarrays there are background and cross-hybridization issues that reduce sensitivity and can confuse results. This difference in background appears to be a very important factor in the superiority of RNA-Seq over microarrays. Third, digital gene expression readouts using RNA-Seq have a virtually unlimited dynamic range, estimated to cover about five orders of magnitude when using 40 million reads with the mouse genome (Mortazavi et al. 2008). Microarrays, on the other hand, have detection problems with very low abundance transcripts because of background noise, and can saturate causing loss of linear response at high transcript levels. Thus, microarrays are limited to an effective dynamic range of only a few hundred fold. Fourth, RNA-Seq is extremely accurate in measuring transcript levels, as determined by qPCR validation (Nagalakshmi et al. 2008) and RNA spike control (Mortazavi et al. 2008). Fifth, RNA-Seq gives an excellent view of alternative RNA processing events. It does this in two ways, by producing a digital quantitative expression level for each exon, and by providing sequence spanning exon junctions. Recent estimates are that more than 90% of genes undergo alternative processing. In some cases, the differentially processed products actually have opposite biological functions. Sixth, promoter start sites can be identified. Finally, RNA-Seq results are highly reproducible (Nagalakshmi et al. 2008; Cloonan et al. 2008).

18.7.1 RNA-Seq Analysis of Early Nephrogenesis

As an example of the power of RNA-Seq, let us consider the conversion of capping mesenchyme progenitor cells to renal vesicles. The earliest stages of nephrogenesis, as the progenitor cells of the capping mesenchyme are induced and begin to differentiate into renal vesicles and then nephrons, are of particular interest. During kidney development there is a continuous outer nephrogenic zone where the forming collecting duct undergoes branching morphogenesis and induces nephron formation. The delicate balance between progenitor renewal and differentiation determines final nephron count, which has important medical consequences (Keller et al. 2003; Poladia et al. 2006). In addition, the induction of capping mesenchyme to form renal vecicles is the remarkable first step in the conversion of an amorphous mesenchyme cloud into the intricate complexity of the nephron.

Genetic approaches, primarily using gene knockouts, have identified many key regulators of early kidney development (Costantini and Kopan 2010). Induction by WNT9b (Carroll et al. 2005), coupled with decreased *Six2* expression (Self et al. 2006), is required for nephrogenesis. Other genes, including *Sall1*, *Fgfr1*, *Fgfr2*, *Fgf8*, *Wt1* and *Bmp7* also play important roles in regulating the renewal/differentiation balance (Costantini and Kopan 2010). In addition, the LCM/microarray approach identified over a thousand genes that undergo differential expression as the capping mesenchyme forms the renal vesicles (Brunskill et al. 2008).

Nevertheless, despite this rather considerable prior analysis, notable new insights were derived from a further study of this process using RNA-Seq (Brunskill and Potter 2012). Of interest, when the gene lists resulting from microarray and RNA-Seq analyses were compared, it was observed that over 90% of genes called differentially expressed by microarrays were validated by RNA-Seq data. Indeed, among other things, RNA-Seq provides a high throughput, global technology for the validation of microarray data. It was surprising, however, that RNA-Seq found many more differentially expressed genes than microarrays, even when more stringent P-value/fold change screening criteria were used. This dramatically improved detection rate appears to be related to the absence of background and the greater dynamic range for RNA-Seq data compared to microarray data.

In this RNA-Seq study of capping mesenchyme differentiation into renal vesicles, the data were primarily analyzed using Avadis NGS software (Brunskill and Potter 2012). The workflow is very similar to that of GeneSpring microarray analysis software, allowing filtering to remove low level expressed genes, statistical testing for differential gene expression, clustering, and functional analysis of resulting gene lists. It is interesting to note that statistical analysis of microarray data is dependent on replicate samples, performed at least in triplicate. This is not the case for RNA-Seq data, however, where the statistical analysis is based on deviation from predicted Poisson distributions, and is dependent on the total number of sequence reads rather than the number of samples analyzed. One such statistical tool for RNA-Seq data analysis is the Audic Claverie test. It is therefore possible to pool biological samples, to average out biological variability, and to perform RNA-Seq on only a single pooled sample of each type to be analyzed. There is, however, some advantage to be gained from performing RNA-Seq on multiple independent biological samples of each type, as this gives a better view of biological variability.

RNA-Seq can be performed using biochemistries that detect only polyadenylated RNAs, or all RNAs, including those without polyadenylation. For example, the Illumina TruSeq method (www.illumina.com/truseq.ilmn) first selects for polyA + RNA and then uses random primers to generate cDNA from their entire lengths. In contrast, the NuGEN Ovation RNA-Seq System V2 method (http://www.nugeninc.com/nugen/index.cfm) has no step for selection of polyadenylated RNAs. Instead random primers are immediately used for cDNA synthesis. The NuGEN method avoids the generation of overwhelming amounts of cDNA from ribosomal RNA by using a "semi random" primer that is depleted in sequences homologous to rRNAs. In the RNA-Seq analysis of the capping mesenchyme transition to renal vesicles, both of these methods were used. This allows the two independent technologies to provide cross validation of each other, at least for the polyadenylated RNAs, and it also permits a dataset comparison that determines whether transcripts are polyadenylated or not.

RNA-Seq data is primarily used to give a digital readout of gene expression levels and to define RNA processing patterns, but RNA-Seq data can also be used to map the positions of enhancers, many of which are themselves transcribed. For example, a study of neuronal enhancers found that about half are transcribed, giving rise to short (under 2 kb) bidirectional transcripts (Kim et al. 2010). Another study of the response to macrophage activation found that PolII transcription of enhancers produced unstable transcripts (De Santa et al. 2010). Therefore, by performing the capping mesenchyme/renal vesicle RNA-Seq with both TruSeq polyA+ and NuGEN random primer technologies, it was possible to map the positions of "non-canonical" (not associated with standard exons) transcripts and to determine their polyadenylation status. Transcribed enhancer maps were made for both capping mesenchyme and renal vesicles, and it was observed that transcripts from most were not polyadenylated (Brunskill and Potter 2012).

Another interesting result of the capping mesenchyme/renal vesicles RNA-Seq study was the striking view of the Hox clusters that emerged. First, almost all of the 39 genes of the four Hox clusters were transcribed in both capping mesenchyme and renal vesicles, with only the genes at the extreme ends of the clusters not expressed. For example, all of the genes of the HoxA cluster were expressed in both capping mesenchyme and renal vesicles, except for *Hoxa1* and *Hoxa13*. Second, not only were the standard Hox genes expressed, but the intergenic regions were also extensively transcribed, giving rise to mostly nonpolyadenylated transcripts. This widespread transcription was remarkable, and could not be attributed to repeat sequences, such as SINES and LINES, which are almost entirely absent in Hox clusters. Third, there were novel exons that were used extensively, and these often dramatically altered the function of the encoded proteins. For example, in the capping mesenchyme a noncanonical alternative 5′ exon was used for *Hoxd11* that gave a frameshift and as a result was noncoding, while in the renal vesicles the standard two exons were used (Fig. 18.4). In addition there was extensive intergenic RNA processing and opposite strand transcription. Indeed, the picture that emerged suggested that the Hox clusters should be considered single supergenes with a multitude of transcriptional and processing possibilities rather than clusters of simple two exon genes.

In summary, RNA-Seq is revolutionizing our view of the genome. With RNA-Seq we see changing RNA processing patterns, enhancer transcription, new genes, long intergenic noncoding RNA transcripts, antisense transcripts, and more. As the price of DNA sequencing continues to plummet, RNA-Seq will continue to improve in power, allowing more reads and a more exhaustive analysis of genome wide transcription.

18.8 Single Cell Analysis

The RNA-Seq analysis of pure populations of cells of a specific type, such as capping mesenchyme, represents a major step forward in the genomics analysis of kidney development. Nevertheless, even the thorough RNA-Seq study of each specific cell type of the developing kidney would still leave significant holes in our understanding. Some questions simply cannot be answered by the ensemble analysis of collections of cells. They require a still higher resolution.

As mentioned earlier, development begins with the fertilized egg, which gives rise to an adult with amazing cellular variety and multiple organ systems. Underlying the enormous complexity of this process are simplifying strategies that make repeated use of individual genes and signaling systems to drive distinct developmental decisions in different contexts. At many stages in development, we observe histologically uniform groups of cells that produce varied progeny. We

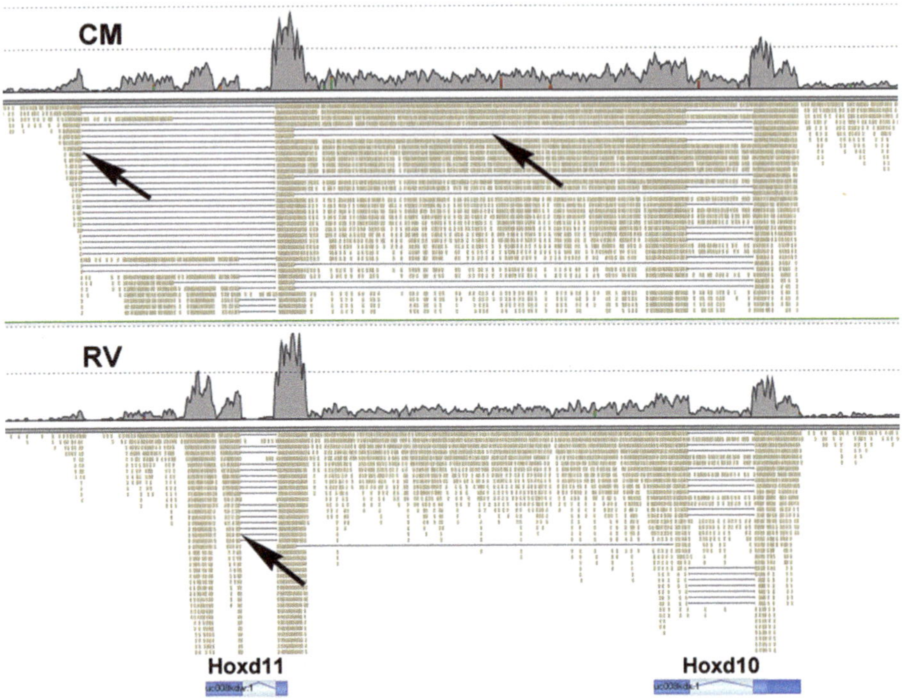

Fig. 18.4 RNA-Seq analysis of transcription of the*Hoxd10-Hoxd11*genomic region. RNA-Seq analysis of capping mesenchyme (*CM*) progenitor cell and renal vesicle (*RV*) differentiated cell RNAs. The TruSeq polyA + selection method of RNA-Seq was used for this illustration. The *blue rectangles* at the *bottom* show the positions of the *Hoxd10* and *Hoxd11* genes, with the *light shaded regions* representing introns. Transcription is *left* to *right*. Each *small rectangle* represents an aligned RNA-Seq sequence read of about 50 bases. Where the sequence aligns to two separated sequences the intervening intronic region is represented as a horizontal line. Note in the capping mesenchyme that there is an alternate 5′ exon often used for *Hoxd11*, which results in a frame shift and a noncoding RNA (*upper left arrow*). Also note the many reads aligning to the intergenic region, indicating the presence of polyadenylated intergenic transcripts. The *top right arrow* highlights introns signifying intergenic splicing between the sequences of *Hoxd10* and *Hoxd11*. The *bottom arrow* shows the canonical intron used in the RV for *Hoxd11*

know that morphogen gradients, flanking cell crosstalk, and even stochastic gene expression can regulate decisions that generate diversity from apparent uniformity. Nevertheless, to date we have not created a detailed blueprint of this process, at single cell resolution, for any developing organ system.

For example, early in kidney development there are primarily two structures, the metanephric mesenchyme, and the ureteric bud. The metanephric mesenchyme is a histologically uniform cloud of mesenchymal cells that will eventually give rise to almost all of the varied cell types of the nephron. When do the first signs of differentiation into distinct developmental lineages appear within the mesenchyme? Within the early mesenchyme are some cells already lineage primed, predetermined to make specific cell types? Indeed, how varied are the gene expression profiles of individual cells of the early metanephric mesenchyme? These kinds of questions can only be answered by the analysis of the transcriptional profiles of single cells.

While the genomics analysis of single cells is clearly highly desirable, there are serious challenges to consider. First there is the technical difficulty of producing an accurate gene expression profile from the extremely small quantity of RNA present in a single cell. Total RNA content per cell depends on cell type, but is generally in the range of 5–30 pg (Copois et al. 2003; Kane et al. 2000). Several methods have been used for the gene expression profiling of single cells. Multiple rounds of *in vitro* transcription have been used for the profiling of single neurons (Kamme et al. 2003). We have found that the RiboSpia displacement DNA polymerase system (Kurn et al. 2005), using a single isothermal round of amplification, the NuGEN OneDirect system, is quite effective. In addition there are PCR based systems that appear to offer an excellent combination of sensitivity (detecting the most expressed genes), and cost (Iscove et al. 2002; Brady and Iscove 1993; Cherry et al. 2009; Brunskill et al. 2011b). None of these systems are perfect, however, meaning that technical noise is a much greater concern than it is for standard microarray/RNA-Seq studies.

In addition to technical noise, there is biological noise to consider. There are extremely small numbers of transcripts per gene in a single cell. Some simple calculations that assume about 10 pg of total RNA per cell, with 2% of this mRNA, determine that there are approximately 160,000 molecules of mRNA per cell. This might seem like a lot, but in fact about 10,000 genes are expressed per cell on average, so this works out to only 16 mRNA transcripts per gene. This also assumes that all mRNAs are of similar abundance, but in fact a typical cell has around a hundred or so genes that are expressed at very high levels, with 1–10,000 copies per cell, accounting in total for about half of the mass of mRNA. The end result is that for the majority of genes expressed, the transcripts are present at an abundance level below ten copies per cell.

These are such small numbers that one would expect significant random fluctuation, even if genes were expressed in a constant steady state manner.

An abundance of data collected over several decades, however, tells us that genes are actually expressed in a pulsatile fashion, further contributing to transcript level variation. Early work in both bacteria and eukaryotes showed that gene expression is largely an on/off process, with gradual induction increasing the percentage of cells with expression, rather than giving incremental increase of expression in each cell (Novick and Weiner 1957; Ko et al. 1990). More recent landmark studies include the use of two copies of the same promoter in a single bacterial, or yeast cell, driving expression of two different fluorescent proteins. The results elegantly demonstrate striking variation, or noise, in the expression levels of two genes with identical promoters in single cells (Elowitz et al. 2002; Ozbudak et al. 2002).

Gene expression varies from cell to cell, or within a single cell as a function of time, as a result of sporadic short bursts of active transcription (Takasuka et al. 1998; Ross et al. 1994; Chubb et al. 2006; Raj et al. 2008; Golding et al. 2005). The causes of the pulsatile nature of gene expression are not fully defined, but one model states that transcription occurs within a limited number of transcription factories in cells (Wansink et al. 1993; Jackson et al. 1993; Osborne et al. 2004). Genes would compete for occupancy of sites within a factory, where they would be highly transcribed. It has been proposed that there are relatively few of these factories, on the order of hundreds per cell, which account for the bulk of the transcription of the approximately 10,000 genes that are expressed.

There are important biological consequences to the noisy nature of gene expression. Each gene is present in only two copies per cell. As mentioned, most expressed genes have on average fewer than ten transcripts per cell. These are very small numbers, statistically speaking. The observed burst mode of gene expression dictates that variations in transcript levels are far greater than would be predicted by a simple Poisson distribution. Indeed there are so many genes with so much variation that it makes one wonder how things ever turn out correctly, especially during development, when correct combinations of transcription factors are thought to drive decisions about the appropriate developmental destiny of a cell. One strategy for biological success is to employ genetic functional redundancy. Indeed, such redundancy is generally acknowledged to be responsible for the surprisingly mild phenotypes often observed in mice with targeted mutations. In addition, the levels of the encoded proteins can be more constant, especially if they have a long half-life. Gene noise is probably an underappreciated cause of incomplete penetrance and variable expressivity, which can persist even on isogenic genetic backgrounds.

Remarkably, it appears that in some cases the noisy nature of gene expression has actually been harnessed to drive developmental decisions. One example is the selection of specific odorant receptors during development of the olfaction system. There are over a thousand receptors and they are activated using a random "Monte Carlo" strategy in a mutually exclusive fashion (Vassar et al. 1993; Tsuboi et al. 1999). There is also evidence that during hematopoiesis the differentiation of stem cells could be regulated by stochastic gene expression events (Chang et al. 2008). Another elegant example is the development of photoreceptors in the *Drosophila* eye, where stochastic variations in *spineless* gene expression drive decisions about photoreceptor differentiation (Wernet et al. 2006).

Despite the many challenges, it is clear that it is now possible to perform genomics analysis of single cells. One must be careful to distinguish biological from technical noise, however, through the performance of many technical replicates of single cell equivalent RNA samples. Because of the biological noise, it is necessary to examine large numbers of single cells before drawing conclusions. Nevertheless, there is great potential power in single cell analysis. By examining large numbers of histologically homogeneous cells from developing systems, it will be possible to define the level of transcriptional heterogeneity present. Evidence of lineage priming, with predilection of specific developmental direction, can be defined. In addition by connecting the overlapping gene expression states that connect the progenitor cell to the fully differentiated cell, it will be possible to more completely define the gene expression blueprint that drives the progression. In the end it will be possible to create single cell resolution blueprints of the gene expression programs that drive organogenesis.

18.9 Craniofacial Development

While most of the focus of this chapter has been on kidney development, the same principles apply to craniofacial development. The face forms primarily from a mix of neural crest and paraxial mesoderm, although there is important interaction with the underlying neuroepithelium (Chai and Maxson 2006). The neural crest cells begin their epithelium to mesenchymal transformation and migration from the neural tube at around E8.5 in the mouse. There are important signaling centers, including the anterior neural ridge and the olfactory placode, that produce growth factors directing migration and survival. The face then forms from a number of buds of tissue, the facial prominences, which consist of mesenchyme and overlying epithelium. These include the mandibular arch, maxillary arch, medial nasal process and lateral nasal process. During the period from around E10 to E12.5 these processes grow, fuse, and form the lower and upper jaw as well as the other major elements of the face.

Trevor Williams and colleagues carried out a particularly extensive microarray study of early craniofacial development (Feng et al. 2009). Three mouse facial prominences, the maxillary, mandibular and frontonasal (which includes both the lateral and medial) were manually dissected. Embryos were examined at five time points, E10.5, E11.0, E11.5, E12.0 and E12.5, and seven independent biological replicates were examined for each sample type, giving very strong statistical power for the data analysis. This excellent work began to define the sets of genes with expression restricted to each prominence. That is, it provided a novel set of biological markers. In addition, it identified key molecular functions and biological processes that were enriched in each prominence as a function of developmental time. For example, the frontonasal prominence was distinguished by expression of neurogenesis genes associated with the formation of the olfactory epithelium.

The FACEBASE consortium (www.facebase.org) is currently working to create an LCM/microarray-RNA-Seq atlas of craniofacial development that will significantly improve the resolution of the genomics analysis. Gene expression profiles are being defined for key developmental compartments driving craniofacial development for mouse embryos from E8.5 to E16.5. By using laser capture microdissection it is possible to subdivide the frontonasal prominence, for example, into the lateral eminence, and the medial eminence, and to separate the mesenchymal and overlying epithelial components (Fig. 18.5). In addition the underlying neuroepithelial compartments are being profiled to better define the cross inductive interactions taking place. As this atlas evolves, from LCM based compartments, to single cell types, to single cells, it will be particularly interesting to determine the heterogeneity of the neural crest cells at the transcriptional level as they migrate from the neural tube to their final destinations, differentiating along the way into the varied cell types they contribute.

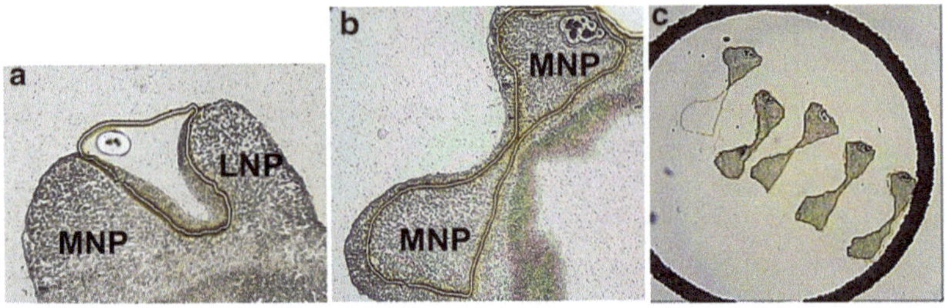

Fig. 18.5 LCM isolation of compartments of craniofacial development. Laser capture microdissection of the medial nasal prominence (*MNP*) and lateral nasal prominence (*LNP*) regions of the E10.5 embryo craniofacial region. (**a**) The laser has been used to cut out the olfactory pit region of epithelium separating the LNP and MNP. (**b**) The laser has cut out the two medial MNP regions. (**c**) A cap with several purified sets of flanking MNPs

Acknowledgement This work was supported in part by NIH grant number RC4-DK090891.

References

Bafico A, et al. Novel mechanism of Wnt signalling inhibition mediated by Dickkopf-1 interaction with LRP6/arrow. Nat Cell Biol. 2001;3(7):683–6.

Brady G, Iscove NN. Construction of cDNA libraries from single cells. Methods Enzymol. 1993;225:611–23.

Brunskill EW, Potter SS. Gene expression programs of mouse endothelial cells in kidney development and disease. PLoS One. 2010;5(8):e12034.

Brunskill EW, Potter SS. RNA-Seq defines novel genes, RNA processing patterns and enhancer maps for the early stages of nephrogenesis: Hox supergenes. Dev Biol. 2012;368:4–17.

Brunskill EW, et al. Atlas of gene expression in the developing kidney at microanatomic resolution. Dev Cell. 2008;15(5):781–91.

Brunskill EW, et al. Defining the molecular character of the developing and adult kidney podocyte. PLoS One. 2011a;6(9):e24640.

Brunskill EW, et al. Genes that confer the identity of the renin cell. J Am Soc Nephrol. 2011b;22(12):2213–25.

Carroll TJ, et al. Wnt9b plays a central role in the regulation of mesenchymal to epithelial transitions underlying organogenesis of the mammalian urogenital system. Dev Cell. 2005;9(2):283–92.

Chai Y, Maxson Jr RE. Recent advances in craniofacial morphogenesis. Dev Dyn. 2006;235(9):2353–75.

Challen G, et al. Temporal and spatial transcriptional programs in murine kidney development. Physiol Genomics. 2005;23(2):159–71.

Chang HH, et al. Transcriptome-wide noise controls lineage choice in mammalian progenitor cells. Nature. 2008;453(7194):544–7.

Cherry TJ, et al. Development and diversification of retinal amacrine interneurons at single cell resolution. Proc Natl Acad Sci U S A. 2009;106(23):9495–500.

Chubb JR, et al. Transcriptional pulsing of a developmental gene. Curr Biol. 2006;16(10):1018–25.

Cloonan N, et al. Stem cell transcriptome profiling via massive-scale mRNA sequencing. Nat Methods. 2008;5(7):613–19.

Copois V, et al. Assessment of RNA quality extracted from laser-captured tissues using miniaturized capillary electrophoresis. Lab Invest. 2003;83(4):599–602.
Costantini F, Kopan R. Patterning a complex organ: branching morphogenesis and nephron segmentation in kidney development. Dev Cell. 2010;18(5):698–712.
De Santa F, et al. A large fraction of extragenic RNA pol II transcription sites overlap enhancers. PLoS Biol. 2010;8(5):e1000384.
Elowitz MB, et al. Stochastic gene expression in a single cell. Science. 2002;297(5584):1183–6.
Feng W, et al. Spatial and temporal analysis of gene expression during growth and fusion of the mouse facial prominences. PLoS One. 2009;4(12):e8066.
Golding I, et al. Real-time kinetics of gene activity in individual bacteria. Cell. 2005;123(6): 1025–36.
Iscove NN, et al. Representation is faithfully preserved in global cDNA amplified exponentially from sub-picogram quantities of mRNA. Nat Biotechnol. 2002;20(9):940–3.
Jackson DA, et al. Visualization of focal sites of transcription within human nuclei. EMBO J. 1993;12(3):1059–65.
Kamme F, et al. Single-cell microarray analysis in hippocampus CA1: demonstration and validation of cellular heterogeneity. J Neurosci. 2003;23(9):3607–15.
Kane MD, et al. Assessment of the sensitivity and specificity of oligonucleotide (50mer) microarrays. Nucleic Acids Res. 2000;28(22):4552–7.
Keller G, et al. Nephron number in patients with primary hypertension. N Engl J Med. 2003;348(2):101–8.
Kim TK, et al. Widespread transcription at neuronal activity-regulated enhancers. Nature. 2010;465(7295):182–7.
Ko MS, Nakauchi H, Takahashi N. The dose dependence of glucocorticoid-inducible gene expression results from changes in the number of transcriptionally active templates. EMBO J. 1990;9(9):2835–42.
Kobayashi A, et al. Distinct and sequential tissue-specific activities of the LIM-class homeobox gene Lim1 for tubular morphogenesis during kidney development. Development. 2005;132(12):2809–23.
Kurn N, et al. Novel isothermal, linear nucleic acid amplification systems for highly multiplexed applications. Clin Chem. 2005;51(10):1973–81.
Mao B, et al. LDL-receptor-related protein 6 is a receptor for Dickkopf proteins. Nature. 2001;411(6835):321–5.
McGinnis N, Kuziora MA, McGinnis W. Human Hox-4.2 and Drosophila deformed encode similar regulatory specificities in Drosophila embryos and larvae. Cell. 1990;63(5):969–76.
Mortazavi A, et al. Mapping and quantifying mammalian transcriptomes by RNA-Seq. Nat Methods. 2008;5(7):621–8.
Nagalakshmi U, et al. The transcriptional landscape of the yeast genome defined by RNA sequencing. Science. 2008;320(5881):1344–9.
Novick A, Weiner M. Enzyme induction as an all-or-none phenomenon. Proc Natl Acad Sci U S A. 1957;43(7):553–66.
Osborne CS, et al. Active genes dynamically colocalize to shared sites of ongoing transcription. Nat Genet. 2004;36(10):1065–71.
Ozbudak EM, et al. Regulation of noise in the expression of a single gene. Nat Genet. 2002;31(1):69–73.
Poladia DP, et al. Link between reduced nephron number and hypertension: studies in a mutant mouse model. Pediatr Res. 2006;59(4 Pt 1):489–93.
Potter SS, et al. Laser capture-microarray analysis of Lim1 mutant kidney development. Genesis. 2007;45(7):432–9.
Raj A, et al. Imaging individual mRNA molecules using multiple singly labeled probes. Nat Methods. 2008;5(10):877–9.
Ross IL, Browne CM, Hume DA. Transcription of individual genes in eukaryotic cells occurs randomly and infrequently. Immunol Cell Biol. 1994;72(2):177–85.

Schmidt-Ott KM, et al. Novel regulators of kidney development from the tips of the ureteric bud. J Am Soc Nephrol. 2005;16(7):1993–2002.

Schwab K, et al. A catalogue of gene expression in the developing kidney. Kidney Int. 2003;64(5):1588–604.

Schwab K, et al. Comprehensive microarray analysis of Hoxa11/Hoxd11 mutant kidney development. Dev Biol. 2006;293(2):540–54.

Self M, et al. Six2 is required for suppression of nephrogenesis and progenitor renewal in the developing kidney. EMBO J. 2006;25(21):5214–28.

Shawlot W, Behringer RR. Requirement for Lim1 in head-organizer function. Nature. 1995;374(6521):425–30.

Stark K, et al. Epithelial transformation of metanephric mesenchyme in the developing kidney regulated by Wnt-4. Nature. 1994;372(6507):679–83.

Stuart RO, Bush KT, Nigam SK. Changes in global gene expression patterns during development and maturation of the rat kidney. Proc Natl Acad Sci U S A. 2001;98(10):5649–54.

Stuart RO, Bush KT, Nigam SK. Changes in gene expression patterns in the ureteric bud and metanephric mesenchyme in models of kidney development. Kidney Int. 2003;64(6): 1997–2008.

Takasuka N, et al. Dynamic changes in prolactin promoter activation in individual living lactotrophic cells. Endocrinology. 1998;139(3):1361–8.

Tsuboi A, et al. Olfactory neurons expressing closely linked and homologous odorant receptor genes tend to project their axons to neighboring glomeruli on the olfactory bulb. J Neurosci. 1999;19(19):8409–18.

Vassar R, Ngai J, Axel R. Spatial segregation of odorant receptor expression in the mammalian olfactory epithelium. Cell. 1993;74(2):309–18.

Wansink DG, et al. Fluorescent labeling of nascent RNA reveals transcription by RNA polymerase II in domains scattered throughout the nucleus. J Cell Biol. 1993;122(2):283–93.

Wernet MF, et al. Stochastic spineless expression creates the retinal mosaic for colour vision. Nature. 2006;440(7081):174–80.

Chapter 19
From SNP Genotyping to Improved Pediatric Healthcare

Jacek W. Biesiada, Senthilkumar Sadhasivam, Michael Wagner, and Jaroslaw Meller

Abstract SNP genotyping arrays have become an important tool for cohort identification and stratification, phenotype-genotype association studies, discovery of disease markers, prediction of molecular phenotypes, and clinical decision support. In this chapter, large-scale SNP genotyping and the resulting informatics challenges are discussed in the context of basic as well as translational studies. Tailored research informatics solutions and integration with clinical informatics systems are illustrated using several specific examples of applications, including: (i) cohort stratification analysis; (ii) prediction of classical HLA alleles from SNP data in the context of pediatric autoimmune diseases; and (iii) predictive decision models for the management of surgical pain and avoidance of opioid-related adverse outcomes in children.

J.W. Biesiada, Ph.D.
Division of Biomedical Informatics, Cincinnati Children's Hospital Medical Center, 3333 Burnet Avenue, ML-7024, Cincinnati, OH 45229-3039, USA

Division of Management and Informatics, Technical University of Silesia, Gliwice, Poland

S. Sadhasivam, M.D., MPH
Department of Pediatrics, University of Cincinnati College of Medicine, Cincinnati, OH, USA

Perioperative and Acute Pain Service, Division of Pediatric Anesthesia, Cincinnati Children's Hospital Medical Center, 3333 Burnet Avenue, ML-7024, Cincinnati, OH 45229-3039, USA

M. Wagner, Ph.D.
Division of Biomedical Informatics, Cincinnati Children's Hospital Medical Center, 3333 Burnet Avenue, ML-7024, Cincinnati, OH 45229-3039, USA

Department of Pediatrics, University of Cincinnati College of Medicine, Cincinnati, OH, USA

J. Meller, Ph.D. (✉)
Division of Biomedical Informatics, Cincinnati Children's Hospital Medical Center, 3333 Burnet Avenue, ML-7024, Cincinnati, OH 45229-3039, USA

Department of Environmental Health, University of Cincinnati College of Medicine, Cincinnati, OH, USA
e-mail: jarek.meller@cchmc.org

J.J. Hutton (ed.), *Pediatric Biomedical Informatics: Computer Applications in Pediatric Research*, Translational Bioinformatics 2, DOI 10.1007/978-94-007-5149-1_19,

19.1 Introduction

Continuing rapid advances in DNA sequencing, imaging and other techniques underpin the generation of increasingly large streams of complex data that need to be integrated with clinical information in order to realize the promise of personalized clinical interventions (Holmes et al. 2009; Chan and Ginsburg 2011; The International HapMap Consortium 2003). As a result, research and clinical centers, including those focused on pediatric health, have been investing heavily in the development of necessary informatics infrastructure, including scalable databases that combine different types of information, analytical and data mining capabilities, as well as better interactive platforms for cross-disciplinary basic and translational research. This chapter focuses on the use of Single Nucleotide Polymorphism (SNP) and re-sequencing arrays (LaFramboise 2009; Schaaf et al. 2011), which readily provide information about genetic variation that can subsequently be combined with demographic, phenotypic and other data to facilitate both basic and translational clinical research (Chou et al. 1987, 2006a; Diatchenko et al. 2005; Wellcome Trust Case Control Consortium 2007; de Bakker et al. 2006). Informatics challenges and solutions pertaining to the use of SNP arrays are discussed, with several case studies to illustrate applications in research and clinical settings.

We start by describing the development and use of an integrated database and analysis platform for joint analysis of SNP genotype and clinical data. This database has been developed in our institution to address the limitations of existing systems and to enable further integration with clinical informatics systems. Emphasis is placed on informatics challenges related to: the evolving nature and increasing complexity of the data, as new re-sequencing platforms are being adopted; data are generated by multiple centers and must be combined to increase the statistical power of analyses; and ongoing development of specialized SNP chips that may present data in new formats or with different attributes.

The first application (Sect. 19.3) concerns the prediction of Human Leukocyte Antigen (HLA) alleles (Lechler and Warrens 2000; Marsh et al. 2010a, 2010b; Aureli et al. 2008; Orozco et al. 2008; Sampaio-Barros et al. 2008; Robinson et al. 2003, 2009) from SNP data. Reliable HLA allele prediction (or imputation) methods have recently been developed (Leslie et al. 2008; Dilthey et al. 2011), offering significant advantages in on-going studies of pediatric autoimmune disorders. At the same time, it adds to the complexity of informatics solutions that are required to fully capitalize on these developments. A number of additional layers, including carefully assessed and unified quality control protocols, population stratification and haplotype reconstruction (de-phasing) approaches, and multiple filters and predictors must be integrated to achieve robust accuracy and high confidence in predictions. Possible applications to donor matching with further progress in this field are also discussed.

The second application (Sect. 19.4) deals with decision support systems that are being developed to minimize opioid-related adverse effects, while managing perioperative pain in children. Children are at higher risk of inadequate pain relief and serious side effects from morphine, the gold standard opioid, in pediatric surgery pain management (Caldas et al. 2004; Duedahl and Hansen 2007). The ability to personalize the use of morphine in order to maximize pain relief while minimizing its adverse effects would have significant public health relevance. Efforts to develop decision support systems for such personalized intervention provide a case study for applications of informatics in translational research. Here, we focus on challenges related to developing robust and accurate decision rules that combine SNP genotype and clinical data in the context of population stratification and gene-gene interactions.

19.2 Managing Large Genotype-Phenotype Data Sets

Advances in commercially available genotyping technology (e.g., by companies such as Affymetrix and Illumina) have facilitated the use of genome-wide association studies (GWAS) to link specific genetic variants with human disease. These studies use allele frequencies in populations under considerations to find correlations between a phenotype (disease status) and individual genomic variants, which typically consist of single nucleotide polymorphisms (SNPs), but also copy number variants (CNVs) and insertion-deletions (indels) (The International HapMap Consortium 2003; LaFramboise 2009; Schaaf et al. 2011).

The large scale of the source data (the most recent Illumina Omni-5 platform assays over 4.3 million SNPs on a single chip) and the need for data on very large cohorts (often tens of thousands of individuals), require efficient data management and pose challenges in data analyses. While a number of very efficient and high-quality GWAS analysis software packages are available (e.g., PLINK, SNPTEST, LAMP (Purcell et al. 2007; Marchini and Howie 2010; Li et al. 2005)), the challenge of transparently managing genome-wide variant information in conjunction with demographic and phenotypic variables is less easily solved with off-the-shelf tools.

19.2.1 Storing and Accessing Raw Genotypes

Genotyping data typically must undergo iterative rounds of filtering and quality control, which usually results in many versions of the same data, which in turn create challenges in documentation and tracking. This motivated us to improve processes for managing large (genome-wide) genotype data sets. In order to provide a central, web-accessible repository for the typically large *raw* genotyping data (e.g., CEL files for Affymetrix and idat files for Illumina platforms) we developed an application (accessible to registered users at http://research.cchmc.org/genotyping), which allows investigators to perform the following functions:

- Upload new data files that conform to a certain file format. All files are parsed, any metadata contained in the file is parsed out by the system.
- Control access to data, based on permissions granted by the owner of the data
- Tag data files with keywords and controlled vocabulary
- Upload text files or spreadsheets with supplemental information about the primary data
- Use keyword, controlled vocabulary and/or free-text searches to retrieve and download primary data
- Use the interface to launch processing (genotype calling) algorithms on a Linux cluster computer, thus enabling access to very high-performance computing power for basic researchers otherwise unfamiliar with computing tools or languages. In particular, we have implemented the CRLMM approach to genotype calling, and we make the Affy Power Tools available to our users as well.

19.2.2 Genotype Phenotype Data Flow

A second application, which we call gwadb (Genome-Wide Association Database), is accessible to registered users at http://research.cchmc.org/gwadb2 and consists of a user-friendly, web-based interface to a relational database holding all relevant data (in particular, genotype calls) and metadata for GWAS analyses. Our aim is to enable clinical and translational researchers to perform GWAS analyses and related tasks on well-curated, well-documented data without requiring knowledge of Linux command lines. We accomplish this by storing raw or minimally processed genotyping data in the same database as basic demographics and phenotypic information. We enable our users not only to download any data in a number of commonly used file formats, but also to process these data through a number of predefined, parameterized workflows which run on a Linux cluster. The capability of executing these complex and often CPU-time and memory-intensive processes as a batch job on a cluster is essential for GWAS-type analyses, as the data sizes often make local processing close to impossible.

Currently the processing workflows that are directly callable from gwadb's web interface include commonly used tools such as PLINK (for quality control, case-control analyses, basic data management etc.), Eigenstrat (for cohort variability visualization) (Fig. 19.1), IMPUTE2 (for imputation of genotypes to the 1000 Genomes reference set), KING (for automated pedigree imputation based on genotypes) (Price et al. 2006; Manichaikul et al. 2010; Howie et al. 2009). All of these are essential to perform GWAS studies. PLINK is widely used in the community and provides a large number of functions to summarize, manage, filter and analyze large-scale SNP data, it has de facto become a standard tool in any analysis. Association p-values can be uploaded directly into a track of a local instance of the UCSC Genome Browser and further processed and visualized.

The analysis of ancestry with tools such as Eigenstrat is equally essential, not only for cohort stratification, but also as a quality control tool since it in effect

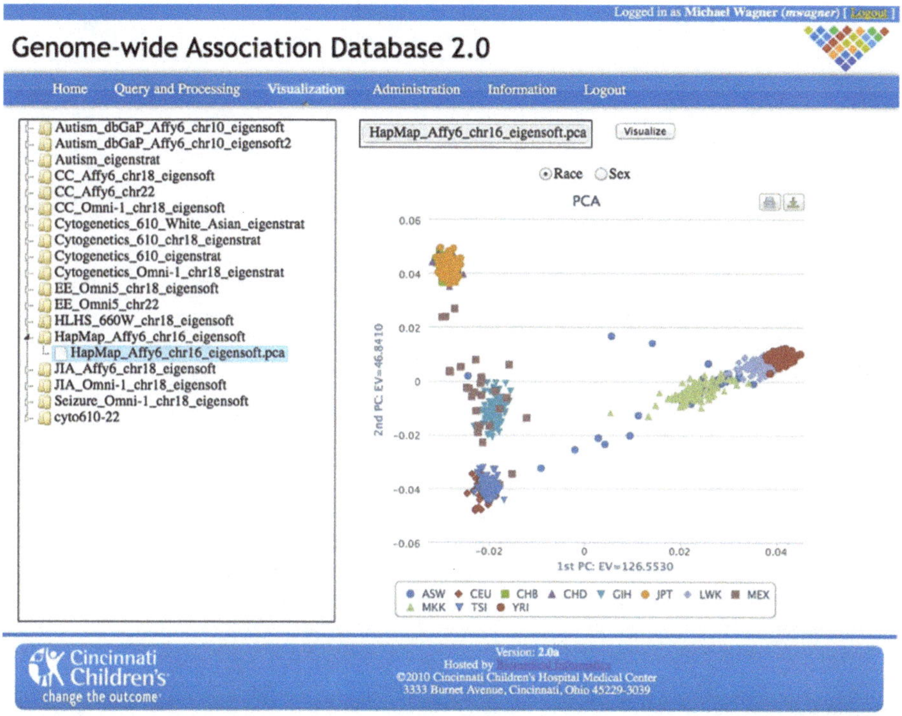

Fig. 19.1 Visualization of population sub-structure using Eigenstrat on HapMap3 data. All SNPs on chromosome 16 were used on this Affymetrix 6 data set. HapMap3 populations (including European ancestry CEU and TSI cohorts) are shown using two main PCA components. In-browser visualization is facilitated by a library from HighCharts.com

allows for verification of self-reported ancestries. Imputation is a crucial tool when performing analyses across different genotyping platforms, and because of its computational cost, it is especially convenient for our end-users to run IMPUTE2 on our computational cluster from gwadb. Finally, GWAS studies generally assume that all data used come from unrelated subjects, and KING is a convenient, fast computational tool that can verify kinships and uncover previously unknown familial relationships. This is especially useful when merging different cohorts in order to verify that no one is enrolled in both cohorts at the same time. As new applications and extensions are considered, more software tools can easily be incorporated into gwadb due its flexible design, modular architecture, and the integration with a Linux computational cluster, which enables parallel large scale computation in the batch mode, as well as on-the-fly analysis whenever applicable.

Gwadb is implemented using open-source tools such as php/mysql. Furthermore all communication to the server is encrypted with SSL and strict guidelines are enforced concerning user names and passwords, thus ensuring security and compliance with security standards. These portals represent complete suites that

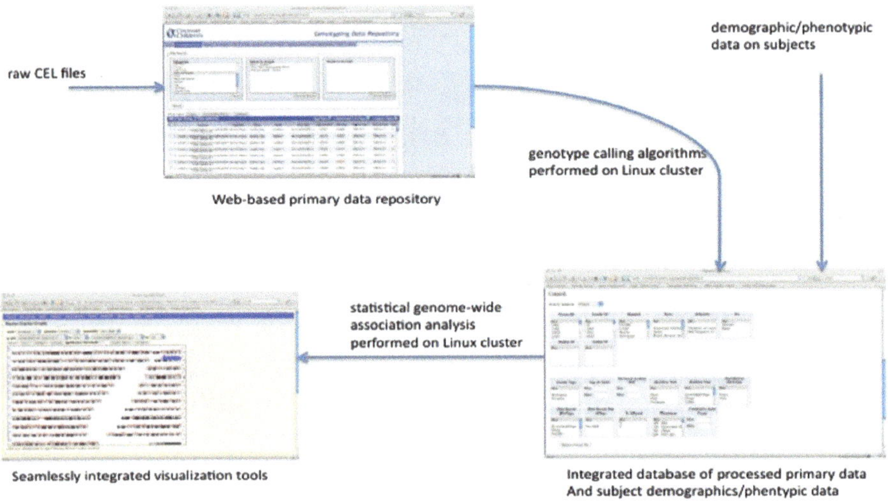

Fig. 19.2 Overview of the genotype/phenotype data flow through web-based repositories, including the raw data repository (*top left*) and gwadb (*bottom right*). We emphasize the ability to spawn computational analysis jobs (e.g., for genotype calling and GWAS analysis) on a high-performance Linux cluster

integrate the management of disk-based storage for primary data, the management and integration of metadata about the primary data with clinical data points, as well as the processing of these data using state-of-the-art algorithms and computational equipment in a user-friendly web interface (Fig. 19.2).

19.3 Prediction of HLA Alleles from SNP Data

Human Leukocyte Antigen (HLA) locus on chromosome 6 encodes a number of highly polymorphic genes that play an essential role in immune system responses (de Bakker et al. 2006; Lechler and Warrens 2000; Marsh et al. 2010a, b; Aureli et al. 2008; Orozco et al. 2008; Sampaio-Barros et al. 2008; Robinson et al. 2003, 2009). In particular, Major Histocompatibility Complex (MHC) class I and class II genes, including HLA-A, -B, -C and DRB1, DQB1, DPB1, respectively, are well known components of the immune system. The function and medical relevance of these genes stem from their ability to bind and selectively recognize peptides derived from pathogens as well as self-peptides. The latter aspect of their function contributes to the role of MHC region as an important susceptibility locus in autoimmune diseases (de Bakker et al. 2006; Lechler and Warrens 2000). In particular, multiple associations between HLA alleles and childhood autoimmune disorders, including type 1 diabetes, celiac disease and Juvenile Idiopathic Arthritis

(JIA), have been observed. For example, HLA-B gene variant (allele) HLA-B35 and HLA-DRB1 gene variants (alleles) DRB1*1101/1104 have been associated with susceptibility to some JIA subtypes, whereas HLA-DR4 appears to be a protective allele. Consequently, HLA loci and their haplotypes are an important component of disease association studies in the context of JIA and other autoimmune diseases (Aureli et al. 2008; Orozco et al. 2008; Sampaio-Barros et al. 2008).

HLA alleles can be identified by direct DNA sequencing of variable regions of HLA genes that encode parts of the protein involved in peptide recognition (e.g., exon 2 of DRB1). Recently, to overcome limitations of this method (including its cost), an alternative approach has been proposed, which takes advantage of single nucleotide polymorphism (SNP) microarrays. These "SNP chips" can be used to re-sequence single letter variants in the whole genome, or they can be tailored to include a specific panel of SNPs, e.g., sampling the HLA locus (LaFramboise 2009; Schaaf et al. 2011).

Efficient methods for SNP genotyping have significantly impacted studies on autoimmunity. Large-scale genotyping projects using SNP arrays are providing rich data on variation within the extended MHC region in different populations. In particular, the existence of strong linkage disequilibrium between MHC SNPs and HLA alleles suggested that SNP tagging (or SNP-based imputing) could be an attractive (and inexpensive) alternative to traditional HLA typing (de Bakker et al. 2006; Leslie et al. 2008; Dilthey et al. 2011).

19.3.1 Linkage Disequilibrium

Extensive LD present within the MHC region is illustrated in Fig. 19.3, using data on a cohort of about 500 healthy controls of Caucasian descent from the midwestern United States. This and similar data sets are made available to researchers through the gwadb database described in Sect. 19.2. SNP genotypes in the region were obtained using the Affymetrix 6 platform, whereas high resolution (4 digit) HLA genotypes were obtained by direct DNA sequencing of the variable exons, following standard protocols. PHASE version 2.1 (Marchini et al. 2006; Stephens and Donnelly 2003; Stephens et al. 2001) was used to reconstruct the haplotypes, which was facilitated by the use of gwadb and the Linux computational cluster. The strength of correlation between individual (biallelic) SNPs and (multiallelic) HLA was then quantified using the relative information, following de Bakker et al. (2006). Different colors represent the extent of LD for each of 8 HLA genes considered here.

As can be seen in the figure, there is a broad peak of LD around each gene. Consequently, SNPs located near (and in majority of cases outside of) the respective HLA genes can be used to impute their alleles based on these strong LD patterns. Such predictions for cohorts of Caucasian ancestry have been demonstrated to reach greater than 90% accuracy at the level of 4 digit alleles for all except one gene considered here, the exception being DRB1 for which greater than 85% accuracy

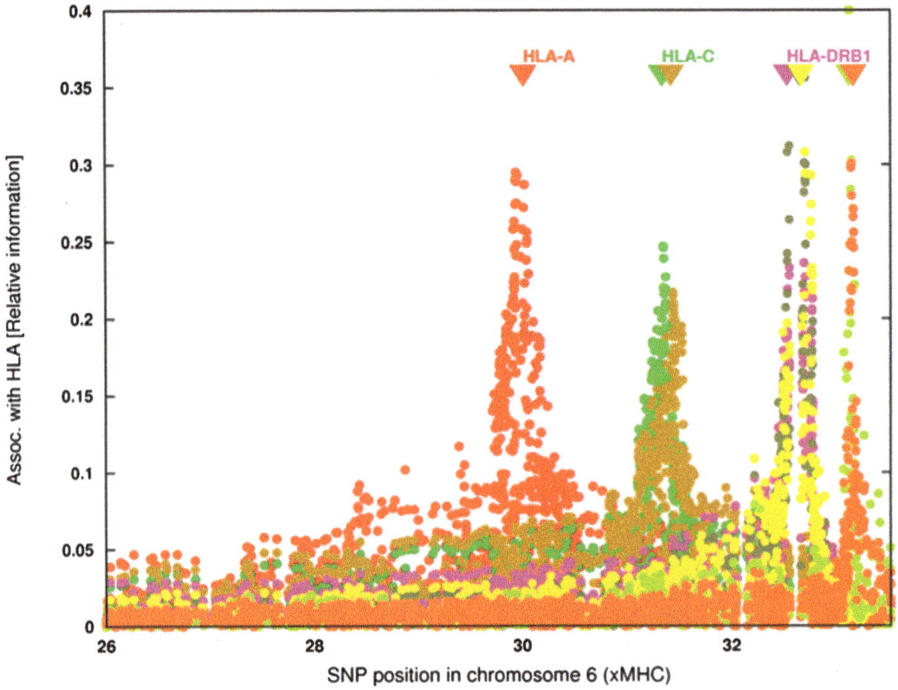

Fig. 19.3 Extensive linkage disequilibrium within the MHC region

has been reported (Leslie et al. 2008; Dilthey et al. 2011). It should be noted that HLA alleles are typically typed with 4-digit resolution for association studies and transplant donor matching.

19.3.2 Prediction of HLA Alleles

Reliable prediction of HLA alleles from SNP data can be obtained by exploiting the presence of linkage disequilibrium (non-random associations) between SNPs within MHC and particular HLA alleles, and by applying advanced statistical and machine learning methods (see Fig. 19.5). Integrating such predictions with platforms for SNP data analyses, such as gwadb, can greatly streamline GWAS studies, especially those on autoimmune disorders. Using this approach, analysis of associations with HLA alleles can be performed by applying SNP-based imputation, without additional sequencing of HLA genes. A number of studies successfully utilizing this approach have been published recently. Applications of SNP-based imputation of HLA alleles should take into account:

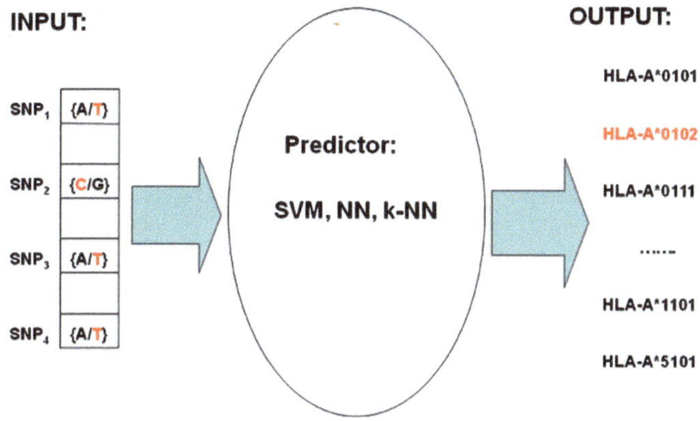

Fig. 19.4 HLA allele prediction from SNP data can be improved by using statistical and machine learning methods, such as Support Vector Machines (*SVM*) or neural network (*NN*); it requires carefully designed, properly stratified and sufficiently large high quality training and validation sets

- Differences between SNP arrays, such as Affymetrix 6.0 or Immunochip, which may require between platform imputation and/or careful selection of SNP subsets to be used for the prediction
- Limitations in accuracy of current methods that at present cannot fully replace HLA typing (especially for cases predicted with lower confidence)
- Careful assessment of population stratification and exclusion of individuals other than those of Caucasian descent (to which most current predictors are limited).

With regards to population stratification, extensions to other than White/CEU populations may require data on specific cohorts and (re-) training of HLA allele predictors. This observation can be extended to relatively rare (overall) classes that nevertheless can play in important role in some specific autoimmune disorder (e.g., as a high risk allele). Representative cohorts for that particular disease (and risk allele) may be required to enable robust extrapolation from training examples and sufficient accuracy for the allele of interest (Fig. 19.4).

Prediction of HLA alleles from SNP data starts from haplotype reconstruction, which allows one to assign individual HLA alleles to chromosomes reconstructed from SNP genotypes (see Fig. 19.5). Two different options are available for that purpose within our platform, integrating two widely used and well benchmarked packages for population-based haplotype reconstruction: Impute2 (Manichaikul et al. 2010) and Phase (version 2.1) (Marchini et al. 2006; Stephens and Donnelly 2003; Stephens et al. 2001). In case of the latter, which is in general more accurate, but also computationally more demanding, reconstruction can be performed within blocks of up to 400 SNPs (depending on the size of the cohort), spawning multiple jobs on the cluster for multiple blocks. Tagging SNPs (Li and Stephens 2003; Marchini et al. 2007; Scheet and Stephens 2006) across the whole region can be included to enable subsequent "stitching" to obtain partial haplotypes.

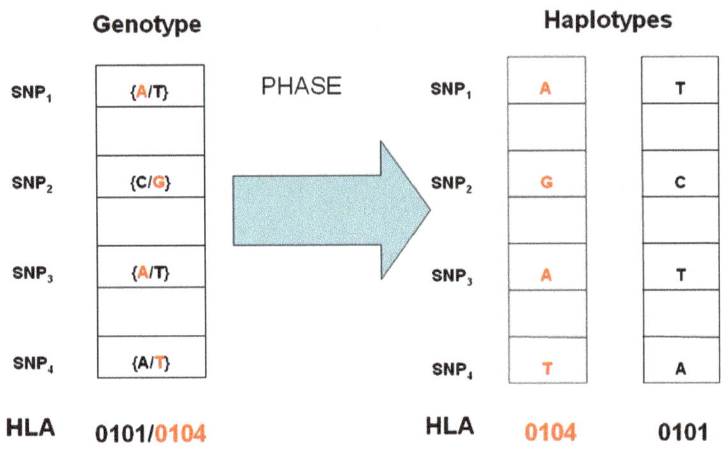

Fig. 19.5 Haplotype reconstruction for HLA prediction

In summary, integrated informatics solutions for storing, processing and analyzing SNP and other sequencing data from multiple clinical studies can greatly facilitate efforts to extend and improve HLA allele prediction. Such databases and tools provide consistent quality control criteria and protocols, as well as access to data potentially pertaining to cohorts of different ancestries that can be used to assess, refine and re-train current methods. In particular, further improvements in terms of the accuracy can be obtained by combining publicly available data with local cohorts being collected and studied at multiple clinical centers. These efforts will further contribute to association studies by providing HLA allele assignment directly from SNP chip data, with impact on diagnosis, outcomes and personalized therapeutic interventions.

19.4 SNP-Based Models for Clinical Decision Rules

We have chosen to present an example of translational research that aims to use SNP genotyping to develop clinical decision rules for management of pediatric surgical pain. Safe and effective analgesia is an important medical and economic problem (Caldas et al. 2004; Duedahl and Hansen 2007). A significant fraction of approximately five million children, who undergo a painful surgery in the US each year, experience inadequate pain relief and serious opioid-related side-effects (Cepeda et al. 2003; Sadhasivam et al. 2012; Esclamado et al. 1989). This part of the chapter focuses on efforts to improve quality of care in the context of pediatric surgery. One of the goals is to develop accurate models for the prediction of clinical responses to opioids, using genotype, clinical and demographic data. It is readily apparent that informatics challenges and solutions for this problem share commonalities with those discussed in the context of HLA allele prediction from

SNP data. At the same time, this application illustrates other issues related to the use of clinical decision support systems. See Chap. 11 for a more extensive discussion of clinical decision support, an active area of translational research in pediatrics.

The efforts to personalize pain management in children build on the growing understanding of how genes and their specific variants influence postoperative pain and the occurrence of serious side effects, such as respiratory depression. Multiple genes involved in opioid metabolism, receptor signaling and transport have been associated with inter-individual variability in clinical response to morphine in small-scale studies of adults (Chou et al. 2006b; Diatchenko et al. 2006, 2007; Klepstad et al. 2004; Nackley and Diatchenko 2010; Nackley et al. 2007; Oertel et al. 2006; Rakvåg et al. 2005, 2008; Coller et al. 2006). Much of this variability in pain perception can be explained by SNPs in catechol-O-methyltransferase (COMT), which is a key regulator of pain perception. Genetic variants of mu opioid receptor (OPRM1), the main target for many opioids including morphine, have also been associated with increased pain sensitivity. Adult carriers of the GG variant of OPRM1 SNP A118G require a 2–4 times higher dose of morphine than AA variants. In addition, specific variants of the ATP Binding Cassette B1 (ABCB1) gene that encodes a transporter protein have been shown to modulate cerebral pharmacokinetics of morphine, thereby to change its analgesic and side effects (Chou et al. 2006b; Diatchenko et al. 2006, 2007; Klepstad et al. 2004; Nackley and Diatchenko 2010; Nackley et al. 2007; Oertel et al. 2006; Rakvåg et al. 2005, 2008; Coller et al. 2006).

With the goal of improving and personalizing the postoperative care and pain management in children, we have initiated a systematic effort to identify genetic variants underlying clinical responses to opioids. Our initial cohort included 199 children undergoing tonsillectomy, of which 39 were African-American and 99 were classified as suffering from obstructive sleep apnea (Esclamado et al. 1989). A standardized protocol allowed for unambiguous assignment and accurate quantification of the observed phenotypes. Genotype data were collected using a specialized SNP panel. The candidate genes included ABCB1, COMT, OPRM1, FAAH, ADRB2 and a number of other genes that were chosen based on their allele frequencies and clinical evidences of important associations in adults with opioid analgesic and adverse effects (Chou et al. 2006b; Diatchenko et al. 2006, 2007; Klepstad et al. 2004; Nackley and Diatchenko 2010; Nackley et al. 2007; Oertel et al. 2006; Rakvåg et al. 2005, 2008; Coller et al. 2006).

Preliminary statistical analyses revealed several significant associations between genetic and non-genetic factors and postoperative opioid adverse effects and inadequate analgesia. In particular, TT genotype of the ABCB1 SNP rs1045642 (C3435T) was found (after adjusting for obstructive sleep apnea) to be associated with a higher risk of morphine induced respiratory depression than CC genotype. Specifically, in ABCB1 TT genotype, resting minute ventilation (MV) after morphine decreased by 47.5% compared to only 18.4% in CC and CT genotypes ($p < 0.05$). A number of other, relatively weak associations were found with SNPs in FAAH, COMT and other genes, as well as indications of epistatic interactions between ABCB1 and FAAH and ABCB1 and ADRB2. We have initiated a systematic multivariate analysis of associations between gene-gene interactions and other confounding factors (such as race, age, body mass index, etc.) and the respective outcomes, using standard CART and C4.5 decision trees (Therneu and Atkinson 2009; Quinlan 1996), as well as other machine learning approaches.

19.4.1 Decision Trees

Decision trees are a standard machine learning technique for multivariate data analysis and classification (Hastie et al. 2009; Witten and Frank 2005; Hothorn 2010). Decision trees can be viewed as a recursive partitioning approach, in which data is hierarchically divided into strata by simple logical rules. The advantage of decision trees is their simplicity, ability to handle both categorical and numerical variables as well as missing values, robustness to outliers and scaling, and the ability to combine feature selection with stratification and classification. Decision trees can also be used to derive easy to interpret and intuitive rules for decision support systems. Here, we use decision trees to select and combine the most predictive SNPs with demographic, clinical and other input features into simple logical rules that enable robust and accurate point-of-care prediction of inadequate pain relief and opioid-related adverse effects.

Different strata with distinct patterns of such interactions (with ABCB1 playing prominent role in some strata) were identified. Our preliminary data, which will guide the design of larger studies in the future as is typical in translational research, suggest that African American children had inadequate pain control and Caucasian children had a higher incidence of adverse effects from similar doses of morphine (Sadhasivam et al. 2012). Concordant differences in allelic frequency of ABCB1 (and other genes) were observed. For example, TT genotype of ABCB1 SNP rs1045642 that predisposes children to opioid induced respiratory depression (>4 fold higher incidence than CC genotype) was found with 27% frequency in Caucasian children, as compared to 2–3% in African-American children.

Here, we briefly describe the results obtained using standard CART and C4.5 decision trees, as implemented in R and Weka, respectively (Therneu and Atkinson 2009; Quinlan 1996; Frank et al. 2004). These methods were first used to identify and analyze potential patterns of gene-gene interactions and other factors predictive of pain sensitivity and inadequate pain relieve. The increased pain sensitivity phenotype is defined here by the need to administer a post-operative analgesic (PA). By contrast, the low pain sensitivity class is defined by no need for post-operative intervention in the form of additional analgesic. This analysis is summarized as an "efficacy" tree shown in Fig. 19.6, panel A, for the post-operative analgesic use vs. no intervention (noi) classification problem, using SNP data and covariates such as race and obstructive sleep apnea (OSA). Leaves (nodes) of the tree that represent strata with increased risk of inadequate pain relieve are highlighted in red, whereas

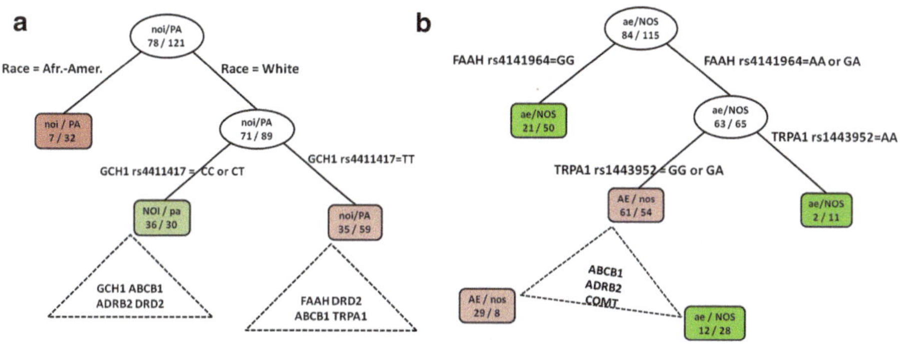

Fig. 19.6 Decision tree based prediction models for inadequate pain relief (efficacy tree, *panel A*) and adverse effects (safety tree, *panel B*); simple decision rules to be applied in clinical settings are discussed in the text. (**a**) Efficacy Tree: no intervenation (*noi*) vs. the need for an additional post-operative analgesic (*pa*). (**b**) Safety Tree: adverse effects (*ae*) vs. no adverse effects, referred to as no symptoms (*nos*)

those with relatively lower risk in green. The number of patients in each of the two classes is shown in each node of the tree to illustrate the results of recursive partitioning of the overall data set into subsequent subsets (strata).

Consistent with known associations between African-Americans, OSA and higher pain sensitivity, race is found to be the most discriminating feature, providing the first split of the overall cohort of 199 patients into two branches. Among Caucasian children, further strata are defined by specific polymorphisms in GCH1 (which was implicated as potential modifier of pain sensitivity and persistence) and interactions involving ABCD1/MC1R, ADRB2, and DRD2, or FAAH, DRD2, ABCD1/MC1R and TRPA1, in the two main sub-branches shown in Fig. 19.6 as dotted triangles. The top part of the "efficacy" decision tree included in Fig. 19.6 panel A can be interpreted in terms of logical rules as follows:

IF (Race=African-American)

THEN ***high risk*** *of increased pain sensitivity and inadequate pain relieve*;

IF ((Race=White) AND (GCH1 rs441417=TT))

THEN ***moderate risk*** *of increased pain sensitivity and inadequate pain relieve*;

IF ((Race=White) AND (GCH1 rs441417=CC OR CT))

THEN ***low risk*** *of increased pain sensitivity and inadequate pain relieve.*

The structure of the tree and the resulting rules that could be used for point of care decision support systems capture a relatively strong correlation between race and increased pain sensitivity (82% of African-American children included in the cohort required additional post-operative analgesic). At the same time, Caucasians could be further stratified by GCH1 and other genes, indicating epistatic effects.

The C allele of GCH1 is associated with somewhat decreased pain sensitivity: about 63% of Caucasian children with homozygous TT genotype required additional post-operative analgesic.

Additional analysis of factors predictive of adverse effects are summarized in the form of a "safety" tree (Fig. 19.6 panel B) for the classification of adverse effects, including respiratory depression, nausea and vomiting, and over-sedation (referred jointly to as AE) vs. the other (no adverse effects) class. Leaves of the tree that correspond to strata with increased risk of AE are highlighted in red. Only two main leaves that can be classified more easily by combinations of ABCB1, ADRB2, COMT and FAAH polymorphisms are shown explicitly within the middle branch. Consistent with previous univariate analysis, GG genotype at FAAH rs4141964 SNP was found to be protective, whereas the other two genotypes in interaction with TRPA1 rs1443952 genotypes GG or GA (and specific further interactions with ABCB1, ADRB2, COMT and FAAH) carry increased risk of adverse effects.

Obstructive sleep apnea is over-represented among African-American patients, who in turn are more likely to be at a higher risk of inadequate pain relief, as also indicated by "efficacy" tree. OSA and race represent important covariates to be further studied using larger cohorts. This is further highlighted by the finding that ADRB2 and FAAH gene polymorphisms can predict race with approximately 80% accuracy. However, OSA itself seems to have a relatively strong, although poorly understood, genetic component. We found that OSA can be predicted using SNPs in just two genes (ADRB2 and ABCB1) with greater than 70% accuracy, compared to 50% for a baseline classifier. It should be emphasized that further analysis of classification trees, stability of feature selection and the observed strata, as well as the accuracy of the resulting decision rules must be carefully assessed by study of larger numbers of patients from diverse populations. Preliminary data, such as presented here, guide the design of larger, more definitive, and more costly studies, as is typical in translational research. As with other machine learning methods, a common strategy to assess the overall accuracy and stability of decision trees, and to control the risk of overfitting is to use a comprehensive cross-validation approach, in which data is randomly split into training and validation subsets (Hastie et al. 2009; Witten and Frank 2005; Hothorn 2010). This approach works well with sufficiently large sample sizes so that different strata are sufficiently well represented in both training and tests sets. Large data sets are required to validate decision rules involving multiple genes and epistatic effects on distinct ethnic backgrounds. It is critical that delivery of personalized and predictive care be based on robust and accurate decision rules.

19.5 Molecular Modeling to Gain Mechanistic Insights

Mechanistic insights into relevant pathways that are affected by genetic variations with consequent phenotypic variation can be provided by molecular modeling of SNP effects on proteins and by network analyses. Such extensions require broad interdisciplinary expertise and an array of modeling and bioinformatics tools (see Chaps. 15 – Network Analysis in Translational Research and 16 – Bioinformatics and Orphan Diseases). However, many such tools can be integrated with informatics solutions for the development and validation of SNP-based predictors and clinical decision support systems. Therefore, we briefly illustrate how molecular modeling can be used to guide sequencing efforts and potentially lead to identification of the most informative variants.

While mechanistic aspects of receptor signaling and metabolism are undoubtedly equally if not more important, we focus here on opioid transport through the ATP-binding cassette B1 (ABCB1) protein-mediated opioid transport. Membrane associated proteins of the ABC family transport various molecules across extra- and intra-cellular membranes. ABCB1 is a member of the multi-drug resistance (MDR) subfamily and is also known as MDR1 (or P-glycoprotein 1, P-gp). This family of proteins has been extensively studied using biochemical, functional and structural approaches, providing a wealth of data for modeling the effects of polymorphisms in ABCB1 (Aller et al. 2009; Loo and Clarke 2005; Pajeva et al. 2006; Klepsch et al. 2011; Yin et al. 2006). ABCB1 encoded protein is an ATP-dependent drug efflux pump for xenobiotic compounds with broad substrate specificity. It is responsible for decreased drug accumulation in multidrug-resistant cells and also functions as an efflux transporter in the blood-brain barrier. In particular, ABCB1 is known to specifically transport morphine through the blood-brain barrier, effectively regulating its concentration in cerebral fluid (Aller et al. 2009; Loo and Clarke 2005; Pajeva et al. 2006; Klepsch et al. 2011; Yin et al. 2006). Inhibition of ABCB1 and gene knock-outs in mice lead to elevated brain levels of morphine. In our preliminary studies, ABCB1 was strongly implicated in mediation of respiratory response to inspired 5% carbon dioxide (CO_2) and was associated with potentially life threatening respiratory depression in children. However, some of the SNPs used in our study appear to be markers of polymorphisms in genes linked to the SNP containing gene. Phenotypes associated with the SNP might then be secondary to functional variation in the linked gene. For example, C3435T is synonymous variant that does not affect the amino acid sequence of ABCB1, but it may mark a variant in a tightly linked gene that does affect opioid binding, transport, or metabolism.

Structural analysis and modeling of the human ABCB1 protein has been greatly facilitated by the availability of resolved crystal structures of other ABC family members, including the mouse abcb1 protein that is highly homologous to the human protein (sequence identity of 87%, difference in length of only four residues). Availability of such highly homologous resolved structural templates (e.g., PDB 3G61 structure of the "inward" mouse transporter conformation – Fig. 19.7), allows one to build reliable, high quality models the human counterpart that can be used for molecular simulation, including molecular dynamics and normal mode analysis, as well as docking simulations to identify putative morphine binding sites. Incidentally, specialized molecular modeling and docking packages (Rosenfeld et al. 2003; Zavodszky et al. 2002; Seco et al. 2009; Lindahl et al. 2006) are available on the same computational cluster and many aspects of potential targets, such as ABCB1, can be analyzed within the same highly integrated environment.

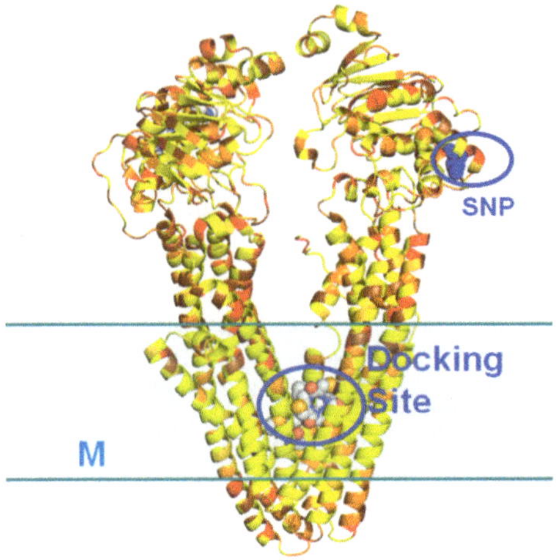

Fig. 19.7 Inward conformation of ABCB1 transporter: polymorphic position SNP C3435T and a docking site within the cavity shown in *blue*; approximate membrane (*M*) orientation indicated by a *rectangular shade*; hydrophobic residues shown in *yellow*, whereas hydrophilic in *red*, *brown* and *orange*

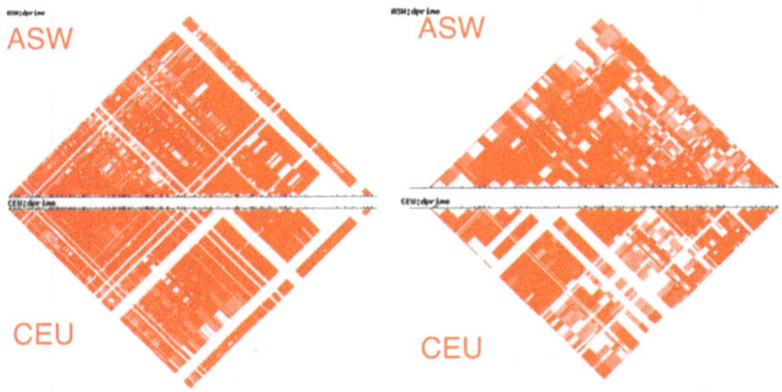

Fig. 19.8 Patterns of linkage disequilibrium within ABCB1 (*left panel*) and FAAH (*right panel*) genes using HapMap3 Caucasian (*CEU*) and African-American (*ASW*) cohorts and D prime measure of LD: shades of *red* indicate the strength of correlation (http://www.hapmap.org) (Note high level of LD, especially across ABCB1; note also differences between CEU and ASW cohorts)

Molecular simulations can provide mechanistic insights and help map truly functional sites, as opposed to markers that may simply be correlated with structural and functional hot spots. However, such analysis must take into account the differences in linkage disequilibrium (LD) patterns in Caucasian and African-American populations, which are illustrated in Fig. 19.8 for two genes considered here. Linkage disequilibrium is extensively discussed in Sect. 19.3. Careful analysis of LD and structural data can be very helpful not only in interpreting the results of association studies, but also in facilitating the selection of SNPs for further sequencing and the development of robust decision support systems. Moreover, imputing SNP alleles at potentially functional sites based on high LD (in analogy to HLA prediction from SNPs around genes of interest) could be used to map relevant alleles even without further sequencing.

Combining statistical and machine learning analysis of associations with systematic analysis of patterns of LD and mechanistic insights, one may be able to optimize the extent of sequencing to identify potential novel polymorphisms. Mapping functional hot spots, such as opioid docking sites, conformational switches or protein-protein interaction interfaces that can regulate the function can be also used to guide the search for the most informative SNPs (or other markers). The influence of these SNPs on opioid pharmacokinetics can be examined, eventually providing a better understanding of the factors which influence morphine sensitivity and adverse reactions to opioids.

19.6 Summing Up

Biomedical research in the post-genomic era is characterized by an exponentially growing body of sequence and other data that need to be mined and analyzed in to move from data to knowledge to clinically meaningful applications. In particular, translational and clinical studies increasingly rely on the ability to combine sequence and other molecular data with clinical and demographic information via clinical information systems. This chapter provides several case studies to illustrate this growing trend in the context of SNP data. In the future, data from sequencing of individual genomes will be applied in similar ways, but will be even more challenging because of the huge data sets and extensive normal genetic variation among individuals. A critical advantage of *in silico* genomic and biomedical studies is that computational methods can take advantage of current computing power to enable complex analyses, which shed light on the intricate networks of molecular interactions, hubs and pathways that underlie observed phenotypes. Understanding these networks can lead to better therapeutic interventions as discussed in Chaps. 16 and 17.

Interpreting the vast (and rapidly growing) amounts of data leads to problems of size and complexity that cannot be solved in routine ways; new algorithms and tools are required. Additionally, the simultaneous consideration of data from many different sources of input, such as expression profiles and polymorphisms, clinical and demographic data etc. poses significant challenges for these methods. It also requires that a new generation of highly integrated and versatile informatics systems be developed that combine databases with analytical and visualization tools.

Specific applications were discussed in this chapter to illustrate how such integrated systems can take advantage of both sequencing and clinical data in the context of translational and fundamental research. The first application dealt with the prediction of HLA alleles form SNP data. Many specific variants (alleles) of classical HLA genes have been implicated as risk or protective factors in pediatric autoimmune disorders, including juvenile rheumatoid arthritis, type 1 diabetes and celiac disease. Matching classical HLA genes is also commonly used to match transplant donors and recipients. Integrated informatics solutions for storing, processing and analyzing SNP and other sequencing data from multiple clinical studies can greatly facilitate efforts to extend and improve prediction of HLA alleles.

Such databases and tools provide consistent quality control criteria and protocols, as well as access to data potentially pertaining to cohorts of different ancestries that can be used to assess, refine and re-train current methods.

Postoperative care and pain management in children provided a second example of application of SNP data. The use of genetic polymorphisms in conjunctions with clinical data can guide therapeutic decisions that personalize and improve care. However, informatics and other challenges must be addressed by researchers and clinicians aiming to optimize such personalized interventions. Multi-center studies are generally necessary in pediatrics to accrue sufficient numbers of patients to assure adequate statistical power to reach significant, reliable conclusions. Such studies pose complex challenges in data capture, warehousing, sharing, and analysis. Data systems of different institutions participating in a multi-center study, especially their electronic health records, frequently lack semantic and syntactic interoperability, which adds to the difficulties. These issues are discussed in greater detail in Chaps. 6 and 8. We believe that approaches similar to those outlined in this chapter can be applied in other translational studies that aim to develop decision rules and to implement evidence based decision support systems for personalized therapeutic interventions.

References

Aller SG, et al. Structure of P-glycoprotein reveals a molecular basis for poly-specific drug binding. Science. 2009;323(5922):1718–22.

Aureli A, et al. Identification of a novel HLA-B allele, HLA-B*3580, with possible implication in transplantation and CTL response. Tissue Antigens. 2008;71(1):90–1.

Caldas JC, Pais-Ribeiro JL, Carneiro SR. General anesthesia, surgery and hospitalization in children and their effects upon cognitive, academic, emotional and sociobehavioral development – a review. Paediatr Anaesth. 2004;14(11):910–15.

Cepeda MS, et al. Side effects of opioids during short-term administration: effect of age, gender, and race. Clin Pharmacol Ther. 2003;74(2):102–12.

Chan IS, Ginsburg GS. Personalized medicine: progress and promise. Annu Rev Genomics Hum Genet. 2011;12:217–44.

Chou CK, et al. Human insulin receptors mutated at the ATP-binding site lack protein tyrosine kinase activity and fail to mediate postreceptor effects of insulin. J Biol Chem. 1987;262(4):1842–7.

Chou W-Y, et al. Human opioid receptor A118G polymorphism affects intravenous patient-controlled analgesia morphine consumption after total abdominal hysterectomy. Anesthesiology. 2006a;105(2):334–7.

Chou W-Y, et al. Association of mu-opioid receptor gene polymorphism (A118G) with variations in morphine consumption for analgesia after total knee arthroplasty. Acta Anaesthesiol Scand. 2006b;50(7):787–92.

Coller JK, et al. ABCB1 genetic variability and methadone dosage requirements in opioid-dependent individuals. Clin Pharmacol Ther. 2006;80(6):682–90.

de Bakker PI, et al. A high-resolution HLA and SNP haplotype map for disease association studies in the extended human MHC. Nat Genet. 2006;38(10):1166–72.

Diatchenko L, et al. Genetic basis for individual variations in pain perception and the development of a chronic pain condition. Hum Mol Genet. 2005;14(1):135–43.

Diatchenko L, et al. Catechol-O-methyltransferase gene polymorphisms are associated with multiple pain-evoking stimuli. Pain. 2006;125(3):216–24.
Diatchenko L, et al. Genetic architecture of human pain perception. Trends Genet. 2007;23(12):605–13.
Dilthey AT, et al. HLA*IMP – an integrated framework for imputing classical HLA alleles from SNP genotypes. Bioinformatics. 2011;27(7):968–72.
Duedahl TH, Hansen EH. A qualitative systematic review of morphine treatment in children with postoperative pain. Paediatr Anaesth. 2007;17(8):756–74.
Esclamado RM, et al. Perioperative complications and risk factors in the surgical treatment of obstructive sleep apnea syndrome. Laryngoscope. 1989;99(11):1125–9.
Frank E, et al. Data mining in bioinformatics using Weka. Bioinformatics. 2004;20(15):2479–81.
Hastie T, et al. The elements of statistical learning. 2nd ed. Dordrecht: Springer; 2009.
Holmes MV, et al. Fulfilling the promise of personalized medicine? Systematic review and field synopsis of pharmacogenetic studies. PLoS One. 2009;4(12):e7960. doi:10.1371/journal.pone.0007960.
Hothorn T. Unbiased recursive partioning: a conditional inference framework. J Comput Graph Stat. 2010;15(3):651–74.
Howie BN, Donnelly P, Marchini J. A flexible and accurate genotype imputation method for the next generation of genome-wide association studies. PLoS Genet. 2009;5(6):e1000529.
Klepsch F, Chiba P, Ecker GF. Exhaustive sampling of docking poses reveals binding hypotheses for propafenone type inhibitors of P-glycoprotein. PLoS Comput Biol. 2011;7(5):e1002036.
Klepstad P, et al. The 118 A > G polymorphism in the human mu-opioid receptor gene may increase morphine requirements in patients with pain caused by malignant disease. Acta Anaesthesiol Scand. 2004;48(10):1232–9.
LaFramboise T. Single nucleotide polymorphism arrays: a decade of biological, computational and technological advances. Nucl Acids Res. 2009;37(13):4181–93.
Lechler R, Warrens A. HLA in health and disease. San Diego: Academic; 2000.
Leslie S, Donnelly P, McVean G. A statistical method for predicting classical HLA alleles from SNP data. Am J Hum Genet. 2008;82(1):48–56.
Li N, Stephens M. Modeling linkage disequilibrium and identifying recombination hotspots using single-nucleotide polymorphism data. Genetics. 2003;165(4):2213–33.
Li M, Boehnke M, Abecasis GR. Joint modeling of linkage and association: identifying SNPs responsible for a linkage signal. Am J Hum Genet. 2005;76(6):934–49.
Lindahl E, et al. NOMAD-Ref: visualization, deformation and refinement of macromolecular structures based on all-atom normal mode analysis. Nucleic Acids Res. 2006;34(Web Server issue):W52–6.
Loo TW, Clarke DM. Recent progress in understanding the mechanism of P-glycoprotein-mediated drug efflux. J Membr Biol. 2005;206(3):173–85.
Manichaikul A, et al. Robust relationship inference in genome-wide association studies. Bioinformatics. 2010;26(22):2867–73.
Marchini J, Howie B. Genotype imputation for genome-wide association studies. Nat Rev Genet. 2010;11:499–511.
Marchini J, et al. A comparison of phasing algorithms for trios and unrelated individuals. Am J Hum Genet. 2006;78(3):437–50.
Marchini J, et al. A new multipoint method for genome-wide association studies by imputation of genotypes. Nat Genet. 2007;39(7):906–13.
Marsh SGE, et al. Nomenclature for factors of the HLA system, 2010. Tissue Antigens. 2010a;75(4):291–455.
Marsh SGE, et al. An update to HLA nomenclature, 2010. Bone Marrow Transplant. 2010b;45(5):846–8.
Nackley AG, Diatchenko L. Assessing potential functionality of catechol-O-methyltransferase (COMT) polymorphisms associated with pain sensitivity and temporomandibular joint disorders. Methods Mol Biol. 2010;617:375–93.

Nackley AG, et al. Catechol-O-methyltransferase inhibition increases pain sensitivity through activation of both beta2- and beta3-adrenergic receptors. Pain. 2007;128(3):199–208.

Oertel BG, et al. The mu-opioid receptor gene polymorphism 118A > G depletes alfentanil-induced analgesia and protects against respiratory depression in homozygous carriers. Pharmacogenet Genomics. 2006;16(9):625–36.

Orozco G, et al. Auto-antibodies, HLA and PTPN22: susceptibility markers for rheumatoid arthritis. Rheumatology (Oxford). 2008;47(2):138–41.

Pajeva IK, Globisch C, Wiese M. Comparison of the inward- and outward-open homology models and ligand binding of human P-glycoprotein. FEBS J. 2006;276(23):7016–26.

Price AL, et al. Principal components analysis corrects for stratification in genome-wide association studies. Nat Genet. 2006;38(8):904–9.

Purcell S, et al. PLINK: a tool set for whole-genome association and population-based linkage analyses. Am J Hum Genet. 2007;81(3):559–75.

Quinlan JR. Improved use of continuous attributes in c4.5. J Artif Intell Res. 1996;4:77–90.

Rakvåg TT, et al. The Val158Met polymorphism of the human catechol-O-methyltransferase (COMT) gene may influence morphine requirements in cancer pain patients. Pain. 2005; 116(1–2):73–8.

Rakvåg TT, et al. Genetic variation in the catechol-O-methyltransferase (COMT) gene and morphine requirements in cancer patients with pain. Mol Pain. 2008;4:64.

Robinson J, et al. IMGT/HLA and IMGT/MHC: sequence databases for the study of the major histocompatibility complex. Nucleic Acids Res. 2003;31(1):311–14.

Robinson J, et al. The IMGT/HLA database. Nucleic Acids Res. 2009;37(Database issue): D1013–17.

Rosenfeld RJ, et al. Automated docking of ligands to an artificial active site: augmenting crystallographic analysis with computer modeling. J Comput Aided Mol Des. 2003;17(8): 525–36.

Sadhasivam S, et al. Race and unequal burden of perioperative pain and opioid related adverse effects in children. Pediatrics. 2012;129(5):832–8.

Sampaio-Barros PD, et al. Frequency of HLA-B27 and its alleles in patients with Reiter syndrome: comparison with the frequency in other spondyloarthropathies and a healthy control population. Rheumatol Int. 2008;28(5):483–6.

Schaaf CP, Wiszniewska J, Beaudet AL. Copy number and SNP arrays in clinical diagnostics. Annu Rev Genomics Hum Genet. 2011;12:25–51.

Scheet P, Stephens M. A fast and flexible statistical model for large-scale population genotype data: applications to inferring missing genotypes and haplotypic phase. Am J Hum Genet. 2006;78(4):629–44.

Seco J, Luque FJ, Barril X. Binding site detection and druggability index from first principles. J Med Chem. 2009;52(8):2363–71.

Stephens M, Donnelly P. A comparison of Bayesian methods for haplotype reconstruction from population genotype data. Am J Hum Genet. 2003;73(5):1162–9.

Stephens M, Smith NJ, Donnelly P. A new statistical method for haplotype reconstruction from population data. Am J Hum Genet. 2001;68(4):978–89.

The International HapMap Consortium. The international HapMap project. Nature. 2003;426: 789–96.

Therneu TM, Atkinson B. rpart: recursive partitioning. 2009. http://cran.r-project.org/web/packages/rpart

Wellcome Trust Case Control Consortium. Genome-wide association study of 14,000 cases of seven common diseases and 3,000 shared controls. Nature. 2007;447(7145):661–78.

Witten IH, Frank E. Data mining: practical machine learning tools and techniques. 2nd ed. San Francisco: Morgan Kaufmann; 2005.

Yin Y, et al. Structure of the multidrug transporter EmrD from *Escherichia coli*. Science. 2006;312(5774):741–4.

Zavodszky MI, et al. Distilling the essential features of a protein surface for improving protein-ligand docking, scoring, and virtual screening. J Comput Aided Mol Des. 2002;16(12): 883–902.

Chapter 20
Genetic Variation and Gene Discovery

John J. Hutton, Phillip Dexheimer, and Gregory A. Grabowski

Abstract There has been remarkable progress in understanding genetic variation and its impact on human health. This chapter describes the variety of genomic technologies now available to advance our understanding of heritable diseases that afflict children. Sequencing of the whole genome or exome (targeted exome capture) by Next-Gen sequencing technology is becoming the standard in translational research to identify novel genetic variants that underlie risk of developing a disease. The process of analysis of raw data from Next-Gen sequencing is complex, computationally intensive, and uses many different software applications. Integrated sets of tools (software applications) are available to compare the DNA sequences of genes in an affected patient with the reference sequence of a normal control and with sequences of other patients with the same disease. Sites in the DNA are identified where there is statistical evidence for a rare or novel variant in the patient, including single or multi-nucleotide substitutions, insertions and deletions (indels), sequence rearrangements, and copy number variations. There can be considerable phenotypic variation among patients with defects in the same gene and conversely patients with similar phenotypes may or may not share variants in the same gene. There is also variation in types of changes in DNA sequences that can cause loss

J.J. Hutton, M.D. FACP (✉)
Department of Pediatrics, University of Cincinnati College of Medicine, Cincinnati, OH, USA

Division of Biomedical Informatics, Cincinnati Children's Hospital Medical Center,
3333 Burnet Avenue, MLC-7024, Cincinnati, OH 45229-3039, USA
e-mail: John.hutton@cchmc.org

P. Dexheimer, M.S.
Divisions of Biomedical Informatics and Human Genetics, Cincinnati Children's Hospital Medical Center, 3333 Burnet Avenue, MLC-7024, Cincinnati, OH 45229-3039, USA

G.A. Grabowski, M.D.
Department of Pediatrics, University of Cincinnati College of Medicine, Cincinnati, OH, USA

Division of Human Genetics, Cincinnati Children's Hospital Medical Center,
3333 Burnet Avenue, MLC-4006, Cincinnati, OH 45229-3039, USA

J.J. Hutton (ed.), *Pediatric Biomedical Informatics: Computer Applications in Pediatric Research*, Translational Bioinformatics 2, DOI 10.1007/978-94-007-5149-1_20,

of function. Stratification of sequencing data at both the phenotypic and genotypic level is critical for effective discovery of sequence variants that cause disease. This chapter includes several illustrations of use of Next-Gen sequencing technology in genetics research.

20.1 Introduction

There has been remarkable progress in understanding genetic variation and its impact on human health. An up-to-date, authoritative online catalog of human genes and genetic disorders is available (OMIM, http://omim.org/). The major focus of genetics research initially was on single gene disorders that are inherited in Mendelian fashion. Most of the single gene disorders are rare (e.g. cystic fibrosis, muscular dystrophy) and it is now clear that most common diseases (e.g. diabetes, obesity, hypertension) are caused by variation in multiple genes, i.e. they are polygenic. This chapter describes the variety of genomic technologies now available to advance our understanding of heritable diseases that afflict children. These tools can be applied to study of both mono- and polygenic disorders.

Monogenic disorders are generally studied in families. Whole–genome sequencing of a small number of members of a family can permit identification of disease causing variants with high degrees of certainty (Roach et al. 2010). In contrast, study of polygenic disorders requires the identification of large cohorts of patients with well-defined phenotypes and characterization of most or all of their genomes. Genetic variation, often measured as SNPs that form haplotypes across the entire genome, is compared between people with a particular condition and unaffected individuals. Detection and use of SNPs to study inheritance are discussed in Chap. 19 – From SNP Genotyping to Improved Pediatric Healthcare. These studies are called Genome-Wide Association Studies (GWAS) and are used to identify common genetic factors that influence health and disease (Manolio 2010), also known as the common disease, common variant hypothesis. If one type of a SNP variant is more frequent in people with a disease, the SNP is said to be "associated" with the disease. The SNPs themselves are not likely to be the genetic change that is causal to the disease; rather they mark the vicinity of the human genome that influences the risk of disease. These regions can span a large segment of the genome that includes many genes and structural variants such as deletions, insertions, inversions, translocations, and copy number variants. To find the causative genetic variant, regions must be examined in great detail using other technologies such as DNA sequencing. A catalog of published genome-wide association studies is maintained by the US National Human Genome Research Institute (http://www.genome.gov/gwastudies/).

In the past, genomic studies of populations have been extremely expensive and logistically complex because they required the recruitment of appropriate patients one by one. High-throughput genotyping was also very expensive or not available. This process has now become much simpler and less expensive because of the widespread adoption of electronic health records (EHRs), the availability of relatively low cost genotyping, and the development of biobanks (Kohane 2011;

Murphy et al. 2009). The EHR includes information such as age, sex, race, ethnicity, height, weight, temperature, blood pressure, medications, diagnoses, laboratory values, procedures, and clinical notes (notes usually include family and social history). To support research and facilitate integration of electronic health record data with other data, information from the EHR is frequently extracted, transformed and loaded into a research patient data warehouse such as i2b2, as described in Chaps. 6 – Research Patient Data Warehousing and 8 – EHR-linked Registries for Studies of Populations. Information about biospecimens, specifically DNA, in the biobank is included in the research patient data warehouse (see Chap. 7 – Biobanking in Pediatric Research). Because translational research, including genetic studies, can require thousands of DNA samples to identify biologically or clinically relevant findings, many institutions systematically collect and store some or all residual clinical samples for future unspecified research (Murphy et al. 2009). Typically these are residual blood samples discarded by the hospital laboratory after all clinical tests have been completed. Use of these samples is easier and safer for patients than samples obtained exclusively for research, as well as being more time and cost effective. Obtaining patient consent for use of residual clinical samples in research is complex and is also discussed in Chap. 7. At the authors' institution approximately 20,000 patients per year consent to have their residual clinical samples stored in the biorepository and made available for research.

Patients can be phenotyped by analysis of data from their EHRs, including both discrete elements such as ICD-9 codes for diagnoses and variables from the text of clinical notes (see Sect. 10.3 – Natural Language Processing: Applications in Pediatric Research – Phenotype Identification). DNA samples from patients with specific phenotypes can then be obtained from the biobank and genotyped. The process of phenotyping, genotyping, storing, integrating, and analyzing the huge amounts of data that is generated in genomics research is very demanding of resources and expertise in biomedical informatics.

20.2 Core Services

A Genetic Variation and Gene Discovery Core requires expensive equipment, extensive hardware and software resources for management of large quantities of data (gigabytes, terabytes, and sometimes petabytes), high-performance computing, high levels of quality control, an experienced staff, and supervision by knowledgeable professionals. It is best managed as a coherent centralized unit that supports clinical and research programs of the entire medical center. Examples of services that may be required to support pediatric research include:

- DNA extraction from blood, saliva, or other tissue
- Sequencing: DNA-Seq, exome sequencing, ChIP-Seq, RNA-Seq
- Microsatellite genotyping
- Analysis of copy number variation
- Detection of insertions and deletions

- High throughput custom SNP genotyping
- Whole genome genotyping
- Mutation discovery
- Identification of transcription factor binding sites
- Profiles of microRNA expression
- Sequencing cDNA libraries to obtain gene expression profiles of every RNA isoform

The technical team should have training in:

- Data quality control
- Sequencing/genotyping project management
- Primer design and PCR optimization
- Genomic database searches
- Genome assembly and annotation
- Bioinformatics for handling and analyzing large datasets
- Regulatory compliance

It is important to note that genomics data generated in a clinical setting may support both research and clinical care. In the United States, the Centers for Medicare & Medicaid Services (CMS) regulates all laboratory testing (except research) performed on humans through the Clinical Laboratory Improvement Amendments (CLIA). Laboratory data cannot be entered into a patient's medical record or provide information for the diagnosis, prevention, or treatment of disease unless the laboratory that generates the data is CLIA certified. Because of the mixing of research and clinical care in academic health centers, core laboratories usually seek CLIA certification so they can provide results for both research and molecular diagnostic purposes. The Core Services Laboratory generates data, not data interpretation. Individual researchers incorporate the results of the analyses into their overall research projects and may or may not consult with faculty colleagues with expertise in genetics and genomics. If the results of research studies are relevant to the health of the patient, how the information is handled must comply with the wishes of the patient, as documented in the informed consent the patient signed when agreeing to participate in the study. Legal and ethical issues relating to informed consent and use of biosamples from patients are discussed in Chap. 7 – Biobanking in Pediatric Research.

20.3 Variant Discovery

20.3.1 DNA Sequencing

Sequencing of the whole genome and selective sequencing of the exome (targeted exome capture) are becoming the standards in translational research to identify genetic variants that underlie risk of developing a heritable disease. Even for monogenic disorders, the chromosomal location of the causative gene is not generally known so that all or a large portion of the genome must be sequenced in efforts to locate the genetic variation underlying a disease. Ng et al. pointed out that the protein coding regions constitute approximately 1% of the human genome (~30 million bases across ~180,000 exons), but are the sites of approximately 85% of variants that cause disease (Ng et al. 2009, 2010a). They developed methods for targeted capture and massively parallel sequencing of exomes and showed that candidate genes for monogenic disorders can be identified by exome sequencing of a small number of individuals. The exome sequences are captured by hybridizing shotgun libraries of genomic DNA to oligonucleotide probes on microarrays or beads that represent mRNAs and miRNAs of ~30,000 genes and ~300,000 exons. Systems for exome capture are commercially available. While exome sequencing is efficient in terms of cost and size of data sets, it can only identify variants in coding regions of genes. Approximately 99% of the genome is not examined by exome sequencing and it does not support detection of structural variants or variants in introns and intergenic regions, where important regulatory elements might reside.

The chain termination method can only sequence 100–1,000 base pairs, so today's commonly used technology for whole genome sequencing requires that the long strands of DNA from chromosomes be broken into millions of short pieces (shotgun sequencing or massively parallel short-read sequencing technology). After sequencing the fragments, they must be reassembled piece-by-piece into a whole strand.

A first step in this so-called Next-Gen DNA sequencing is isolating genomic DNA, shearing it into appropriate sized fragments, processing to add sequencing adapters, and then amplifying fragments by PCR to create genomic libraries (shotgun library). Similar processes can be used to create libraries to sequence mRNAs and microRNAs. Genomic libraries can be enriched for specific tasks such as: resequencing all coding regions (exomes) in the human genome or resequencing a specific region using on-chip DNA probe baits or bead bound RNA probes. To prevent errors, all samples are barcoded and all of the complex, largely automated, steps in sequencing and analysis utilize the barcodes to track samples for unique identification.

The genomic library of amplified, short DNA fragments with adapters is loaded into the Next-Gen sequencing system, where the individual fragments are sequenced base by base. Output varies depending on the instrument, but generally ranges from several million reads for small "bench-top" sequencers up to about three billion reads for the largest system currently available. The fragment read length is generally 100–200 base pairs, so that a single run produces somewhere between 0.5 and 600 "gigabases" (billion bases). The sequencing system assigns quality scores to each base call using various predictors of sequencing errors. They are a measure of the uncertainty of base calls, i.e. the probability of a base call being wrong. This probability is usually based on the ability of the sequencer to distinguish among the four nucleotides at a given position, and is assigned in a very instrument-specific fashion. These scores reveal how much of the data from a given run is useful in a sequencing or assembly experiment.

Because of the random nature of fragmentation and sequencing, there is no guarantee that every region of the genome is adequately represented in the raw data. Coverage of the genome can be improved by sequencing larger numbers of fragments so the chance that one particular fragment is represented is increased. Ajay et al. (2011) performed a depth-of-coverage analysis to determine the relationship between the amount of sequence data generated and the proportion of the genome where genotypes could be called with confidence. Technology is rapidly advancing, but using standard methods available in 2010, these authors concluded that an average mapped depth of 50X with high quality bases was required to produce confident genotype calls for >94% of the genome and >80% of the coding exome (this is what they minimally defined as comprehensive). They also pointed out that this level of completeness is relevant to study of personal genomes, but is not the standard for population scale sequencing.

20.3.2 Calling Variants

Analysis of the raw reads to derive any meaningful information is a complex process. Algorithms are available (Table 20.1) to support the process of moving from raw sequences to meaningful information about the genome. Separating true variation from machine artifact is a critical challenge. The process of analysis is computationally intensive and uses many different software applications. To optimize efficiency and avoid errors, a structured framework for analysis of Next-Gen sequencing data must be in place. There are many ways to do this, but one well-described, widely used example is the Genome Analysis Toolkit (GATK) developed at the Broad Institute by DePristo (2011). GATK is a publically available, unified analytic framework to move from raw sequence reads to variant discovery and evaluation among multiple samples simultaneously. It includes a wide variety of software applications, with a strong emphasis on data quality assurance to support variant discovery and genotyping (Fig. 20.1).

Phase 1 in the GATK pipeline focuses on preprocessing the raw reads to remove common sources of noise before attempting to find variants. The first two steps, aligning to the genome and marking duplicate reads, take place outside the GATK framework. Sequence aligners (see Table 20.1 for examples) must accurately and quickly place each 100–200 bp read on the reference sequence (such as those available from the Genome Reference Consortium, see Table 20.2). Once the reads are aligned, three major sources of noise are removed. The first source of noise is biased sampling of the source library. This can occur during PCR amplification, if particular fragments are amplified with higher efficiency or a library is overamplified. To eliminate this noise, multiple reads that appear to originate from the same fragment are identified as "duplicates" and one representative is chosen. Also in this stage, the base qualities assigned by the sequencer are recalibrated

Table 20.1 Examples of tools useful in analyses of genomic sequences

Tool	URL	Description
Illumina applications	http://www.illumina.com/	Multiple applications to support sequencing, SNP genotyping, analysis of copy number variants
FindPeaks	http://bcgsc.ca/platform/bioinfo/software/findpeaks	Open source, ChIP-Seq analysis
Partek Genomics Suite	http://partek.com	Supports all microarray and Next-Gen sequencing technologies. ChIP-chip, ChIP-Seq, and microRNA in a single software package
GenoLogics	http://genologics.com	Laboratory information management system for Next-Gen sequencing; automate workflows, integrate instruments, manage samples and associated information
Applied Biosystems applications	http://www.appliedbiosystems.com	Multiple applications to support Next-Gen sequencing, genotyping and genetic variation. Sequence Scanner software is free
Sequencher	http://genecodes.com	DNA assembly; analysis of data sets from Next-Gen and capillary electrophoresis
IMPUTE and other GWA tools	http://www.stats.ox.ac.uk/~marchini/software/gwas/gwas.html	Free; package of genome-wide association tools
Broad Institute GenePattern	http://www.broadinstitute.org/cancer/software/genepattern	Free; genomic analysis platform; access to more than 200 tools for gene expression analysis, proteomics, SNP analysis, flow cytometry, RNA-Seq analysis
Genome Analysis Toolkit, GATK	http://www.broadinstitute.org/gatk/	Free; software package to analyze Next-Gen resequencing data with a primary focus on variant discovery and genotyping
Phred/Phrap/Consed	http://www.phrap.org/phredphrapconsed.html	Free; DNA assembly software
Plink/gPlink	http://pngu.mgh.harvard.edu/~purcell/plink/	Free; whole genome association analysis toolset
DNAStar Lasergene	http://dnastar.com	Comprehensive DNA and protein sequence analysis software suite; functions range from sequence assembly and SNP detection to automated virtual cloning and primer design

(continued)

Table 20.1 (continued)

Tool	URL	Description
Integrative Genomics Viewer	http://www.broadinstitute.org/igv/	Free; Desktop viewer for a variety of genomic data, especially suited for viewing alignments from Next-Gen sequencing
Bowtie	http://bowtie-bio.sourceforge.net	Ultrafast sequence aligner, aligns short DNA sequence reads to large genomes.
Burrows-Wheeler Aligner	http://bio-bwa.sourceforge.net	Aligns short sequences against a large sequence database, e.g. the human genome
TopHat	http://tophat.cbcb.umd.edu	Splice-site aware sequence aligner, for use in aligning RNA-seq reads
Samtools	http://samtools.sourceforge.net	Specifications and utilities for handling Next-Gen data, the SAM format is the *de facto* standard for storing and sharing reads and alignments
Picard	http://picard.sourceforge.net/index.shtml	Collection of utilities for advanced handling of SAM files, includes utility to mark duplicate reads

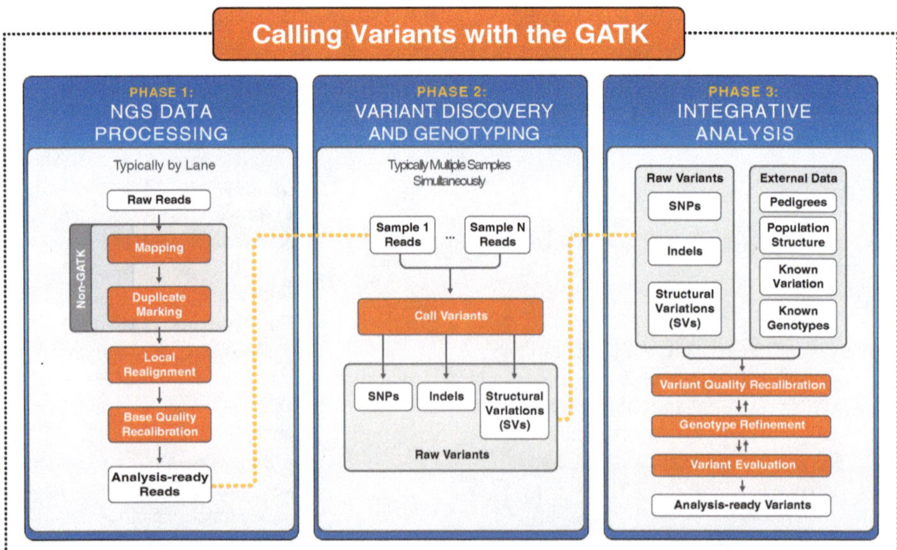

Fig. 20.1 Workflow to move from raw sequence reads to variant discovery and evaluation using the Genome Analysis Toolkit (DePristo 2011; Used with permission of the Broad Institute)

Table 20.2 Examples of databases useful in analyses of genomic variation and gene discovery

Database	URL	Description
OMIM	http://www.ncbi.nlm.nih.gov/omim/	Compendium of human genes and genetic phenotypes
GWAS Studies	http://www.genome.gov/gwastudies/	Catalog of published genome-wide association studies
Ensembl	http://www.ensembl.org/index.html	Annotated genome databases for humans and other eukaryotes
Entrez	www.ncbi.nlm.nih.gov/Entrez/	Search engine for multiple databases
1000 Genomes	http://www.1000genomes.org/home	Deep catalog of human genetic variation
dbSNP	http://www.ncbi.nlm.nih.gov/snp/	Database of SNPs, insertions/deletions, microsatellites, non-polymorphic variants
dbVar	www.ncbi.nlm.nih.gov/dbvar/	Database of large scale genomic variants such as insertions, deletions and inversions
HapMap	http://hapmap.ncbi.nlm.nih.gov/	Catalogs common SNPs and other genetic variants, how they are organized/linked on chromosomes, how they are distributed in populations
UCSC Genome Browser	http://genome.ucsc.edu/	Reference sequence and working draft assemblies for a large collection of genomes.
Genome Reference Consortium	http://www.ncbi.nlm.nih.gov/projects/genome/assembly/grc/index.shtml#	Genome reference assemblies, consensus representation of the genome
Cancer Genome Atlas	http://cancergenome.nih.gov/	Comprehensive sequencing, characterization, and analysis of the genomic changes in more than 20 types of cancer

based on the observed mismatch rate. Finally, reads that appear to span indels are re-aligned with a more sensitive aligner in order to reduce the false mappings used by the genotyper.

Phase 2 of the analysis focuses on variant discovery and genotyping. The sequence of the sample is compared with the reference sequence. Tools in GATK analyze the sequences and identify those sites in the sample DNA with statistical evidence for an alternative allele, including single nucleotide polymorphisms (SNPs) and insertions and deletions (indels). At this stage of analysis, the program does not have enough information to reliably distinguish true variants from sequencing or alignment artifacts. Therefore, the variant discovery process is very sensitive, even at the cost of low specificity. The putative variants are stored, together with various metrics calculated during discovery, and these become the raw variants.

Phase 3 The aim of Phase 3 is to separate true sequence variants in the sample DNA from machine artifacts. In order to accomplish this, known variations from external sources such as 1000 Genomes or dbSNP (see Table 20.2) are used to train a machine-learning algorithm. This algorithm uses the metrics collected during Phase 2 and the high-quality external data to learn what real variation "looks like". Each of the raw variants is scored by how well they match that ideal, and a threshold is chosen by adjusting the sensitivity to the high-quality set. The sensitivity chosen depends both on the external data and the goals of the experiment.

Ideally, all variants that appear to be relevant to the specific project will be verified by standard Sanger sequence analysis to generate a list of analysis-ready variants, one of which might, for example, be associated with a disease. Verification is critical and particularly important for the return of results of clinical utility. All of these analyses are ideally completed by single integrated computational environments with little need for human intervention, which is the aim of the Genome Analysis Toolkit.

20.3.3 Other Measures of Genome Variation

- **Single Nucleotide Polymorphism (SNP) and re-sequencing arrays** provide information about genetic variation. Data from arrays can be used for cohort identification and stratification, phenotype-genotype association studies, discovery of disease markers, prediction of molecular phenotypes, and clinical decision support, as discussed in Chap. 19, From SNP Genotyping to Improved Pediatric Healthcare.
- **RNA-Seq** technology is rapidly replacing gene expression microarrays. RNA is first reverse transcribed to cDNA, which is then sequenced using high throughput Next-Gen sequencing. The results provide a digital measure of gene expression levels. More abundant RNAs give rise to more copies of cDNA, which then give more DNA sequence reads. It also provides information on differently spliced transcripts, non-coding RNAs, post-transcriptional mutations or editing, and gene fusions. RNA-Seq is discussed in Sects. 18.7 and 18.8, Defining Genetic Blueprints – Kidney and Craniofacial Development, RNA-Seq.
- **Chromatin Immunoprecipitation (ChIP)** followed by genomic tiling microarray (ChIP-Chip) or massive parallel sequencing (ChIP-Seq) are two of the most widely used approaches for genome-wide identification of physical interactions between transcription factors and the regulatory DNA sequences to which they bind. Such studies provide direct evidence of regulatory relationships. The integration of ChIP-Seq and mRNA microarray results enables further dissection of the direct vs. indirect effects of selected transcription factors *in vivo*. Differentially expressed genes identified from mRNA microarray analysis with positive binding sites from ChIP-Seq analysis are considered to be direct transcriptional targets. ChIP-Seq is discussed in Sect. 17.4.7, Transcriptional Networks – Control of Lung Maturation, mRNA Microarray and ChIP-Seq Integration.

20.3.4 Human Genome Variation

Variation in genomic sequences among humans is very great and makes it difficult to identify variants that cause (monogenic) or are one of the causes (polygenic) of a disease. There are approximately three billion bases in the haploid human genome. A first challenge, when analyzing the sequence of a genome, is to be certain that there is sufficient coverage of the genome to call variants. As noted in Sect. 20.3.1, an average mapped depth of 50X with high quality bases is required to produce confident genotype calls for >94% of the genome. For clinical work where results could affect diagnosis, counseling, and treatment of patients, 75X coverage is becoming the standard. In its 2010 report on variants of functional consequence in the human, the 1000 Genomes Project Consortium reported that an individual typically differs from the reference human DNA sequence at 10,000–11,000 non-synonymous sites, 10,000–12,000 synonymous sites, 190–210 in-frame indels, 80–100 premature stop codons, 10–50 splice-site disrupting variants and 220–250 deletions that shift the reading frame (The 1000 Genomes Project Consortium 2010). The Consortium estimated that the genome of each individual carries approximately 250–300 loss-of-function variants in annotated genes and is heterozygous for 50–100 variants that are known to cause human disorders. The per generation base pair mutation rate was estimated to be $1.0–1.2 \times 10^{-8}$, based on study of two mother-father-child trios. Because of this extensive genomic variation in every human being, it is clear that merely sequencing the DNA of one human afflicted with a disease cannot identify the causative gene, even if the disease is monogenic.

20.4 Human Genome Variation and Human Disease

Gene mapping and sequencing technologies are being widely applied to studies of heritable human disease. Several examples will be discussed to illustrate general principles and strategies.

20.4.1 Genetic Mapping and Exome Sequencing

Puffenberger et al. combined genetic mapping of recessive conditions by SNP microarrays with Next-Gen exome sequencing to identify sequence variants associated with five novel diseases (Puffenberger et al. 2012). Their work was highly productive in identifying causative variants of recessive diseases because of the care they took to select appropriate individuals for study, including families and the population from which they were derived. Subjects were members of the genetically isolated Plain people (Old Order Amish and Mennonite). Recessive disorders occur with high frequency in this population and most affected individuals are homozygous for a pathogenic variant. For the past decade or so, Strauss et al. (Strauss et al. 2005) have carried out genome-wide homozygosity mapping studies on affected individuals with a variety of diseases using high-density SNP arrays. They have localized causative genes to homozygous haplotype blocks with a mean size of 4.4 Mb (range 1.6–8.4 Mb), which is too large to be sequenced by conventional technology, e.g. Sanger sequencing. The mean number of genes in a block was estimated to be 79. The basic strategy to discover causative genes was to sequence exomes within the haplotype block associated with each disorder to identify novel variants that were homozygous in affected individuals. While Next-Gen sequencing and the workflow for variant discovery shown in Fig. 20.1 were employed, it is important to note that the whole genome was not being sequenced, limited numbers of genes/exons were being studied, and the condition was monogenic recessive so both alleles at the causative locus would be novel, i.e. associated with a rare disease, not a polymorphism found in the normal population. In most cases, both parents and siblings were available for study. Novelty of DNA sequence variants found in patients was determined by their absence from lists of variants maintained by dbSNP and the 1000 Genomes Project (Table 20.2). When available, siblings, parents, and affected patients from other Old Order Amish/Mennonite demes were sequenced to confirm appropriate segregation of alleles and/or presence of homozygous variant alleles in other patients with the same disease phenotype. Nucleotide sequences of pathogenic variants were confirmed by Sanger sequencing.

The authors' study of Infantile Parkinsonism-dystonia syndrome provides a specific example of the types of data and analyses required to conclude that variation in a specific gene is associated with a disease (Puffenberger et al. 2012). Exons of the haplotype block from an affected patient were sequenced. Known variants were excluded and a list of all novel homozygous variants in the block was tabulated. Nineteen novel homozygous variants were identified in an affected patient: eight synonymous, nine missense, one 5′-UTR, and one splice site. The nine missense and one splice site variant were considered potentially pathogenic. Linkage analyses and direct sequencing of parents, an affected sib, and six unaffected sibs identified variation in the *SLC6A3* gene (*SLC6A3 IVS9 + 1 G > T*) as most likely to be causative. This gene encodes the dopamine transporter, which functionally is known to be associated with Parkinson's Disease. There are no statistical methods to assign a probability to each of the 10 or so potentially pathogenic variants in terms of its being causative, so the selection of one variant as causative is a best, educated guess.

Identification of causative genes in other disorders was made possible by the availability of family members and in some cases by patients with the same disease from other Old Amish/Mennonite demes, where because of genetic isolation, there is sharing of rare variants. This assumption of homogeneity of rare alleles that cause disease greatly simplifies interpretation of results of exome sequencing. As the authors of this study point out, identification of homozygosity for a rare allele does not prove the allele causes the disease. While the effect of a mutation on the gene product, i.e. protein, can be predicted, it is important to show that changes in function of the affected gene product can reasonably be expected to cause the observed phenotype.

20.4.2 Genetic Modifiers of the Course of Disease

Gaucher disease is an autosomal recessive lysosomal storage disorder with systemic accumulation of lysosomal glucocerebrosides that can be treated effectively with enzyme replacement therapy (Grabowski et al. 2010). There is an increased risk of cancer and recurrent cancer in Gaucher patients, but it is not known whether the accumulation of glucocerebrosides is causative. Lo et al. (2012) carried out a genome analysis of two sibs with Gaucher disease who, as children, developed new onset T-cell lymphoblastic lymphoma. The aim was to identify the genetic cause of cancer in these Gaucher patients.

Sequencing of the coding region of the Gaucher gene in the family revealed homozygosity for the D137N mutation in the affected sibs and heterozygosity of mutant and normal alleles in the parents. Genomic DNA was extracted from normal peripheral blood cells of the parents and children, whole exomes were captured, and sequenced using Next-Gen technology (mean coverage 82X) coupled with a variant discovery workflow similar to Fig. 20.1. Novel variants were identified by exclusion of polymorphic variants listed in the dbSNP and 1000 Genomes databases. Because the two sibs were identically affected by T-cell lymphoma, the investigators postulated that the disease causing variants would be homozygous by descent from the parents. Six rare mutations were found that were homozygous in the two sibs and heterozygous in the parents. Two of the six were felt to be relevant to the patients, based on known functions of the six genes. One was a missense variant in the Gaucher gene and the second was a frame-shift mutation with premature termination in the gene *MSH6.* Neither of the two mutations was found in 5,000 control chromosomes. Both mutations were confirmed by Sanger sequencing and were found to be heterozygous in the parents. The *MSH6* gene is a member of a family of genes that maintain integrity of the genome by correcting errors that arise during DNA replication. Defective DNA repair is associated with cancer causation. The authors conclude that genes other that the Gaucher gene underlie the increased susceptibility of some Gaucher patients to cancer and that these genetic modifiers can be identified by individual genome analysis.

20.4.3 Discovering a Gene that Causes an Autosomal Dominant Disorder

Discovering a gene that causes a monogenic autosomal dominant disorder by whole genome sequencing is conceptually more difficult that discovering a gene that causes an autosomal recessive disorder. Search for the cause of a recessive disorder is based on the fact that the affected individual must be homozygous for one rare, novel variant or heterozygous for two different, but rare, variants in the same gene. Dominant inheritance requires the affected individual to have only a single copy of the rare novel variant, i.e. typically the individual has one rare causative variant and one wild-type allele. Searching whole genome or exome sequences for homozygous rare variants (recessive disorder) generally limits the list of possible causative variants more quickly than listing all rare variants in an exome with no requirement that they be homozygous (dominant disorder).

Kabuki syndrome is a rare multiple malformation disorder with considerable variation in phenotype. Most cases are sporadic, but parent to child transmission has occasionally been observed, suggesting that it is an autosomal dominant disease. Because of the very small number of cases, lack of availability of families with multiple affected individuals, and phenotypic heterogeneity, conventional approaches to gene discovery had not been successful in identifying causative variants. Ng et al. successfully applied Next-Gen sequencing of exomes to identify a candidate gene and mutations that are a major cause of Kabuki syndrome (Ng et al. 2010b), but success required modifications of the standard approaches to gene discovery in heritable diseases. Exomes of ten unrelated probands were sequenced and examined for a candidate gene with at least one novel variant in each proband, using a workflow similar to that shown in Fig. 20.1. The search identified only one candidate gene, but it was considered to be a false positive so the search was unsuccessful.

The investigators postulated three possible reasons for their unsuccessful search, reasons that should be considered by other investigators seeking causes of heritable diseases: (1) the syndrome is phenotypically heterogeneous and probands might not share the same rare mutation; (2) not all mutations in the targeted exome were identified; (3) the causative mutation might not be in an exome or might be in a gene not fully represented in the exome library. They reanalyzed their data set with two changes: First, patients were assigned a categorical rank by degree of compliance with the canonical Kabuki phenotype, such as presence of specific facial characteristics and developmental delay. Novel variants in the same gene that were shared by two or more probands (rather than all 10) were then sought in subsets of patients beginning with the patients with the most typical Kabuki syndrome and adding probands incrementally to identify multiple candidate genes. Second, a loss of function filter was applied to categorize the impact of each novel variant, including, for example, nonsense versus nonsynonymous substitution, splice site disruption, frameshift versus in frame indel. The strategy was successful. Seven of the 10 probands had a novel nonsense or frameshift mutation in the *MLL2* gene, a histone methyltransferase. Each of these 7 was initially identified by Next-Gen sequencing of exomes and was then confirmed by Sanger sequencing. Sanger sequencing of the *MLL2* gene identified frameshift indels in two of the three probands in whom mutations had not been identified by Next-Gen sequencing. Ng et al. concluded that stratification of data at both the phenotypic and genotypic level will be critical for exome and whole genome sequencing to reach their full potential as tools for discovery of genes underlying heritable diseases (Ng et al. 2010b).

References

Ajay SS, et al. Accurate and comprehensive sequencing of personal genomes. Genome Res. 2011;21:1498–505.

DePristo MA. A framework for variation discovery and genotyping using next-generation DNA sequencing data. Nat Genet. 2011;43:491–8.

Grabowski GA, Petsko GA, Kolodny EH. Gaucher disease. In: Valle D, Beaudet A, Vogelstein B, Kinzler KW, Antonarakis SE, et al., editors. The on-line metabolic and molecular bases of inherited diseases. New York: McGraw-Hill; 2010. Chapter 146.1. www.ommbid.com

Kohane IS. Using electronic health records to drive discovery in disease genomics. Nat Rev Genet. 2011;12:417–28.

Lo SM, et al. Phenotype diversity in type 1 Gaucher disease: discovering the genetic basis of Gaucher disease/hematologic malignancy phenotype by individual genome analysis. Blood. 2012;119:4731–40.

Manolio TA. Genomewide association studies and assessment of the risk of disease. N Engl J Med. 2010;363:166–76.

Murphy S, et al. Instrumenting the health care enterprise for discovery research in the genomic era. Genome Res. 2009;19:1675–81.

Ng SB, et al. Targeted capture and massively parallel sequencing of twelve human exomes. Nature. 2009;461:272–6.

Ng SB, et al. Exome sequencing identifies the cause of a mendelian disorder. Nat Genet. 2010a;42:30–5.

Ng SB, et al. Exome sequencing identifies MLL2 mutations as a cause of Kabuki syndrome. Nat Genet. 2010b;42:790–3.

Puffenberger EG, et al. Genetic mapping and exome sequencing identify variants associated with five novel diseases. PLoS One. 2012;7(1):e28936. doi:10.1371/journal.pone.0028936.

Roach JC, et al. Analysis of genetic inheritance in a family quartet by whole-genome sequencing. Science. 2010;328:636–9.

Strauss KA, et al. Genome-wide arrays as a diagnostic tool: clinical description, genetic mapping, and molecular characterization of Salla disease in an Old Order Mennonite Population. Am J Med Genet. 2005;138A:262–7.

The 1000 Genomes Project Consortium. A map of human genome variation from population-scale sequencing. Nature. 2010;467:1061–73.

Index

J.J. Hutton (ed.), *Pediatric Biomedical Informatics: Computer Applications in Pediatric Research*, Translational Bioinformatics 2, DOI 10.1007/978-94-007-5149-1,

FSC
www.fsc.org
MIX
Papier | Fördert gute Waldnutzung
FSC® C083411

Kolibri 360 GmbH
Ferdinand-Jühlke-Straße 7
99095 Erfurt, Deutschland
produktsicherheit@kolibri360.de